STUDY GUIDE
AND
SOLUTIONS MANUAL
TO ACCOMPANY

ORGANIC
CHEMISTRY

TWELFTH EDITION

T. W. GRAHAM SOLOMONS

University of South Florida

CRAIG B. FRYHLE

Pacific Lutheran University

SCOTT A. SNYDER

University of Chicago

ROBERT G. JOHNSON

Xavier University

JON ANTILLA

University of South Florida

WILEY

ACKNOWLEDGMENTS

We are grateful to those people who have made many helpful suggestions for various editions of this study guide. These individuals include: George R. Jurch, George R. Wenzinger, and J. E. Fernandez at the University of South Florida; Darell Berlin, Oklahoma State University; John Mangravite, West Chester State College; J. G. Traynham, Louisiana State University; Desmond M. S. Wheeler, University of Nebraska; Chris Callam, The Ohio State University; Sean Hickey, University of New Orleans; and Neal Tonks, College of Charleston.

We are especially grateful to R.G. (Bob) Johnson (Xavier University) for his dedication and many contributions to this Study Guide over the years.

T. W. Graham Solomons; Craig B. Fryhle; Scott A. Snyder; Jon Antilla

Cover Image   Structure image from the RCSB PDB (www.rcsb.org) of 1FKB (Van Duyne, G. D., Standaert, R. F., Schreiber, S. L., Clardy, J. C. (1992) Atomic Structure of the Ramapmycin Human Immunophilin Fkbp-12 Complex, *J. Amer. Chem. Soc.* 1991, **113**, 7433.) created with JSMol.

This book was set in 10/12 Times Roman by Aptara Noida, UP.

ISBN 978-1-119-07732-9

Binder-Ready version ISBN 978-1-119-07733-6

Printed in the United States of America

10 9 8 7 6 5 4 3 2 1

CONTENTS

Special Topics are available only in *WileyPlus*,
www.wileyplus.com. Solutions to problems in
the Special Topics are found on the following
pages:

To the Student

Contrary to what you may have heard, organic chemisty does not have to be a difficult course. It will be a rigorous course, and it will offer a challenge. But you will learn more in it than in almost any course you will take—and what you learn will have a special relevance to life and the world around you. However, because organic chemistry can be approached in a logical and systematic way, you will find that with the right study habits, mastering organic chemistry can be a deeply satisfying experience. Here, then, are some suggestions about how to study:

1. **Keep up with your work from day to day-never let yourself get behind.** Organic chemistry is a course in which one idea almost always builds on another that has gone before. It is essential, therefore, that you keep up with, or better yet, be a little ahead of your instructor. Ideally, you should try to stay one day ahead of your instructor's lectures in your own class preparations. The lecture, then, will be much more helpful because you will already have some understanding of the assigned material. Your time in class will clarify and expand ideas that are already familiar ones.

2. **Study material in small units, and be sure that you understand each new section before you go on to the next.** Again, because of the cumulative nature of organic chemistry, your studying will be much more effective if you take each new idea as it comes and try to understand it completely before you move on to the next concept.

3. **Work all of the in-chapter and assigned problems.** One way to check your progress is to work each of the in-chapter problems when you come to it. These problems have been written just for this purpose and are designed to help you decide whether or not you understand the material that has just been explained. You should also carefully study the Solved Problems. If you understand a Solved Problem and can work the related in-chapter problem, then you should go on; if you cannot, then you should go back and study the preceding material again. Work all of the problems assigned by your instructor from the end of the chapter, as well. Do all of your problems in a notebook and bring this book with you when you go to see your instructor for extra help.

4. **Write when you study.** Write the reactions, mechanisms, structures, and so on, over and over again. Organic chemistry is best assimilated through the fingertips by writing, and not through the eyes by simply looking, or by highlighting material in the text, or by referring to flash cards. There is a good reason for this. Organic structures, mechanisms, and reactions are complex. If you simply examine them, you may think you understand them thoroughly, but that will be a misperception. The reaction mechanism may make sense to you in a certain way, but you need a deeper understanding than this. You need to know the material so thoroughly that you can explain it to someone else. This level of understanding comes to most of us (those of us without photographic memories) through writing. Only by writing the reaction mechanisms do we pay sufficient attention to their details, such as which atoms are connected to which atoms, which bonds break in a reaction and which bonds form, and the three-dimensional aspects of the structures. When we write reactions and mechanisms, connections are made in our brains that provide the long-term memory needed for success in organic chemistry. We virtually guarantee that your grade in the course will be directly proportional to the number of pages of paper that you fill with your own writing in studying during the term.

5. **Learn by teaching and explaining.** Study with your student peers and practice explaining concepts and mechanisms to each other. Use the *Learning Group Problems* and other exercises your instructor may assign as vehicles for teaching and learning interactively with your peers.

6. **Use the answers to the problems in the *Study Guide* in the proper way.** Refer to the answers only in two circumstances: (1) When you have finished a problem, use the Study Guide to check your answer. (2) When, after making a real effort to solve the problem, you find that you are completely stuck, then look at the answer for a clue and go back to work out the problem on your own. The value of a problem is in solving it. If you simply read the problem and look up the answer, you will deprive yourself of an important way to learn.

7. **Use molecular models when you study.** Because of the three-dimensional nature of most organic molecules, molecular models can be an invaluable aid to your understanding of them. When you need to see the three-dimensional aspect of a particular topic, use the Molecular Visions™ model set that may have been packaged with your textbook, or buy a set of models separately. An appendix to the *Study Guide* that accompanies this text provides a set of highly useful molecular model exercises.

8. **Make use of the rich online teaching resources in** *WileyPLUS* including ORION's adaptive learning system.

INTRODUCTION

"Solving the Puzzle" *or* "Structure Is Everything (Almost)"

As you begin your study of organic chemistry it may seem like a puzzling subject. In fact, in many ways organic chemistry is like a puzzle—a jigsaw puzzle. But it is a jigsaw puzzle with useful pieces, and a puzzle with fewer pieces than perhaps you first thought. In order to put a jigsaw puzzle together you must consider the shape of the pieces and how one piece fits together with another. In other words, solving a jigsaw puzzle is about **structure.** In organic chemistry, molecules are the pieces of the puzzle. Much of organic chemistry, indeed life itself, depends upon the fit of one molecular puzzle piece with another. For example, when an antibody of our immune system acts upon a foreign substance, it is the puzzle-piece-like fit of the antibody with the invading molecule that allows "capture" of the foreign substance. When we smell the sweet scent of a rose, some of the neural impulses are initiated by the fit of a molecule called geraniol in an olfactory receptor site in our nose. When an adhesive binds two surfaces together, it does so by billions of interactions between the molecules of the two materials. Chemistry is truly a captivating subject.

As you make the transition from your study of general to organic chemistry, it is important that you solidify those concepts that will help you understand the structure of organic molecules. A number of concepts are discussed below using several examples. We also suggest that you consider the examples and the explanations given, and refer to information from your general chemistry studies when you need more elaborate information. There are also occasional references below to sections in your text, Solomons, Fryhle, and Snyder *Organic Chemistry,* because some of what follows foreshadows what you will learn in the course.

SOME FUNDAMENTAL PRINCIPLES WE NEED TO CONSIDER

What do we need to know to understand the structure of organic molecules? First, we need to know where electrons are located around a given atom. To understand this we need to recall from general chemistry the ideas of **electron configuration** and **valence shell electron orbitals,** especially in the case of atoms such as carbon, hydrogen, oxygen, and nitrogen. We also need to use **Lewis valence shell electron structures.** These concepts are useful because the shape of a molecule is defined by its constituent atoms, and the placement of the atoms follows from the location of the electrons that bond the atoms. Once we have a Lewis structure for a molecule, we can consider **orbital hybridization** and **valence shell electron pair repulsion (VSEPR) theory** in order to generate a three-dimensional image of the molecule.

Secondly, in order to understand why specific organic molecular puzzle pieces fit together we need to consider the attractive and repulsive forces between them. To understand this we need to know how electronic charge is distributed in a molecule. We must use tools such as **formal charge** and **electronegativity.** That is, we need to know which parts of a molecule are relatively positive and which are relatively negative—in other words, their **polarity.** Associations between molecules strongly depend on both shape and the complementarity of their electrostatic charges (polarity).

When it comes to organic chemistry it will be much easier for you to understand why organic molecules have certain properties and react the way they do if you have an appreciation for the structure of the molecules involved. Structure is, in fact, almost everything, in that whenever we want to know why or how something works we look ever more deeply into its structure. This is true whether we are considering a toaster, jet engine, or an organic reaction. If you can visualize the shape of the puzzle pieces in organic chemistry (molecules), you will see more easily how they fit together (react).

SOME EXAMPLES

In order to review some of the concepts that will help us understand the structure of organic molecules, let's consider three very important molecules—water, methane, and methanol (methyl alcohol). These three are small and relatively simple molecules that have certain similarities among them, yet distinct differences that can be understood on the basis of their structures. Water is a liquid with a moderately high boiling point that does not dissolve organic compounds well. Methanol is also a liquid, with a lower boiling point than water, but one that dissolves many organic compounds easily. Methane is a gas, having a boiling point well below room temperature. Water and methanol will dissolve in each other, that is, they are miscible. We shall study the structures of water, methanol, and methane because the principles we learn with these compounds can be extended to much larger molecules.

Water

HOH

Let's consider the structure of water, beginning with the central oxygen atom. Recall that the atomic number (the number of protons) for oxygen is eight. Therefore, an oxygen atom also has eight electrons. (An ion may have more or less electrons than the atomic number for the element, depending on the charge of the ion.) Only the valence (outermost) shell electrons are involved in bonding. Oxygen has six valence electrons—that is, six electrons in the second principal shell. (Recall that the number of valence electrons is apparent from the group number of the element in the periodic table, and the row number for the element is the principal shell number for its valence electrons.) Now, let's consider the electron configuration for oxygen. The sequence of atomic orbitals for the first three shells of any atom is shown below. Oxygen uses only the first two shells in its lowest energy state.

$$1s, 2s, 2p_x, 2p_y, 2p_z, 3s, 3p_x, 3p_y, 3p_z$$

The p orbitals of any given principal shell (second, third, etc.) are of equal energy. Recall also that each orbital can hold a maximum of two electrons and that each equal energy orbital must accept one electron before a second can reside there (Hund's rule). So, for oxygen we place two electrons in the $1s$ orbital, two in the $2s$ orbital, and one in each of the $2p$ orbitals, for a subtotal of seven electrons. The final eighth electron is paired with another in one of the $2p$ orbitals. The ground state configuration for the eight electrons of oxygen is, therefore

$$1s^2\ 2s^2\ 2p_x^{\,2}\ 2p_y^{\,1}\ 2p_z^{\,1}$$

where the superscript numbers indicate how many electrons are in each orbital. In terms of relative energy of these orbitals, the following diagram can be drawn. Note that the three $2p$ orbitals are depicted at the same relative energy level.

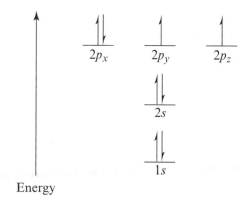

Energy

Now, let's consider the shape of these orbitals. The shape of an **s orbital** is that of a sphere with the nucleus at the center. The shape of each **p orbital** is approximately that of a dumbbell or lobe-shaped object, with the nucleus directly between the two lobes. There is one pair of lobes for each of the three p orbitals (p_x, p_y, p_z) and they are aligned along the x, y, and z coordinate axes, with the nucleus at the origin. Note that this implies that the three p orbitals are at 90° angles to each other.

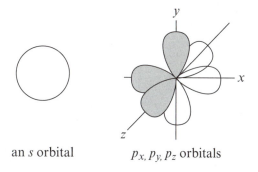

an s orbital p_x, p_y, p_z orbitals

Now, when oxygen is bonded to two hydrogens, bonding is accomplished by the sharing of an electron from each of the hydrogens with an unpaired electron from the oxygen. This type of bond, involving the sharing of electrons between atoms, is called a **covalent bond.** The formation of covalent bonds between the oxygen atom and the two hydrogen atoms is advantageous because each atom achieves a full valence shell by the sharing of these electrons. For the oxygen in a water molecule, this amounts to satisfying the octet rule.

A **Lewis structure** for the water molecule (which shows only the valence shell electrons) is depicted in the following structure. There are two nonbonding pairs of electrons around the oxygen as well as two bonding pairs.

In the left-hand structure the six valence electrons contributed by the oxygen are shown as dots, while those from the hydrogens are shown as x's. This is done strictly for bookkeeping purposes. All electrons are, of course, identical. The right-hand structure uses the convention that a bonding pair of electrons can be shown by a single line between the bonded atoms.

This structural model for water is only a *first approximation,* however. While it is a proper Lewis structure for water, it is not an entirely correct three-dimensional structure. It might appear that the angle between the hydrogen atoms (or between any two pairs of electrons in a water molecule) would be 90°, but this is not what the true angles are in a water molecule. The angle between the two hydrogens is in fact about 105°, and the nonbonding electron pairs are in a different plane than the hydrogen atoms. The reason for this arrangement is that groups of bonding and nonbonding electrons tend to repel each other due to the negative charge of the electrons. Thus, the ideal angles between bonding and nonbonding groups of electrons are those angles that allow maximum separation in three-dimensional space. This principle and the theory built around it are called the **valence shell electron pair repulsion (VSEPR) theory.**

VSEPR theory predicts that the ideal separation between four groups of electrons around an atom is 109.5°, the so-called **tetrahedral angle.** At an angle of 109.5° all four electron groups are separated equally from each other, being oriented toward the corners of a regular tetrahedron. The exact tetrahedral angle of 109.5° is found in structures where the four groups of electrons and bonded groups are identical.

In water, there are two different types of electron groups—pairs bonding the hydrogens with the oxygen and nonbonding pairs. Nonbonding electron pairs repel each other with greater force than bonding pairs, so the separation between them is greater. Consequently, the angle between the pairs bonding the hydrogens to the oxygen in a water molecule is compressed slightly from 109.5°, being actually about 105°. As we shall see shortly, the angle between the four groups of bonding electrons in methane (CH_4) *is* the ideal tetrahedral angle of 109.5°. This is because the four groups of electrons and bound atoms are identical in a methane molecule.

Orbital hybridization is the reason that 109.5° is the ideal tetrahedral angle. As noted earlier, an *s* orbital is spherical, and each *p* orbital is shaped like two symmetrical lobes aligned along the *x, y,* and *z* coordinate axes. Orbital hybridization involves taking a weighted average of the valence electron orbitals of the atom, resulting in the same number of new hybridized orbitals. With four groups of valence electrons, as in the structure of water, one *s* orbital and three *p* orbitals from the second principal shell in oxygen are hybridized (the $2s$ and $2p_x$, $2p_y$, and $2p_z$ orbitals). The result is four new hybrid orbitals of equal energy designated as sp^3 orbitals (instead of the original three *p* orbitals and one *s* orbital). Each of the four sp^3 orbitals has roughly 25% *s* character and 75% *p* character. The geometric result is that the major lobes of the four sp^3 orbitals are oriented toward the corners of a tetrahedron with an angle of 109.5° between them.

sp^3 hybrid orbitals
(109.5° angle between lobes)

In the case of the oxygen in a water molecule, where two of the four sp^3 orbitals are occupied by nonbonding pairs, the angle of separation between them is larger than 109.5° due to additional electrostatic repulsion of the nonbonding pairs. Consequently, the angle between the bonding electrons is slightly smaller, about 105°.

More detail about orbital hybridization than provided above is given in Sections 1.9–1.15 of *Organic Chemistry*. With that greater detail it will be apparent from consideration of orbital hybridization that for three groups of valence electrons the ideal separation is 120° (trigonal planar), and for two groups of valence electrons the ideal separation is 180° (linear). VSEPR theory allows us to come to essentially the same conclusion as by the mathematical hybridization of orbitals, and it will serve us for the moment in predicting the three-dimensional shape of molecules.

Methane

$$CH_4$$

Now let's consider the structure of methane (CH_4). In methane there is a central carbon atom bearing four bonded hydrogens. Carbon has six electrons in total, with four of them being valence electrons. (Carbon is in Group IVA in the periodic table.) In methane each valence electron is shared with an electron from a hydrogen atom to form four covalent bonds. This information allows us to draw a Lewis structure for methane (see below). With four groups of valence electrons the VSEPR theory allows us to predict that the three-dimensional shape of a methane molecule should be tetrahedral, with an angle of 109.5° between each of the bonded hydrogens. This is indeed the case. Orbital hybridization arguments can also be used to show that there are four equivalent sp^3 hybrid orbitals around the carbon atom, separated by an angle of 109.5°.

All H-C-H angles are 109.5°

The structure at the far right above uses the dash-wedge notation to indicate three dimensions. A solid wedge indicates that a bond projects out of the paper toward the reader. A dashed bond indicates that it projects behind the paper away from the viewer. Ordinary lines represent bonds in the plane of the paper. The dash-wedge notation is an important and widely used tool for depicting the three-dimensional structure of molecules.

Methanol

$$CH_3OH$$

Now let's consider a molecule that incorporates structural aspects of both water and methane. Methanol (CH_3OH), or methyl alcohol, is such a molecule. In methanol, a central carbon atom has three hydrogens and an O–H group bonded to it. Three of the four valence electrons of the carbon atom are shared with a valence electron from the hydrogen atoms, forming three C—H bonds. The fourth valence electron of the carbon is shared with a valence electron from the oxygen atom, forming a C–O bond. The carbon atom now has an octet of valence electrons through the formation of four covalent bonds. The angles between these four covalent bonds is very near the ideal tetrahedral angle of 109.5°, allowing maximum separation between them. (The valence orbitals of the carbon are sp^3 hybridized.) At the oxygen atom, the situation is very similar to that in water. The oxygen uses its two unpaired valence electrons to form covalent bonds. One valence electron is used in the bond with the carbon atom, and the other is paired with an electron from the hydrogen to form the O–H bond. The remaining valence electrons of the oxygen are present as two nonbonding pairs, just as in water. The angles separating the four groups of electrons around the oxygen are thus near the ideal angle of 109.5°, but reduced slightly in the C–O–H angle due to repulsion by the two nonbonding pairs on the oxygen. (The valence orbitals of the oxygen are also sp^3 hybridized since there are four groups of valence electrons.) A Lewis structure for methanol is shown below, along with a three-dimensional perspective drawing.

THE "CHARACTER" OF THE PUZZLE PIECES

With a mental image of the three-dimensional structures of water, methane, and methanol, we can ask how the structure of each, as a "puzzle piece," influences the interaction of each molecule with identical and different molecules. In order to answer this question we have to move one step beyond the three-dimensional shape of these molecules. We need to consider not only the location of the electron groups (bonding and nonbonding) but also the distribution of electronic charge in the molecules.

First, we note that nonbonding electrons represent a locus of negative charge, more so than electrons involved in bonding. Thus, water would be expected to have some partial negative charge localized in the region of the nonbonding electron pairs of the oxygen. The same would be true for a methanol molecule. The lower case Greek δ (delta) means "partial."

Secondly, the phenomenon of electronegativity influences the distribution of electrons, and hence the charge in a molecule, especially with respect to electrons in covalent bonds. **Electronegativity** is the propensity of an element to draw electrons toward it in a covalent bond. The trend among elements is that of increasing electronegativity toward the upper right corner of the periodic table. (Fluorine is the most electronegative element.) By observing the relative locations of carbon, oxygen, and hydrogen in the periodic table, we can see that oxygen is the most electronegative of these three elements. Carbon is more electronegative than hydrogen, although only slightly. Oxygen is significantly more electronegative than hydrogen. Thus, there is substantial separation of charge in a water molecule, due not only to the nonbonding electron pairs on the oxygen but also to the greater electronegativity of the oxygen with respect to the hydrogens. The oxygen tends to draw electron density toward itself in the bonds with the hydrogens, leaving the hydrogens partially positive. The resulting separation of charge is called **polarity.** The oxygen–hydrogen bonds are called **polar covalent bonds** due to this separation of charge. If one considers the net effect of the two nonbonding electron pairs in a water molecule as being a region of negative charge, and the hydrogens as being a region of relative positive charge, it is clear that a water molecule has substantial separation of charge, or polarity.

An analysis of polarity for a methanol molecule would proceed similarly to that for water. Methanol, however, is less polar than water because only one O–H bond is present. Nevertheless, the region of the molecule around the two nonbonding electron pairs of the oxygen is relatively negative, and the region near the hydrogen is relatively positive. The electronegativity difference between the oxygen and the carbon is not as large as that between oxygen and hydrogen, however, so there is less polarity associated with the C–O bond. Since there is even less difference in electronegativity between hydrogen and carbon in the three C–H bonds, these bonds contribute essentially no polarity to the molecule. The net effect for methanol is to make it a polar molecule, but less so than water due to the nonpolar character of the CH_3 region of the molecule.

Now let's consider methane. Methane is a nonpolar molecule. This is evident first because there are no nonbonding electron pairs, and secondly because there is relatively little electronegativity difference between the hydrogens and the central carbon. Furthermore, what little electronegativity difference there is between the hydrogens and the central carbon atom is negated by the symmetrical distribution of the C–H bonds in the tetrahedral shape of methane. The slight polarity of each C–H bond is canceled by the symmetrical orientation of the four C–H bonds. If considered as vectors, the vector sum of the four slightly polar covalent bonds oriented at 109.5° to each other would be zero.

Net dipole is zero.

The same analysis would hold true for a molecule with identical bonded groups, but groups having electronegativity significantly different from carbon, so long as there were symmetrical distribution of the bonded groups. Tetrachloromethane (carbon tetrachloride) is such a molecule. It has no net polarity.

Net dipole is zero.

INTERACTIONS OF THE PUZZLE PIECES

Now that you have an appreciation for the polarity and shape of these molecules it is possible to see how molecules might interact with each other. The presence of polarity in a molecule bestows upon it attractive or repulsive forces in relation to other molecules. The negative part of one molecule is attracted to the positive region of another. Conversely, if there is little polarity in a molecule, the attractive forces it can exert are very small [though not completely nonexistent, due to dispersion forces (Section *2.13B* in *Organic Chemistry*)]. Such effects are called **intermolecular forces** (forces between molecules), and strongly depend on the polarity of a molecule or certain bonds within it (especially O—H, N—H, and other bonds between hydrogen and more electronegative atoms with nonbonding pairs). Intermolecular forces have profound effects on physical properties such as **boiling point, solubility,** and **reactivity.** An important manifestation of these properties is that the ability to isolate a pure compound after a reaction often depends on differences in boiling point, solubility, and sometimes reactivity among the compounds present.

Boiling Point

An intuitive understanding of boiling points will serve you well when working in the laboratory. The polarity of water molecules leads to relatively strong intermolecular attraction between water molecules. One result is the moderately high boiling point of water (100 °C, as compared to 65 °C for methanol and –162 °C for methane, which we will discuss shortly). Water has the highest boiling point of these three example molecules because it will strongly associate with itself by attraction of the partially positive hydrogens of one molecule (from the electronegativity difference between the O and H) to the negatively charged region in another water molecule (where the nonbonding pairs are located).

hydrogen bonds

The specific attraction between a partially positive hydrogen atom attached to a heteroatom (an atom with both nonbonding and bonding valence electrons, e.g., oxygen or nitrogen) and the nonbonding electrons of another heteroatom is called **hydrogen bonding.** It is a form of **dipole-dipole attraction** due to the polar nature of the hydrogen–heteroatom bond. A given water molecule can associate by hydrogen bonding with several other water molecules, as shown above. Each water molecule has two hydrogens that can associate with the non-bonding pairs of other water molecules, and two nonbonding pairs that can associate with the hydrogens of other water molecules. Thus, several hydrogen bonds are possible for each water molecule. It takes a significant amount of energy (provided by heat, for example) to give the molecules enough kinetic energy (motion) for them to overcome the polarity-induced attractive forces between them and escape into the vapor phase (evaporation or boiling).

Methanol, on the other hand, has a lower boiling point (65 °C) than water, in large part due to the decreased hydrogen bonding ability of methanol in comparison with water. Each methanol molecule has only one hydrogen atom that can participate in a hydrogen bond with the nonbonding electron pairs of another methanol molecule (as compared with two for each water molecule). The result is reduced intermolecular attraction between methanol molecules and a lower boiling point since less energy is required to overcome the lesser intermolecular attractive forces.

The CH_3 group of methanol does not participate in dipole–dipole attractions between molecules because there is not sufficient polarity in any of its bonds to lead to significant partial positive or negative charges. This is due to the small electronegativity difference between the carbon and hydrogen in each of the C–H bonds.

Now, on to methane. Methane has no hydrogens that are eligible for hydrogen bonding, since none is attached to a heteroatom such as oxygen. Due to the small difference in electronegativity between carbon and hydrogen there are no bonds with any significant polarity. Furthermore, what slight polarity there is in each C–H bond is canceled due to the tetrahedral symmetry of the molecule. [The minute attraction that is present between methane molecules is due to dispersion forces, but these are negligible in comparison to dipole–dipole interactions that exist when significant differences in electronegativity are present in molecules such as water and methanol.] Thus, because there is only a very weak attractive force between methane molecules, the boiling point of methane is very low (−162 °C) and it is a gas at ambient temperature and pressure.

Solubility

An appreciation for trends in solubility is very useful in gaining a general understanding of many practical aspects of chemistry. The ability of molecules to dissolve other molecules or solutes is strongly affected by polarity. The polarity of water is frequently exploited during the isolation of an organic reaction product because water will not dissolve most organic compounds but will dissolve salts, many inorganic materials, and other polar byproducts that may be present in a reaction mixture.

As to our example molecules, water and methanol are miscible with each other because each is polar and can interact with the other by dipole–dipole hydrogen bonding interactions. Since methane is a gas under ordinary conditions, for the purposes of this discussion let's consider a close relative of methane—hexane. Hexane (C_6H_{14}) is a liquid having only carbon—carbon and carbon—hydrogen bonds. It belongs to the same chemical family as methane. Hexane is *not* soluble in water due to the essential absence of polarity in its bonds. Hexane is slightly soluble in methanol due to the compatibility of the nonpolar CH_3 region of methanol with hexane. The old saying "like dissolves like" definitely holds true. This can be extended to solutes, as well. Very polar substances, such as ionic compounds, are usually freely soluble in water. The high polarity of salts generally prevents most of them from being soluble in methanol, however. And, of course, there is absolutely no solubility of ionic substances in hexane. On the other hand, very nonpolar substances, such as oils, would be soluble in hexane.

Thus, the structure of each of these molecules we've used for examples (water, methanol, and methane) has a profound effect on their respective physical properties. The presence of nonbonding electron pairs and polar covalent bonds in water and methanol versus the complete absence of these features in the structure of methane imparts markedly different physical properties to these three compounds. Water, a small molecule with strong intermolecular forces, is a moderately high boiling liquid. Methane, a small molecule with only very weak intermolecular forces, is a gas. Methanol, a molecule combining structural aspects of both water and methane, is a relatively low boiling liquid, having sufficient intermolecular forces to keep the molecules associated as a liquid, but not so strong that mild heat can't disrupt their association.

Reactivity

While the practical importance of the physical properties of organic compounds may only be starting to become apparent, one strong influence of polarity is on the **reactivity** of molecules. It is often possible to understand the basis for a given reaction in organic chemistry by considering the relative polarity of molecules and the propensity, or lack thereof, for them to interact with each other.

Let us consider one example of reactivity that can be understood at the initial level by considering structure and polarity. When chloromethane (CH_3Cl) is exposed to hydroxide ions (HO^-) in water a reaction occurs that produces methanol. This reaction is shown below.

$$CH_3Cl \;+\; HO^- (\text{as NaOH dissolved in water}) \;\rightarrow\; HOCH_3 \;+\; Cl^-$$

This reaction is called a substitution reaction, and it is of a general type that you will spend considerable time studying in organic chemistry. The reason this reaction occurs readily can be understood by considering the principles of structure and polarity that we have been discussing. The hydroxide ion has a negative charge associated with it, and thus should be attracted to a species that has positive charge. Now recall our discussion of electronegativity and polar covalent bonds, and apply these ideas to the structure of chloromethane. The chlorine atom is significantly more electronegative than carbon (note its position in the periodic table). Thus, the covalent bond between the carbon and the chlorine is polarized such that there is partial negative charge on the chlorine and partial positive charge on the carbon. This provides the positive site that attracts the hydroxide anion!

The intimate details of this reaction will be studied in Chapter 6 of your text. Suffice it to say for the moment that the hydroxide ion attacks the carbon atom using one of its nonbonding electron pairs to form a bond with the carbon. At the same time, the chlorine atom is pushed away from the carbon and takes with it the pair of electrons that used to bond it to the carbon.

The result is substitution of OH for Cl at the carbon atom and the synthesis of methanol. By calculating **formal charges** (Section 1.5 in the text) one can show that the oxygen of the hydroxide anion goes from having a formal negative charge in hydroxide to zero formal charge in the methanol molecule. Similarly, the chlorine atom goes from having zero formal charge in chloromethane to a formal negative charge as a chloride ion after the reaction. *The fact that the reaction takes place at all rests largely upon the complementary polarity of the interacting species. This is a pervasive theme in organic chemistry.*

Acid-base reactions are also very important in organic chemistry. Many organic reactions involve at least one step in the overall process that is fundamentally an acid-base reaction. Both Brønsted-Lowry acid-base reactions (those involving proton donors and acceptors) and Lewis acid-base reactions (those involving electron pair acceptors and donors, respectively) are important. In fact, the reaction above can be classified as a Lewis acid-base reaction in that the hydroxide ion acts as a Lewis base to attack the partially positive carbon as a Lewis acid. It is strongly recommended that you review concepts you have learned previously regarding acid-base reactions. Chapter 3 in *Organic Chemistry* will help in this regard, but it is advisable that you begin some early review about acids and bases based on your previous studies. Acid–base chemistry is widely applicable to understanding organic reactions.

JOINING THE PIECES

Finally, while what we have said above has largely been in reference to three specific compounds, water, methanol, and methane, the principles involved find exceptionally broad application in understanding the structure, and hence reactivity, of organic molecules in general. You will find it constantly useful in your study of organic chemistry to consider the electronic structure of the molecules with which you are presented, the shape caused by the distribution of electrons in a molecule, the ensuing polarity, and the resulting potential for that molecule's reactivity. What we have said about these very small molecules of water, methanol, and methane can be extended to consideration of molecules with 10 to 100 times as many atoms. You would simply apply these principles to small sections of the larger molecule one part at a time. The following structure of Streptogramin A provides an example.

A region with trigonal planar bonding

A region with tetrahedral bonding

A few of the partially positive and partially negative regions are shown, as well as regions of tetrahedral and trigonal planar geometry. See if you can identify more of each type.

Streptogramin A

A natural antibacterial compound that blocks protein synthesis at the 70S ribosomes of Gram-positive bacteria.

We have not said much about how overall shape influences the ability of one molecule to interact with another, in the sense that a key fits in a lock or a hand fits in a glove. This type of consideration is also extremely important, and will follow with relative ease if you have worked hard to understand the general principles of structure outlined here and expanded upon in the early chapters of *Organic Chemistry*. An example would be the following. Streptogramin A, shown above, interacts in a hand-in-glove fashion with the 70S ribosome in bacteria to inhibit binding of transfer RNA at the ribosome. The result of this interaction is the prevention of protein synthesis in the bacterium, which thus accounts for the antibacterial effect of Streptogramin A. Other examples of hand-in-glove interactions include the olfactory response to geraniol mentioned earlier, and the action of enzymes to speed up the rate of reactions in biochemical systems.

FINISHING THE PUZZLE

In conclusion, if you pay attention to learning aspects of structure during this initial period of "getting your feet wet" in organic chemistry, much of the three-dimensional aspects of molecules will become second nature to you. You will immediately recognize when a molecule is tetrahedral, trigonal planar, or linear in one region or another. You will see the potential for interaction between a given section of a molecule and that of another molecule based on their shape and polarity, and you will understand why many reactions take place. Ultimately, you will find that there is much less to memorize in organic chemistry than you first thought. You will learn how to put the pieces of the organic puzzle together, and see that structure is indeed almost everything, just applied in different situations!

1

THE BASICS: BONDING AND MOLECULAR STRUCTURE

Another Approach to Writing Lewis Structures

When we write Lewis structures using this method, we assemble the molecule or ion from the constituent atoms showing only the valence electrons (i.e., the electrons of the outermost shell). By having the atoms share electrons, we try to give each atom the electronic structure of a noble gas. For example, we give hydrogen atoms two electrons because this gives them the structure of helium. We give carbon, nitrogen, oxygen, and fluorine atoms eight electrons because this gives them the electronic structure of neon. The number of valence electrons of an atom can be obtained from the periodic table because it is equal to the group number of the atom. Carbon, for example, is in Group IVA and has four valence electrons; fluorine, in Group VIIA, has seven; hydrogen, in Group 1A, has one. As an illustration, let us write the Lewis structure for CH_3F. In the example below, we will at first show a hydrogen's electron as x, carbon's electrons as o's, and fluorine's electrons as dots.

Example A

3 H$^\times$, $\circ\overset{\circ}{\underset{\circ}{C}}\circ$, and $\cdot\overset{\cdot\cdot}{\underset{\cdot\cdot}{F}}\colon$ are assembled as

$$\overset{\displaystyle H}{\underset{\displaystyle H}{H\overset{\times\,\circ}{\underset{\times\,\circ}{\circ C \circ}}\overset{\cdot\cdot}{\underset{\cdot\cdot}{F}}\colon}} \quad \text{or} \quad \overset{\displaystyle H}{\underset{\displaystyle H}{H \colon C \colon \overset{\cdot\cdot}{\underset{\cdot\cdot}{F}}\colon}}$$

If the structure is an ion, we add or subtract electrons to give it the proper charge. As an example, consider the chlorate ion, $ClO_3{}^-$.

Example B

$\colon\overset{\cdot\cdot}{\underset{\cdot\cdot}{Cl}}\cdot$, and $\overset{\circ\circ}{\underset{\circ\circ}{\circ O \circ}}$ and an extra electron $\times$ are assembled as

$$\left[\overset{\overset{\circ\circ}{O}\circ}{\underset{}{\overset{\circ\circ}{\underset{\circ\circ}{O}}\colon \overset{}{\underset{}{Cl}} \times \overset{\circ\circ}{\underset{\circ\circ}{O}}\circ}} \right]^- \quad \text{or} \quad \left[\overset{\colon\overset{\cdot\cdot}{O}\colon}{\underset{}{\colon\overset{\cdot\cdot}{\underset{\cdot\cdot}{O}}\colon \overset{}{\underset{}{Cl}}\colon\overset{\cdot\cdot}{\underset{\cdot\cdot}{O}}\colon}} \right]^-$$

1.1 ^{14}N, 7 protons and 7 neutrons; ^{15}N, 7 protons and 8 neutrons

1.2   (a) one   (b) seven   (c) four   (d) three   (e) eight   (f) five

1.3   (a) O   (b) N   (c) Cl   (d) S

1.4   (a) ionic   (b) covalent   (c) covalent   (d) covalent

1.5   (a) and (b) Lewis structures of CH_3F and $CHCl_3$

1.6   Lewis structure of CH_3OH

1.7   Lewis structure of acetaldehyde CH_3CHO

1.8   (a) $H-\ddot{\underset{..}{F}}:$ (d) $H-\ddot{\underset{..}{O}}-\ddot{N}=\dot{\underset{.}{O}}$ (g) phosphoric acid structure

(b) $:\ddot{\underset{..}{F}}-\ddot{\underset{..}{F}}:$ (e) $H-\ddot{\underset{..}{O}}-\underset{..}{S}-\ddot{\underset{..}{O}}-H$ (h) carbonic acid structure

(c) $H-\underset{H}{\overset{H}{C}}-\ddot{\underset{..}{F}}:$ (f) $\left[H-\underset{H}{\overset{H}{B}}-H \right]^{-}$

1.9   phosphate ion structure

1.10 (a)
$$H-\overset{\overset{\displaystyle H}{|}}{\underset{\underset{\displaystyle H}{|}}{C}}-\ddot{\overset{..}{O}}\!:^{-}$$

(c) $^{-}\!:C\equiv N:$

(e)
$$H-\ddot{\overset{..}{O}}-\overset{\overset{\displaystyle \cdot\ddot{O}\cdot}{\|}}{C}-\ddot{\overset{..}{O}}\!:^{-}$$

(b) $H-\overset{\overset{\displaystyle ..}{}}{\underset{\underset{\displaystyle H}{|}}{\ddot{N}}}\!:^{-}$

(d)
$$H-\overset{\overset{\displaystyle \cdot\ddot{O}\!:}{\|}}{C}-\ddot{\overset{..}{O}}\!:^{-}$$

(f) $H-C\equiv C\!:^{-}$

1.11 (a)
$$H-\overset{\overset{\displaystyle H}{|}}{\underset{\underset{\displaystyle H}{|}}{C}}-\overset{\overset{\displaystyle H}{|}}{\underset{\underset{\displaystyle H}{}}{C}}^{+}$$

(d)
$$H-\overset{\overset{\displaystyle :\ddot{O}:^{-}}{|}}{\underset{\underset{\displaystyle H}{|}}{C}}-H$$

(g)
$$H-\overset{\overset{\displaystyle H}{|}}{\underset{\underset{\displaystyle H}{|}}{C}}-C\equiv N:$$

(b) $H-\overset{\overset{\displaystyle ..+}{}}{\underset{\underset{\displaystyle H}{|}}{\ddot{O}}}-H$

(e)
$$H-\overset{\overset{\displaystyle H}{|}}{\underset{\underset{\displaystyle H}{|}}{C}}-\overset{\overset{\displaystyle H}{|}}{\underset{\underset{\displaystyle H}{|}}{N}}^{+}-H$$

(h)
$$H-\overset{\overset{\displaystyle H}{|}}{\underset{\underset{\displaystyle H}{|}}{C}}-\overset{+}{N}\equiv N:$$

(c)
$$\underset{\underset{\displaystyle H}{}}{\overset{\overset{\displaystyle \cdot\ddot{O}\cdot}{\|}}{C}}-\ddot{\overset{..}{O}}\!:^{-}$$

(f)
$$H-\overset{\overset{\displaystyle ..+}{}}{\underset{\underset{\displaystyle H}{|}}{\ddot{O}}}-H$$
$$H-\overset{\overset{\displaystyle |}{|}}{\underset{\underset{\displaystyle H}{|}}{C}}-H$$

1.12
$$H-\overset{\overset{\displaystyle H}{|}}{\underset{\underset{\displaystyle H}{|}}{C}}-\ddot{\overset{..}{O}}-\overset{\overset{\displaystyle H}{|}}{\underset{\underset{\displaystyle H}{|}}{C}}-\overset{\overset{\displaystyle H}{|}}{\underset{\underset{\displaystyle H}{|}}{C}}-H$$

1.13 $CH_3CHCHCHCH_3$ or $(CH_3)_2CHCH(CH_3)CH(CH_3)_2$

with CH_3 on top and CH_3, CH_3 below

1.14 (a)

(b)

(c) CH_3, CH_3, H, $C=C$, CH_3, CH_2 $\quad = \quad$ (skeletal structure)

(d) CH_3, CH_2, CH_2, CH_2, CH_3 $\quad = \quad$ (skeletal structure)

(e) CH_3, CH_2, CH, CH_2, CH_3, OH $\quad = \quad$ (skeletal structure with OH)

(f) $CH_2=C$, CH_2-CH_3, CH_2-CH_3 $\quad = \quad$ (skeletal structure)

(g) O, C, CH_3, CH_2, CH_2, CH_3 $\quad = \quad$ (skeletal structure with O)

(h) Cl, CH_3, CH, CH_2, CH, CH_3, CH_3 $\quad = \quad$ (skeletal structure with Cl)

1.15 (a) and (d) are constitutional isomers with the molecular formula C_5H_{12}.

(b) and (e) are constitutional isomers with the molecular formula $C_5H_{12}O$.

(c) and (f) are constitutional isomers with the molecular formula C_6H_{12}.

1.16 (a) $H-\overset{\displaystyle H}{\underset{\displaystyle H}{C}}-\overset{\displaystyle Cl}{\underset{\displaystyle H}{C}}-\overset{\displaystyle H}{\underset{\displaystyle H}{C}}-\overset{\displaystyle H}{\underset{\displaystyle H}{C}}-\overset{\displaystyle O}{C}-\overset{\displaystyle H}{\underset{\displaystyle H}{C}}-H$

(c) (ring structure with C=C double bond and branching)

(b) (ring structure with O—H)

1.17 (a)

(Note that the **Cl** atom and the three **H** atoms may be written at any of the four positions.)

(b)

or   and so on

(c)

and others

(d)

and others

1.18 (a)

(b) and (c). Since the two resonance structures are equivalent, each should make an equal contribution to the overall hybrid. The C—O bonds should therefore be of equal length (they should be of bond order 1.5), and each oxygen atom should bear a 0.5 negative charge.

1.19 (a)

(b)

(c)

(d)

1.20 (a)

(b)

(c)

(d)

(e)

(f) $H_2\bar{C}$—C($\overset{:\overset{..}{O}:}{||}$)—$CH_3$ $\longleftrightarrow$ H_2C=C($\overset{:\overset{..}{O}:^-}{|}$)—$CH_3$

$\overset{\delta-}{H_2C}$---C($\overset{O^{\delta-}}{||}$)—$CH_3$

(g) CH_3—$\overset{..}{\underset{..}{S}}$—$\overset{+}{C}H_2$ $\longleftrightarrow$ CH_3—$\overset{..}{S}$=CH_2

CH_3—$\overset{\delta+}{\underset{..}{S}}$---$\overset{\delta+}{C}H_2$

(h) CH_3—$\overset{+}{N}$($\overset{\overset{..}{O}:}{}$)($\overset{}{\underset{..}{O}:^-}$) $\longleftrightarrow$ CH_3—$\overset{+}{N}$($\overset{\overset{..}{O}:^-}{}$)($\overset{}{\underset{..}{O}:}$) $\longleftrightarrow$ CH_3—$\overset{2+}{N}$($\overset{\overset{..}{O}:^-}{}$)($\overset{}{\underset{..}{O}:^-}$)

(minor)

CH_3—$\overset{+}{N}$($\overset{O^{\delta-}}{}$)($\overset{}{O^{\delta-}}$)

1.21 (a)

H—$\overset{+}{\underset{H}{C}}$—$N(\overset{CH_3}{})(CH_3)$ $\longleftrightarrow$ H—$\overset{}{\underset{H}{C}}$=$\overset{+}{N}(\overset{CH_3}{})(CH_3)$ CH_2=$\overset{+}{N}(CH_3)_2$ because all atoms

have a complete octet (rule 3), and there are more covalent bonds (rule 1).

(b) CH_3—C($\overset{:\overset{..}{O}:}{||}$)($\underset{\overset{..}{O}.—H}{}$) $\longleftrightarrow$ CH_3—C($\overset{:\overset{..}{O}:^-}{}$)($\underset{\overset{+}{\overset{..}{O}}—H}{}$) CH_3—C($\overset{:\overset{..}{O}}{||}$)($\underset{:\overset{..}{O}—H}{}$)   because it has no charge separation (rule 2).

(c) $:NH_2$—C≡$\overset{..}{N}:$ $\longleftrightarrow$ $\overset{+}{N}H_2$=C=$\overset{..}{\underset{}{N}}^-$ $:NH_2$—C≡N: because it has no charge separation (rule 2).

1.22 (a) Cis-trans isomers are not possible.

(b) $\overset{CH_3}{\underset{H}{}}$C=C$\overset{CH_3}{\underset{H}{}}$ and $\overset{CH_3}{\underset{H}{}}$C=C$\overset{H}{\underset{CH_3}{}}$

(c) Cis-trans isomers are not possible.

(d) $\overset{CH_3CH_2}{\underset{H}{}}$C=C$\overset{Cl}{\underset{H}{}}$ and $\overset{CH_3CH_2}{\underset{H}{}}$C=C$\overset{H}{\underset{Cl}{}}$

1.23 sp^3

1.24 sp^3

1.25 *sp²*

1.26 *sp*

1.27 (a)

$$\left[\begin{array}{c} H \\ H \overset{B}{\underset{H}{\longrightarrow}} H \end{array} \right]^{-}$$

There are four bonding pairs.
The geometry is tetrahedral.

(b)   :F̈—Be—F̈:   There are two bonding pairs about
the central atom. The geometry is linear.

(c)

$$\left[\begin{array}{c} H \\ H \overset{N}{\underset{H}{\longrightarrow}} H \end{array} \right]^{+}$$

There are four bonding pairs.
The geometry is tetrahedral.

(d)

There are two bonding pairs and two nonbonding pairs.
The geometry is tetrahedral and the shape is angular.

(e)

$$\begin{array}{c} H \\ | \\ H \overset{B}{\diagdown} H \end{array}$$

There are three bonding pairs. The geometry
is trigonal planar.

(f)

$$\begin{array}{c} F \\ F \overset{C}{\underset{F}{\longrightarrow}} F \end{array}$$

There are four bonding pairs around the central
atom. The geometry is tetrahedral.

(g)

$$\begin{array}{c} F \\ F \overset{Si}{\underset{F}{\longrightarrow}} F \end{array}$$

There are four bonding pairs around the central atom.
The geometry is tetrahedral.

(h)

$$\left[\begin{array}{c} \ddot{} \\ Cl \overset{C}{\underset{Cl}{\diagdown}} Cl \end{array} \right]^{-}$$

There are three bonding pairs and one nonbonding pair
around the central atom. The geometry is tetrahedral and
the shape is trigonal pyramidal.

1.28 (a)

trigonal planar at each carbon atom

(b)   CH₃—C≡C—CH₃ linear   (c) H—C≡N: linear

Problems

Electron Configuration

1.29 (a) Na^+ has the electronic configuration, $1s^2 2s^2 2p^6$, of Ne.

(b) Cl^- has the electronic configuration, $1s^2 2s^2 2p^6 3s^2 3p^6$, of Ar.

(c) F^+ and (h) Br^+ do not have the electronic configuration of a noble gas.

(d) H^- has the electronic configuration, $1s^2$, of He.

(e) Ca^{2+} has the electronic configuration, $1s^2 2s^2 2p^6 3s^2 3p^6$, of Ar.

(f) S^{2-} has the electronic configuration, $1s^2 2s^2 2p^6 3s^2 3p^6$, of Ar.

(g) O^{2-} has the electronic configuration, $1s^2 2s^2 2p^6$, of Ne.

Lewis Structures

1.30 (a) (b) (c) (d)

1.31 (a)   (c)

(b)   (d)

1.32

(a)   (b)   (c)

(d)   (e)

Structural Formulas and Isomerism

1.33 (a) $(CH_3)_2CHCH_2\ddot{O}H$

(c)
$$H_2C\text{—}CH_2$$
$$|\qquad|$$
$$HC\text{=}CH$$

(b)
$$\overset{\cdot\ddot{O}\cdot}{\underset{\parallel}{}}$$
$(CH_3)_2CHCCH(CH_3)_2$

(d) $(CH_3)_2CHCH_2CH_2\ddot{O}H$

1.34 (a) $C_4H_{10}O$ (c) C_4H_6

(b) $C_7H_{14}O$ (d) $C_5H_{12}O$

1.35 (a) Different compounds, not isomeric

(b) Same compound

(c) Same compound

(d) Same compound

(e) Same compound

(f) Constitutional isomers

(g) Different compounds, not isomeric

(h) Same compound

(i) Different compounds, not isomeric

(j) Same compound

(k) Constitutional isomers

(l) Different compounds, not isomeric

(m) Same compound

(n) Same compound

(o) Same compound

(p) Constitutional isomers

1.36 (a)

(b)

(c)

(d)

(e)   or

(f)

1.37

1.38

(Other structures are possible.)

Resonance Structures

1.39

1.40

1.41 (a)

(b)

(c)

(d)

(e)

(f)

(g)

(h)

(i)

1.42   (a) While the structures differ in the position of their electrons, they also differ in the positions of their nuclei and thus *they are not resonance structures*. (In cyanic acid the hydrogen nucleus is bonded to oxygen; in isocyanic acid it is bonded to nitrogen.)
(b) The anion obtained from either acid is a resonance hybrid of the following structures:

$$^{-}:\ddot{O}-C\equiv N: \longleftrightarrow \ddot{O}=C=\ddot{N}:^{-}$$

1.43

$$\begin{array}{c} H \\ | \\ H-C \\ | \\ H \end{array}$$

(a) A $+1$ charge. ($F-4-{}^{6}/_{2}-2=+1$)

(b) A $+1$ charge. (It is called a methyl cation.)

(c) Trigonal planar, that is,

$$\begin{array}{c} H \\ | \\ \overset{+}{C} \\ H \quad\quad H \end{array}$$

(d) sp^2

1.44

$$\begin{array}{c} H \\ | \\ H-C: \\ | \\ H \end{array}$$

(a) A -1 charge. ($F = 4 - 6/2 - 2 = -1$)

(b) A -1 charge. (It is called a methyl anion.)

(c) Trigonal pyramidal, that is

$$\begin{array}{c} \overset{\cdot\cdot}{C^{-}} \\ H \quad H \\ H \end{array}$$

(d) sp^3

1.45

$$\begin{array}{c} H \\ | \\ H-C\cdot \\ | \\ H \end{array}$$

(a) No formal charge. ($F = 4 - {}^{6}/_{2} - 1 = 0$)

(b) No charge.

(c) sp^2, that is,

$$\begin{array}{c} H \\ C-H \\ H \end{array}$$

1.46 (a) H_2CO or CH_2O

sp²

(b) $H_2C=CHCH=CH_2$

(c) $H_2C=C=C=CH_2$

1.47 (a) and (b)

(c) Because the two resonance structures are equivalent, they should make equal contributions to the hybrid and, therefore, the bonds should be the same length.

(d) Yes. We consider the central atom to have two groups or units of bonding electrons and one unshared pair.

1.48

$$:N{\equiv}\overset{+}{N}{-}\ddot{N}{:}^{2-} \longleftrightarrow {}^-{:}\ddot{N}{=}\overset{+}{N}{=}\ddot{N}{:}^- \longleftrightarrow {}^{2-}{:}\ddot{N}{-}\overset{+}{N}{\equiv}N{:}$$

A   B   C

Structures **A** and **C** are equivalent and, therefore, make equal contributions to the hybrid. The bonds of the hybrid, therefore, have the same length.

1.49 (a)

(b) $(CH_3)_2NH$ $CH_3CH_2NH_2$

(c) $(CH_3)_3N$ $CH_3CH_2NHCH_3$ $CH_3CH_2CH_2NH_2$ $CH_3\underset{\underset{NH_2}{|}}{C}HCH_3$

(d)

1.50 (a) constitutional isomers   (b) the same

(c) resonance forms   (d) constitutional isomers

(e) resonance forms   (f) the same

Challenge Problems

1.51 (a) $\ddot{O}=\overset{+}{N}=\ddot{O}$

(b) Linear

(c) Carbon dioxide

1.52 Set A:

Set B:

Set C:

[and unstable enol forms of a, b, and c]

Set D:

Set E: (i.e., $CH_3CH_2\overset{\bar{\cdot\cdot}}{C}H_2$ and $CH_3\overset{\bar{\cdot\cdot}}{C}HCH_3$)

1.53 (a) No, a carbon atom in its ground state would have 2 electrons in the $1s$ orbital, 2 electrons in the $2s$ orbital, and only 2 unpaired electrons in the degenerate $2p_x, 2p_y$, and $2p_z$ orbitals. So the two unpaired electrons can pair with only 2 hydrogen atoms with their one unpaired electron, respectively to form the compound CH_2, which would be divalent and have 180 degree bond angles.

(b) In this case 4 unpaired electrons can combine with 4 hydrogen atoms to give CH_4, the correct bonding for methane, a tetravalent compound. However, the tetrahedral geometry known to exist for methane would not result from bonding at the $2s$ and three $2p$ orbitals in the excited state. Hybridized sp^3 orbitals are required for tetrahedral geometry.

1.54 (a) Dimethyl ether   Dimethylacetylene cis-1,2-Dichloro-1,2-difluoroethene

(b)

(c)

1.55 The large lobes centered above and below the boron atom represent the $2p$ orbital that was not involved in hybridization to form the three $2sp^2$ hybrid orbitals needed for the three boron-fluorine covalent bonds. This orbital is not a pure $2p$ atomic orbital, since it is not an isolated atomic p orbital but rather part of a molecular orbital. Some of the other lobes in this molecular orbital can be seen near each fluorine atom.

1.56 The two resonance forms for this anion are $^-$:$\ddot{C}H_2$—CH=$\overset{..}{\overset{.}{O}}$ and CH_2=CH—$\ddot{\overset{..}{O}}$:$^-$. The MEP indicates that the resonance contributor where the negative charge on the anion is on the oxygen is more important, which is what we would predict based on the fact that oxygen is more electronegative than carbon.

Resonance hybrid, $CH_2 \overset{\delta^-}{\text{---}} CH \overset{\delta^-}{\text{---}} O$

QUIZ

$5 + 12 + 1 = 18$

1.1 Which of the following is a valid Lewis dot formula for the nitrite ion (NO_2^-)?

(a) $^-$:$\ddot{\overset{..}{O}}$—$\ddot{N}$=$\ddot{\overset{..}{O}}$:   (b) :$\ddot{\overset{..}{O}}$=$\ddot{N}$—$\ddot{\overset{..}{O}}$:$^-$ (c) :$\ddot{\overset{..}{O}}$—$\ddot{N}$=$\ddot{\overset{..}{O}}$:   (d) Two of these

(e) None of the above

1.2 What is the hybridization state of the boron atom in BF_3?

(a) s (b) p (c) sp (d) sp^2 (e) sp^3

1.3 BF_3 reacts with NH_3 to produce a compound, $F\!-\!B\!-\!N\!-\!H$. The hybridization state of B is

$$\begin{array}{cc} F & H \\ | & | \\ F\!-\!B\!-\!N\!-\!H \\ | & | \\ F & H \end{array}$$

(a) s (b) p (c) sp (d) sp^2 (e) sp^3

1.4 The formal charge on N in the compound given in Problem 1.3 is

(a) -2 (b) -1 (c) 0   (d) $+1$ (e) $+2$

1.5 The correct bond-line formula of the compound whose condensed formula is $CH_3CHClCH_2CH(CH_3)CH(CH_3)_2$ is

(a)

(b)

(c)

(d)

(e)

1.6 Write another resonance structure for the acetate ion.

Acetate ion

1.7 In the boxes below write condensed structural formulas for constitutional isomers of $CH_3(CH_2)_3CH_3$.

$(CH_3)_2 CH_2 CH_3$

$CH_3 (CH_2 CH_3)CH_3$

1.8 Write a three-dimensional formula for a constitutional isomer of compound **A** given below. Complete the partial structure shown.

A

Constitutional isomer of A

1.9   Consider the molecule $(CH_3)_3B$ and give the following:

(a) Hybridization state of boron

(b) Hybridization state of carbon atoms

(c) Formal charge on boron

(d) Orientation of groups around boron

(e) Dipole moment of $(CH_3)_3B$

1.10   Give the formal charge on oxygen in each compound.

(a) $CH_3-\overset{\cdot\cdot}{\underset{|}{O}}-CH_3$ $\boxed{+}$ (c) $\boxed{-}$
  CH_3

(b) $\boxed{0}$

1.11   Write another resonance structure in which all of the atoms have a formal charge of zero.

1.12   Indicate the direction of the net dipole moment of the following molecule.

1.13   Write bond-line formulas for all compounds with the formula C_3H_6O.

$12 + 6 + 6 = 24$

2 FAMILIES OF CARBON COMPOUNDS: FUNCTIONAL GROUPS, INTERMOLECULAR FORCES, AND INFRARED (IR) SPECTROSCOPY

SOLUTIONS TO PROBLEMS

2.1 Although you were only asked to provide two possible constitutional isomers, below are the structures of all the possible products, including some unique forms of alkene bonding that you will encounter in later sections of the text.

1-pentyne

2-pentyne

trans-1,3-pentadiene

1,4-pentadiene

cis-1,3-pentadiene

1,2-pentadiene

(compounds with adjacent double bonds are called allenes: see Section 5.18)

2,3-pentadiene

(and its non-superposable mirror image stereoisomer: see Section 5.3 regarding enantiomers and 5.18 regarding allenes)

2,3-pentadiene

2.2 (a) H—F or $\overset{\delta+ \quad \delta-}{\text{H—F}}$

(b) I—Br or $\overset{\delta+ \quad \delta-}{\text{I—Br}}$

(c) Br—Br
$\mu = 0$ D

(d) F—F
$\mu = 0$ D

2.3   VSEPR theory predicts a trigonal planar structure for BF_3.

$$\mu = 0 \text{ D}$$

The vector sum of the bond moments of a trigonal planar structure would be zero, resulting in a prediction of $\mu = 0$ D for BF_3. This correlates with the experimental observation and confirms the prediction of VSEPR theory.

2.4   The shape of $CCl_2{=}CCl_2$ (below) is such that the vector sum of all of the C—Cl bond moments is zero.

2.5   The fact that SO_2 has a dipole moment indicates that the molecule is angular, not linear.

$$\mu = 1.63 \text{ D} \qquad \left(\text{not} \quad \mu = 0 \text{ D} \right)$$

An angular shape is what we would expect from VSEPR theory, too.

2.6   VSEPR theory predicts the following.

net dipole

2.7   In $CFCl_3$ the large C—F bond moment opposes the C—Cl moments, leading to a net dipole moment in the direction of the fluorine. Because hydrogen is much less electronegative than fluorine, no such opposing effect occurs in $CHCl_3$; therefore, it has a net dipole moment that is larger and in the direction of the chlorine atoms.

Smaller net   Larger net
dipole moment   dipole moment

2.8 (a) [structure] net dipole moment   (b) [structure] $\mu = 0$ D

(c) [structure] net dipole moment   (d) [structure] $\mu = 0$ D

2.9 (a) [structure] net dipole moment

[structure] cis — net dipole moment — [structure] trans $\mu = 0$ D

Cis-trans isomers

(b) [structure] net dipole moment

[structure] cis — net dipole moment — [structure] trans $\mu = 0$ D

Cis-trans isomers

2.10 (a) [structure] Br   and   [structure] Br

(b) [structure with Br]

(c) [structure with Br]

2.11 (a) [structure] F   (b) [structure] Cl

(c) Propyl bromide   (d) Isopropyl fluoride

(e) Phenyl iodide

2.12 (a) [structure] OH   and   [structure] OH

(b) [structure with OH]

(c) [structure] OH

2.13 (a)  OH   (b) OH

2.14 (a) O   (b) O

(c) O   (d) Methyl propyl ether

(e) Diisopropyl ether   (f) Methyl phenyl ether

2.15

ether

OCH₃

alkene

OH

phenol

2.16 (a) Isopropylpropylamine   (b) Tripropylamine

(c) Methylphenylamine   (d) Dimethylphenylamine

(e) NH₂   (f) CH_3-N-CH_3 or $(CH_3)_3N$
                          |
                         CH_3

(g) N—CH₃

2.17 (a) (e) only   (b) (a, c)   (c) (b, d, f, g)

2.18 (a)
$$CH_3-\overset{\displaystyle CH_3}{\underset{\displaystyle CH_3}{N:}} \ + \ H-\ddot{\underset{..}{Cl}}: \ \longrightarrow \ CH_3-\overset{\displaystyle CH_3}{\underset{\displaystyle CH_3}{\overset{+}{N}}}-H \ + \ :\ddot{\underset{..}{Cl}}:^-$$

(b) sp^3

2.19

2.20 (a)

(b)

2.21

2.22

+ others

2.23 $CH_3 - C$ $\overset{..}{O} - CH_2CH_3$ ⟷ $CH_3 - C$ $\overset{+}{O} - CH_2CH_3$

2.24 $CH_3 - C$ $\overset{..}{N}H_2$ ⟷ $CH_3 - C$ $\overset{+}{N}H_2$

2.25 (a) OH would boil higher because its molecules can form hydrogen bonds to each other through the $-\overset{..}{\underset{..}{O}} - H$ group.

(b) would boil higher because its molecules can form hydrogen bonds to each other through the $-\overset{..}{N} - H$ group.

(c) HO OH because by having two $-\overset{..}{\underset{..}{O}} - H$ groups, it can form more hydrogen bonds.

2.26 Cyclopropane would have the higher melting point because its cyclic structure gives it a rigid compact shape that would permit stronger crystal lattice forces.

2.27 d < a < b < c

(c) has the highest boiling point due to hydrogen bonding involving its O—H group.

(b) is a polar molecule due to its C=O group, hence higher boiling than the essentially non-polar (a) and (d).

(a) has a higher boiling point than (d) because its unbranched structure permits more van der Waals attractions.

2.28 If we consider the range for carbon-oxygen double bond stretching in an aldehyde or ketone to be typical of an unsubstituted carbonyl group, we find that carbonyl groups with an oxygen or other strongly electronegative atom bonded to the carbonyl group, as in carboxylic acids and esters, absorb at somewhat higher frequencies. On the other hand, if a nitrogen atom is bonded to the carbonyl group, as in an amide, then the carbonyl stretching frequency is lower than that of a comparable aldehyde or ketone. The reason for this trend is that strongly electronegative atoms increase the double bond character of the carbonyl, while the unshared electron pair of an amide nitrogen atom contributes to the carbonyl resonance hybrid to give it less double bond character.

Functional Groups and Structural Formulas

2.29 (a) Ketone   (b) Alkyne   (c) Alcohol   (d) Aldehyde
(e) Alcohol   (f) Alkene

2.30 (a) Three carbon-carbon double bonds (alkene) and a 2° alcohol

(b) Phenyl, carboxylic acid, amide, ester, and a 1° amine

(c) Phenyl and a 1° amine

(d) Carbon-carbon double bond and a 2° alcohol

(e) Phenyl, ester, and a 3° amine

(f) Carbon-carbon double bond and an aldehyde

(g) Carbon-carbon double bond and 2 ester groups

2.33 (a) 1°   (b) 2°   (c) 3°   (d) 3°   (e) 2°

2.34 (a) 2°   (b) 1°   (c) 3°   (d) 2°   (e) 2°   (f) 3°

2.35 (a) Me—O— …   Me—O— …   —O—

(b) … OH   … OH   … OH   … OH

(c) … OH   (d) … OH

(e) …   …

(f) …

(g) …

(h) …

(i) …

(j) …

(k) …

(l)   (m) Me_3N

(n) ...

2.36 Crixivan has the following functional groups:

2.37 The following formula is for Taxol, a natural compound with anticancer activity. Taxol has the following functional groups.

Physical Properties

2.38 (a) The O—H group of Vitamin A is the hydrophilic portion of the molecule, but the remainder of the molecule is not only hydrophobic but much larger. Attractive dispersion forces between the hydrophobic region of one Vitamin A molecule and another outweighs the effect of hydrogen bonding to water through a single hydroxyl group. Hence, Vitamin A is not expected to be water soluble.

(b) For Vitamin B_3, there are multiple hydrophilic sites. The carbonyl oxygen and the O—H of the acid function as well as the ring nitrogen can all hydrogen bond to water. Since the hydrophobic portion (the ring) of the molecule is modest in size, the molecule is expected to be water soluble.

2.39 The attractive forces between hydrogen fluoride molecules are the very strong dipole-dipole attractions that we call *hydrogen bonds*. (The partial positive charge of a hydrogen fluoride molecule is relatively exposed because it resides on the hydrogen nucleus. By contrast, the positive charge of an ethyl fluoride molecule is buried in the ethyl group and is shielded by the surrounding electrons. Thus the positive end of one hydrogen fluoride molecule can approach the negative end of another hydrogen fluoride molecule much more closely, with the result that the attractive force between them is much stronger.)

2.40 The cis isomer is polar while the trans isomer is nonpolar ($\mu = 0$ D). The intermolecular attractive forces are therefore greater in the case of the cis isomer, and thus its boiling point should be the higher of the two.

2.41 Because of its ionic character—it is a salt—the compound is water-soluble. The organic cation and the bromide ion are well-solvated by water molecules in a fashion similar to sodium bromide. The compound also is soluble in solvents of low polarity such as diethyl ether (though less so than in water). The hydrophobic alkyl groups can now be regarded as lipophilic—groups that seek a nonpolar environment. Attractive forces between the alkyl groups of different cations can be replaced, in part, by attractive dispersion forces between these alkyl groups and ether molecules.

2.42 (a) and (b) are polar and hence are able to dissolve ionic compounds. (c) is non-polar and will not dissolve ionic compounds.

2.43 (a) $H_3C\text{—}F$ type structure (methyl fluoride, with dipole toward F)

(b) difluoromethane structure (dipole)

(c) trifluoromethane structure (dipole)

(d) CF_4 — No dipole moment

(e) chlorofluoromethane CH_2ClF (dipole)

(f) BCl_3 — No dipole moment

(g) $F\text{—}Be\text{—}F$ — No dipole moment

(h) dimethyl ether $(H_3C)_2O$ (dipole)

(i) methanol type $H_3C\text{—}O\text{—}H$ (dipole)

(j) formaldehyde $H_2C{=}O$ (dipole)

2.44 (a) Dimethyl ether: There are four electron pairs around the central oxygen: two bonding pairs and two nonbonding pairs. We would expect sp^3 hybridization of the oxygen with a bond angle of approximately 109.5° between the methyl groups.

$$H_3C\text{—}O\text{—}CH_3 \qquad \mu > 0\ \text{D}$$

(b) Trimethylamine: There are four electron pairs around the central nitrogen: three bonding pairs and one nonbonding pair. We would expect sp^3 hybridization of the nitrogen with a bond angle of approximately 109.5° between the methyl groups.

$$H_3C\text{\tiny{\textbackslash}}\overset{\displaystyle \overset{..}{N}}{\underset{H_3C}{|}}\diagdown_{CH_3} \qquad \mu > 0 \text{ D}$$

(c) Trimethylboron: There are only three bonding electron pairs around the central boron. We would expect sp^2 hybridization of the boron with a bond angle of 120° between the methyl groups.

$$\overset{\displaystyle CH_3}{\underset{H_3C\diagup \overset{|}{B}\diagdown CH_3}{|}} \qquad \mu = 0 \text{ D}$$

(d) Carbon dioxide: There are four bonding electron pairs around the central carbon atom. We would expect sp hybridization of the carbon atom with a bond angle of 180° between the oxygen atoms.

$$O{=}C{=}O \qquad \mu = 0 \text{ D}$$

2.45 Without one (or more) polar bonds, a molecule cannot possess a dipole moment and, therefore, it cannot be polar. If the bonds are directed so that the bond moments cancel, however, the molecule will not be polar even though it has polar bonds.

2.46 (a) $\diagup\diagdown\diagup$OH because its molecules can form hydrogen bonds to each other through its $-\overset{..}{\underset{..}{O}}-$H group.

(b) HO$\diagdown\diagup\diagdown$OH because with two $-\overset{..}{\underset{..}{O}}-$H groups, its molecules can form more hydrogen bonds with each other.

(c) $\diagup\diagdown\diagup\diagdown\diagup$OH because its molecules can form hydrogen bonds to each other.

(d) $\triangleright$—OH [same reason as (c)].

(e) ⬡NH because its molecules can form hydrogen bonds to each other through the

$-\overset{..}{\underset{|}{N}}-$H group.

(f) $\overset{F\diagdown \quad \diagup F}{\diagup \quad \diagdown}$ because its molecules will have a larger dipole moment. (The trans compound will have $\mu = 0$ D.)

(g) [structure: propyl group with C=O and OH — propanoic acid] OH [same reason as (c)].

(h) Nonane, because of its larger molecular weight and larger size, will have larger van der Waals attractions.

(i) [structure] $=O$ because its carbonyl group is far more polar than the double bond of [structure].

IR Spectroscopy

2.47 (a) The alcohol would have a broad absorption from the O—H group in the 3200 to 3500 cm^{-1} region of its IR spectrum. The ether would have no such absorption.

(c) The ketone would have a strong absorption from its carbonyl group near 1700 cm^{-1} in its IR spectrum. The alcohol would have a broad absorption due to its hydroxyl group in the 3200 to 3500 cm^{-1} region of its IR spectrum.

(d) Same rationale as for (a).

(e) The secondary amine would have an absorption near 3300 to 3500 cm^{-1} arising from N—H stretching. The tertiary amine would have no such absorption in this region since there is no N—H group present.

(g) Both compounds would exhibit absorptions near 1710 to 1780 cm^{-1} due to carbonyl stretching vibrations. The carboxylic acid would also have a broad absorption somewhere between 2500 and 3500 cm^{-1} due to its hydroxyl group. The ester would not have a hydroxyl absorption.

(i) The ketone would have a strong absorption from its carbonyl group near 1700 cm^{-1} in its IR spectrum. The alkene would have no such absorption but would have an absorption between 1620 and 1680 cm^{-1} due to C=C stretching.

2.48

Hydrogen bond

$$CH_3CH_2 - C \underset{\overset{\ddots}{O} - H \cdots \overset{\ddots}{O}}{\overset{\overset{\ddots}{O} \cdots H - \overset{\ddots}{O}}{\Big\langle}} C - CH_2CH_3$$

Hydrogen bond

2.49 There are two peaks as a result of the asymmetric and symmetric stretches of the carbonyl groups.

asymmetric   symmetric

Multiconcept Problems

2.50 Any four of the following:

$$\underset{\text{Ketone}}{\underset{\displaystyle CH_3\overset{\displaystyle O}{\overset{\|}{C}}CH_3}{}}$$ $$\underset{\text{Aldehyde}}{CH_3CH_2\overset{\displaystyle O}{\overset{\|}{C}}H}$$ $$\underset{\text{Ether}}{\begin{array}{c} H_2C-CH_2 \\ |\qquad | \\ H_2C-O \end{array}}$$ $$\underset{\text{Ether}}{H_3C-HC\overset{O}{\diagdown\!\!\!\diagup}CH_2}$$

$$\underset{\text{Alkene, alcohol}}{CH_2{=}CHCH_2OH}$$ $$\underset{\text{Alkene, ether}}{CH_2{=}CH-O-CH_3}$$ $$\underset{\text{Alcohol}}{\begin{array}{c} H_2C \\ | \quad\diagdown \\ \quad\;\;CHOH \\ | \quad\diagup \\ H_2C \end{array}}$$

The ketone carbonyl absorption is in the 1680–1750 cm^{-1} range; that for the aldehyde is found in the 1690–1740 cm^{-1} region. The C—O absorption for the ethers is observed at about 1125 cm^{-1}. The C=C absorption occurs at approximately 1650 cm^{-1}. Absorption for the (hydrogen-bonded) O—H group takes the form of a broad band in the 3200–3550 cm^{-1} region.

2.51 (a)

$$\underset{A}{CH_3CH_2\overset{\displaystyle O}{\overset{\|}{C}}NH_2}\qquad \underset{B}{\underset{\overset{\displaystyle |}{H}}{CH_3\overset{\displaystyle O}{\overset{\|}{C}}NCH_3}}\qquad \underset{C}{\underset{\overset{\displaystyle |}{H}}{H\overset{\displaystyle O}{\overset{\|}{C}}NCH_2CH_3}}\qquad \underset{D}{\underset{\overset{\displaystyle |}{CH_3}}{H\overset{\displaystyle O}{\overset{\|}{C}}NCH_3}}$$

(b) **D**, because it does not have a hydrogen that is covalently bonded to nitrogen and, therefore, its molecules cannot form hydrogen bonds to each other. The other molecules all have a hydrogen covalently bonded to nitrogen and, therefore, hydrogen-bond formation is possible. With the first molecule, for example, hydrogen bonds could form in the following way:

$$CH_3CH_2\overset{\displaystyle \ddot{O}{:}\text{----}H-\overset{\displaystyle \overset{H}{|}}{N}{:}}{\underset{\underset{\overset{|}{H}}{\ddot{N}-H\text{----}{:}\ddot{O}}}{C}}CCH_2CH_3$$

(c) All four compounds have carbonyl group absorption at about 1650 cm^{-1}, but the IR spectrum for each has a unique feature.

 A shows two N—H bands (due to symmetrical and asymmetrical stretching) in the 3100–3400 cm^{-1} region.

 B has a single stretching absorption band in that same region since it has only a single N—H bond.

 C has two absorption bands, due to the H—C bond of the aldehyde group, at about 2820 cm^{-1} and 2920 cm^{-1}, as well as one for the N—H bond.

 D does not absorb in the 3100–3500 cm^{-1} region, as the other compounds do, since it does not possess a N—H bond.

2.52 The molecular formula requires unsaturation and/or one or more rings. The IR data exclude the functional groups: $-OH$, $\overset{O}{\underset{\|}{\diagdown C \diagup}}$ and $\diagup C = C \diagdown$. Oxygen (O) must be present in an ether linkage and there can be either a triple bond, a double bond and one ring, or two rings present to account for the low hydrogen-to-carbon ratio. These are the possible structures:

$$HC \equiv CCH_2OCH_3 \qquad HC \equiv COCH_2CH_3 \qquad CH_3C \equiv COCH_3$$

2.53 $-\overset{\displaystyle |}{C}-\overset{\displaystyle O}{\overset{\|}{C}}-O-\overset{\displaystyle |}{C}-$ (a cyclic ester)

with $-(CH_2)_n-$ bridging.

Challenge Problems

2.54

A   B   B′

The 1780 cm^{-1} band is in the general range for C=O stretching so structure B′ is considered one of the possible answers, but only B would have its C=O stretch at this high frequency (B′ would be at about 1730 cm^{-1}).

2.55 (a) HO⋯OH (H H) *cis*   HO⋯H (H OH) *trans*

(b) The cis isomer will have the 3572 cm^{-1} band because only in it are the two hydroxyl groups close enough to permit intramolecular hydrogen-bonding. (Intermolecular hydrogen-bonding is minimal at high dilution.)

2.56 OH, $-CH_3$, $-CH_3$

C

2.57 The helical structure results from hydrogen bonds formed between amide groups—specifically between the carbonyl group of one amide and the N—H group of another.

QUIZ

2.1 Which of the following pairs of compounds is *not* a pair of constitutional isomers?

(a)   and

(b)   and

(c)   and

(d) $CH_3CH_2C\equiv CH$ and $CH_3CH=C=CH_2$

(e) $CH_3CHCH(CH_3)_2$ and $(CH_3)_2CHCH(CH_3)_2$
    $\quad\quad |$
    $\quad CH_3$

2.2 Which of the answers to Problem 2.1 contains an ether group?

2.3 Which of the following pairs of structures represents a pair of isomers?

(a) [structure] and [structure]

(b) [structure] and [structure]

$$
\begin{array}{c}
\qquad\quad CH_3 \\
\qquad\quad | \\
\end{array}
$$

(c) $CH_3CH_2CHCH_2CH_3$ and

$$
\begin{array}{c}
CH_3CH_2CHCH_3 \\
| \\
CH_2CH_3
\end{array}
$$

(d) More than one of these pairs are isomers.

2.4 Give a bond-line formula for each of the following:

(a) A tertiary alcohol with the formula $C_5H_{12}O$

(b) An N,N-disubstituted amide with the formula C_4H_9NO

(c) The alkene isomer of $C_2H_2Cl_2$ that has no dipole moment

(d) An ester with the formula $C_2H_4O_2$

(e) The isomer of $C_2H_2Cl_2$ that cannot show cis-trans isomerism

(f) The isomer of C_3H_8O that would have the lowest boiling point

(g) The isomer of $C_4H_{11}N$ that would
    have the lowest boiling point

2.5 Write the bond-line formula for a constitutional isomer of the compound shown below
that does not contain a double bond.

CH_3CH_2CH=CH_2

2.6 Circle the compound in each pair that would have the higher boiling point.

(a) OH or

(b) N—H or N—CH₃

(c) or OH

(d) CH₃ OH or CH₃ CH₃

(e) CH₃ or CH₃

2.7 Give an acceptable name for each of the following:

(a) C_6H_5

(b) N CH₃ C_6H_5

(c) —NH₂

3 ACIDS AND BASES: AN INTRODUCTION TO ORGANIC REACTIONS AND THEIR MECHANISMS

SOLUTIONS TO PROBLEMS

3.1

(a)

(b)

3.2 (a)

(b)

(c)

3.3 (a) Lewis base

(b) Lewis acid

(c) Lewis base

(d) Lewis base

(e) Lewis acid

(f) Lewis base

34

3.4

$$CH_3 - \overset{H}{\underset{CH_3}{\overset{|}{N:}}} + \quad \overset{\ddot{F}}{\underset{\ddot{F}}{\overset{|}{B}}} - \ddot{F}: \longrightarrow CH_3 - \overset{H}{\underset{CH_3}{\overset{|}{\overset{+}{N}}}} - \overset{:\ddot{F}:}{\underset{:F:}{\overset{|}{\overset{-}{B}}}} - \ddot{F}:$$

Lewis base   Lewis acid

3.5 (a) $K_a = \dfrac{[H_3O^+][HCO_2^-]}{[HCO_2H]} = 1.77 \times 10^{-4}$

Let $x = [H_3O^+] = [HCO_2^-]$ at equilibrium

then, $0.1 - x = [HCO_2H]$ at equilibrium

but, since the K_a is very small, x will be very small and $0.1 - x \simeq 0.1$

Therefore,

$$\frac{(x)(x)}{0.1} = 1.77 \times 10^{-4}$$

$$x^2 = 1.77 \times 10^{-5}$$

$$x = 0.0042 = [H_3O^+] = [HCO_2^-]$$

(b) % Ionized $= \dfrac{[H_3O^+]}{0.1} \times 100$ or $\dfrac{[HCO_2^-]}{0.1} \times 100$

$$= \frac{.0042}{0.1} \times 100 = 4.2\%$$

3.6 (a) $pK_a = -\log 10^{-7} = -(-7) = 7$

(b) $pK_a = -\log 5.0 = -0.7$

(c) Since the acid with a $K_a = 5$ has a larger K_a, it is the stronger acid.

3.7 One way to tackle this problem is to compare the pK_a values for the protonated species of each of the choices (i.e. the molecule that would be formed if each behaved as a base). The resultant compound with the higher pK_a of the two is the weaker acid; hence, its conjugate base (the original choice) is a stronger base.

(a) (c) (b) (d)

3.8 The pK_a of the methylaminium ion is equal to 10.6 (Section 3.6C). Since the pK_a of the anilinium ion is equal to 4.6, the anilinium ion is a stronger acid than the methylaminium ion, and aniline ($C_6H_5NH_2$) is a weaker base than methylamine (CH_3NH_2).

3.9

3.10

3.11 (a) Negative. Because the atoms are constrained to one molecule in the product, they have to become more ordered.

(b) Approximately zero.

(c) Positive. Because the atoms are in two separate product molecules, they become more disordered.

3.12 (a) If $K_{eq} = 1$

then,

$$\log K_{eq} = 0 = \frac{-\Delta G°}{2.303RT}$$
$$\Delta G° = 0$$

(b) If $K_{eq} = 10$

then,

$$\log K_{eq} = 1 = \frac{-\Delta G°}{2.303RT}$$

$\Delta G° = -(2.303)(0.008314 \text{ kJ mol}^{-1} \text{ K}^{-1})(298 \text{ K}) = -5.71 \text{ kJ mol}^{-1}$

(c) $\Delta G° = \Delta H° - T\Delta S°$

$\Delta G° = \Delta H° = -5.71 \text{ kJ mol}^{-1}$ if $\Delta S° = 0$

3.13 Structures **A** and **B** make equal contributions to the overall hybrid. This means that the carbon-oxygen bonds should be the same length and that the oxygens should bear equal negative charges.

3.14 (a) $CHCl_2CO_2H$ would be the stronger acid because the electron-withdrawing inductive effect of *two chlorine atoms* would make its hydroxyl proton more positive. The electron-withdrawing effect of the two chlorine atoms would also stabilize the dichloroacetate ion more effectively by dispersing its negative charge more extensively.

(b) CCl_3CO_2H would be the stronger acid for reasons similar to those given in (a), except here there are three versus two electron-withdrawing chlorine atoms involved.

(c) CH_2FCO_2H would be the stronger acid because the electron-withdrawing effect of a fluorine atom is greater than that of a bromine atom (fluorine is more electronegative).

(d) CH_2FCO_2H is the stronger acid because the fluorine atom is nearer the carboxyl group and is, therefore, better able to exert its electron-withdrawing inductive effect.

(*Remember:* Inductive effects weaken steadily as the distance between the substituent and the acidic group increases.)

(a)   (b)   (c) and (d)

3.15 The inductive effect of an electron-withdrawing group such as a nitro group changes the charge distribution in the molecule so as to decrease the electron density of the ring and the oxygen, causing the proton to be held less strongly; it can also stabilize the phenoxide by dispersing its negative charge. Moreover, through resonance, a positive charge can be placed on the hydroxyl oxygen. These collective effects make the proton to be held much less strongly than in phenol, and hence 4-nitrophenol is more acidic.

Inductive contribution: *Resonance contribution:*

As shown below in resonance structures **A-D** for 2,4,6-trinitrophenol (picric acid), these effects are more pronounced since there are now three nitro groups that can make the hydroxyl oxygen dramatically more positive, and its proton is much more acidic as a result.

A   B

C   D

3.16 Dissolve the mixture in a solvent such as CH_2Cl_2 (one that is immiscible with water). Using a separatory funnel, extract this solution with an aqueous solution of sodium bicarbonate. This extraction will remove the benzoic acid from the CH_2Cl_2 and transfer it (as sodium benzoate) to the aqueous bicarbonate solution. Acidification of this aqueous extract will cause benzoic acid to precipitate; it can then be separated by filtration and purified by recrystallization.

The CH_2Cl_2 solution can now be extracted with an aqueous solution of sodium hydroxide. This will remove the 4-methylphenol (as its sodium salt). Acidification of the aqueous extract will cause the formation of 4-methylphenol as a water-insoluble layer. The 4-methylphenol can then be extracted into ether, the ether removed, and the 4-methylphenol purified by distillation.

The CH_2Cl_2 solution will now contain only toluene and (CH_2Cl_2). These can be separated easily by fractional distillation.

3.17 All compounds containing oxygen and most compounds containing nitrogen will have an unshared electron pair on their oxygen or nitrogen atom. These compounds can, therefore, act as bases and accept a proton from concentrated sulfuric acid. When they accept a proton, these compounds become either oxonium ions or ammonium ions, and having become ionic, they are soluble in the polar medium of sulfuric acid. The only nitrogen compounds that do not have an electron pair on their nitrogen atom are quaternary ammonium compounds, and these, already being ionic, also dissolve in the polar medium of concentrated sulfuric acid.

3.18 (a)

$$CH_3\ddot{O}-H \quad + \quad {}^-:H \quad \xrightarrow{\text{methanol}} \quad CH_3\ddot{O}:^- \quad + \quad H_2$$

Stronger acid   Stronger base   Weaker   Weaker acid
$pK_a = 16$ (from NaH)   base $pK_a = 35$

(b)

$$CH_3CH_2\ddot{O}-H \quad + \quad {}^-:\ddot{N}H_2 \quad \xrightarrow{\text{ethanol}} \quad CH_3CH_2\ddot{O}:^- \quad + \quad :NH_3$$

Stronger acid   Stronger base   Weaker   Weaker acid
$pK_a = 16$ (from NaNH$_2$)   base $pK_a = 38$

(c)

$$H-\underset{\underset{H}{|}}{\ddot{N}}-H \quad + \quad {}^-:CH_2CH_3 \quad \xrightarrow{\text{hexane}} \quad {}^-:\ddot{N}H_2 \quad + \quad CH_3CH_3$$

Stronger acid   Stronger base   Weaker   Weaker acid
$pK_a = 38$ (from CH$_3$CH$_2$Li)   base $pK_a = 50$

(d)

$$H-\overset{\overset{H}{|}}{\underset{\underset{H}{|}}{\overset{+}{N}}}-H \quad + \quad {}^-:\ddot{N}H_2 \quad \xrightarrow{\text{liq. NH}_3} \quad :NH_3 \quad + \quad :NH_3$$

Stronger acid   Stronger base   Weaker   Weaker acid
$pK_a = 9.2$ (from NaNH$_2$)   base $pK_a = 38$
(from NH$_4$Cl)

(e) H—Ö—H + ⁻:ÖC(CH₃)₃ $\xrightarrow{\text{H}_2\text{O}}$ H—Ö:⁻ + HOC(CH₃)₃

Stronger acid   Stronger base   Weaker   Weaker acid
$pK_a = 15.7$ [from (CH₃)₃CONa]   base $pK_a = 18$

(f) No appreciable acid-base reaction would occur because HO⁻ is not a strong enough base to remove a proton from (CH₃)₃COH.

3.19 (a) HC≡CH + NaH $\xrightarrow{\text{hexane}}$ HC≡CNa + H₂

(b) HC≡CNa + D₂O $\longrightarrow$ HC≡CD + NaOD

(c) CH₃CH₂Li + D₂O $\xrightarrow{\text{hexane}}$ CH₃CH₂D + LiOD

(d) CH₃CH₂OH + NaH $\xrightarrow{\text{hexane}}$ CH₃CH₂ONa + H₂

(e) CH₃CH₂ONa + T₂O $\longrightarrow$ CH₃CH₂OT + NaOT

(f) CH₃CH₂CH₂Li + D₂O $\xrightarrow{\text{hexane}}$ CH₃CH₂CH₂D + LiOD

Problems

Brønsted-Lowry Acids and Bases

3.20 (a) ⁻:ṄH₂ (the amide ion)   (d) H—C≡C:⁻ (the ethynide ion)

(b) H—Ö:⁻ (the hydroxide ion)   (e) CH₃Ö:⁻ (the methoxide ion)

(c) :H⁻ (the hydride ion)   (f) H₂Ö (water)

3.21 ⁻:ṄH₂ > :H⁻ > H—C≡C:⁻ > CH₃Ö:⁻ ≈ H—Ö:⁻ > H₂O

3.22 (a) H₂SO₄   (d) NH₃

(b) H₃O⁺   (e) CH₃CH₃

(c) CH₃N⁺H₃   (f) CH₃CO₂H

3.23 H₂SO₄ > H₃O⁺ > CH₃CO₂H > CH₃N⁺H₃ > NH₃ > CH₃CH₃

3.24

OH   OH   OH   O₂C—OH   O₂C—OH

OCH₃   NO₂   CF₃   CF₃   5 = most acidic

2   1   3   5   4

3.25 (a)

Fluorine is electron withdrawing; a methyl (alkyl) group is not

(b)

Nitro is electron withdrawing; a methyl (alkyl) group is not

(c)

Only this nitro group can exert electron withdrawing though resonance

(d)

This compound is a phenol; the other is an alcohol

(e)

Fluorine is more electron withdrawing than bromine

Lewis Acids and Bases

3.26 (a)

$$CH_3CH_2 \text{—} \ddot{\underset{\cdot\cdot}{C}l}: + AlCl_3 \longrightarrow CH_3CH_2 \text{—} \overset{+}{\underset{\cdot\cdot}{C}l} \text{—} \overset{\overset{\displaystyle :\ddot{C}l:}{|}}{\underset{\underset{\displaystyle :\ddot{C}l:}{|}}{Al}} \text{—} \ddot{\underset{\cdot\cdot}{C}l}:$$

Lewis base   Lewis acid

(b)

$$CH_3 \text{—} \ddot{O}H + BF_3 \longrightarrow CH_3 \text{—} \overset{+}{\underset{\underset{\displaystyle H}{|}}{O}} \text{—} \overset{\overset{\displaystyle :\ddot{F}:}{|}}{\underset{\underset{\displaystyle :\ddot{F}:}{|}}{B}} \text{—} \ddot{\underset{\cdot\cdot}{F}}:$$

Lewis base   Lewis acid

(c)

$$CH_3 \text{—} \overset{\overset{\displaystyle CH_3}{|}}{\underset{\underset{\displaystyle CH_3}{|}}{\overset{+}{C}}} + H_2\ddot{O}: \longrightarrow CH_3 \text{—} \overset{\overset{\displaystyle CH_3}{|}}{\underset{\underset{\displaystyle CH_3}{|}}{C}} \text{—} \overset{+}{O}H_2$$

Lewis acid   Lewis base

Curved-Arrow Notation

3.27 (a) $CH_3-\overset{..}{\underset{..}{O}}H$ + $H-\overset{..}{\underset{..}{I}}:$ $\longrightarrow$ $CH_3-\overset{+}{\overset{..}{O}}-H$ + $:\overset{..}{\underset{..}{I}}:^-$
$\qquad\qquad\qquad\qquad\qquad\qquad\qquad\qquad\qquad\quad\overset{|}{H}$

(b) $CH_3-\overset{..}{N}H_2$ + $H-\overset{..}{\underset{..}{Cl}}:$ $\longrightarrow$ $CH_3-\overset{\overset{H}{|}}{\underset{\underset{H}{|}}{\overset{+}{N}}}-H$ + $:\overset{..}{\underset{..}{Cl}}:^-$

(c)

3.28 (a)

(b)

(c)

(d)

3.29 (a) $CH_3CH_2-\overset{\overset{O}{||}}{C}-\overset{..}{\underset{..}{O}}H$ + $^-:\overset{..}{\underset{..}{O}}-H$ $\longrightarrow$ $CH_3CH_2-\overset{\overset{O}{||}}{C}-\overset{..}{\underset{..}{O}}:^-$ + $H-\overset{..}{\underset{..}{O}}-H$

(b) $C_6H_5-\overset{\overset{\overset{..}{O}{:}}{||}}{\underset{\underset{:\overset{..}{O}:}{||}}{S}}-\overset{..}{\underset{..}{O}}H$ + $^-:\overset{..}{\underset{..}{O}}-H$ $\longrightarrow$ $C_6H_5-\overset{\overset{\overset{..}{O}{:}}{||}}{\underset{\underset{.\overset{..}{O}.}{||}}{S}}-\overset{..}{\underset{..}{O}}:^-$ + $H-\overset{..}{\underset{..}{O}}-H$

(c) No appreciable acid-base reaction takes place because CH_3CH_2ONa is too weak a base to remove a proton from ethyne.

(d) $H-C\equiv C-H$ + $^-:CH_2CH_3$ $\xrightarrow{\text{hexane}}$ $H-C\equiv C:^-$ + CH_3CH_3
$\qquad\qquad\qquad\quad$(from
$\qquad\qquad\quad$ $LiCH_2CH_3$)

(e) $CH_3-CH_2-\overset{..}{\underset{..}{O}}-H$ + $^-:CH_2CH_3$ $\xrightarrow{\text{hexane}}$ $CH_3-CH_2-\overset{..}{\underset{..}{O}}:^-$ + CH_3CH_3
(from
$LiCH_2CH_3$)

Acid-Base Strength and Equilibrium

3.30 $CH_3CH_2-\overset{..}{\underset{..}{O}}-H$ + $^-:C\equiv C-H$ $\xrightarrow{\text{liq. NH}_3}$ $CH_3CH_2-\overset{..}{\underset{..}{O}}:^-$ + $H-C\equiv C-H$

3.31 (a) $pK_a = -\log 1.77 \times 10^{-4} = 4 - 0.248 = 3.75$

(b) $K_a = 10^{-13}$

3.32 (a) HB is the stronger acid because it has the smaller pK_a.

(b) Yes. Since A^- is the stronger base and HB is the stronger acid, the following acid-base reaction will take place.

$$A:^- \quad + \quad H-B \quad \longrightarrow \quad A-H \quad + \quad B:^-$$

Stronger	Stronger	Weaker	Weaker
base	acid	acid	base
	$pK_a = 10$	$pK_a = 20$	

3.33 (a) $C_6H_5-C\equiv C-H$ + $NaNH_2$ $\xrightarrow{\text{ether}}$ $C_6H_5-C\equiv C:^- Na^+$ + NH_3
then
$C_6H_5-C\equiv C:^- Na^+$ + T_2O $\longrightarrow$ $C_6H_5-C\equiv C-T$ + $NaOT$

(b) $CH_3-\underset{\underset{CH_3}{|}}{CH}-O-H$ + NaH $\longrightarrow$ $CH_3-\underset{\underset{CH_3}{|}}{CH}-O^- Na^+$ + H_2
then
$CH_3-\underset{\underset{CH_3}{|}}{CH}-O^- Na^+$ + D_2O $\longrightarrow$ $CH_3-\underset{\underset{CH_3}{|}}{CH}-O-D$ + $NaOD$

(c) $CH_3CH_2CH_2OH$ + NaH $\longrightarrow$ $CH_3CH_2CH_2O^- Na^+$ + H_2
then
$CH_3CH_2CH_2O^- Na^+$ + D_2O $\longrightarrow$ $CH_3CH_2CH_2OD$ + $NaOD$

3.34 (a) $CH_3CH_2OH > CH_3CH_2NH_2 > CH_3CH_2CH_3$

Oxygen is more electronegative than nitrogen, which is more electronegative than carbon. The O-H bond is most polarized, the N-H bond is next, and the C-H bond is least polarized.

(b) $CH_3CH_2O^- < CH_3CH_2\bar{N}H < CH_3CH_2\bar{C}H_2$

The weaker the acid, the stronger the conjugate base.

3.35 (a) $CH_3C{\equiv}CH > CH_3CH{=}CH_2 > CH_3CH_2CH_3$

(b) $CH_3CHClCO_2H > CH_3CH_2CO_2H > CH_3CH_2CH_2OH$

(c) $CH_3CH_2\overset{+}{O}H_2 > CH_3CH_2OH > CH_3OCH_3$

3.36 (a) $CH_3\overset{+}{N}H_3 < CH_3NH_2 < CH_3\bar{N}H$

(b) $CH_3O^- < CH_3\bar{N}H < CH_3\bar{C}H_2$

(c) $CH_3C{\equiv}C^- < CH_3CH{=}\bar{C}H < CH_3CH_2\bar{C}H_2$

General Problems

3.37 The acidic hydrogens must be attached to oxygen atoms. In H_3PO_3, one hydrogen is bonded to a phosphorus atom:

3.38 (a)

(b)

(c)

(d) $H—\overset{..}{\underset{..}{O}}:^- + \ ^+CH_3—\overset{..}{\underset{..}{I}}: \quad \longrightarrow \quad H—\overset{..}{\underset{..}{O}}—CH_3 + :\overset{..}{\underset{..}{I}}:^-$

(e) $H—\overset{..}{\underset{..}{O}}:^- + \ ^+H—CH_2—\overset{\overset{\textstyle CH_3}{|}}{\underset{\underset{\textstyle CH_3}{|}}{C}}—\overset{..}{\underset{..}{Cl}}: \quad \longrightarrow \quad CH_2=\overset{\overset{\textstyle CH_3}{\diagup}}{\underset{\underset{\textstyle CH_3}{\diagdown}}{C}} \ + :\overset{..}{\underset{..}{Cl}}:^- + H—\overset{..}{\underset{..}{O}}—H$

3.39 (a) Assume that the acidic and basic groups of glycine in its two forms have acidities and basicities similar to those of acetic acid and methylamine. Then consider the equilibrium between the two forms:

We see that the ionic form contains the groups that are the weaker acid and weaker base. The equilibrium, therefore, will favor this form.

(b) The high melting point shows that the ionic structure better represents glycine.

3.40 (a) The second carboxyl group of malonic acid acts as an electron-withdrawing group and stabilizes the conjugate base formed (i.e., $HO_2CCH_2CO_2{}^-$) when malonic acid loses a proton. [Any factor that stabilizes the conjugate base of an acid always increases the strength of the acid (Section 3.11C).] An important factor here may be an entropy effect as explained in Section 3.10.

(b) When $^-O_2CCH_2CO_2H$ loses a proton, it forms a dianion, $^-O_2CCH_2CO_2{}^-$. This dianion is destabilized by having two negative charges in close proximity.

3.41 HB is the stronger acid.

3.42 $\Delta G° = \Delta H° - T\Delta S°$
$= 6.3 \text{ kJ mol}^{-1} - (298 \text{ K})(0.0084 \text{ kJ mol}^{-1}\text{K}^{-1})$
$= 3.8 \text{ kJ mol}^{-1}$

$\log K_{eq} = \log K_a = -pK_a = -\dfrac{\Delta G°}{2.303RT}$

$pK_a = \dfrac{\Delta G°}{2.303RT}$

$pK_a = \dfrac{3.8 \text{ kJ mol}^{-1}}{(2.303)(0.008314 \text{ kJ mol}^{-1}\text{K}^{-1})(298 \text{ K})}$

$pK_a = 0.66$

3.43 The dianion is a hybrid of the following resonance structures:

If we mentally fashion a hybrid of these structures, we see that each carbon-carbon bond is a single bond in three structures and a double bond in one. Each carbon-oxygen bond is a double bond in two structures and a single bond in two structures. Therefore, we would expect all of the carbon-carbon bonds to be equivalent and of the same length, and exactly the same can be said for the carbon-oxygen bonds.

Challenge Problems

3.44 (a) *A* is $CH_3CH_2S^-$ *B* is CH_3OH

       C is $CH_3CH_2SCH_2CH_2O^-$ *D* is $CH_3CH_2SCH_2CH_2OH$

       E is HO^-

(b)

3.45 (a) $CH_3(CH_2)_8OD + CH_3(CH_2)_8Li \longrightarrow CH_3(CH_2)_8O^- Li^+ + CH_3(CH_2)_8D$

Hexane could be used as solvent. Liquid ammonia and ethanol could not because they would compete with $CH_3(CH_2)_8OD$ and generate mostly non-deuterio-labelled $CH_3(CH_2)_7CH_3$.

(b) $\bar{N}H_2 + CH_3C{\equiv}CH \longrightarrow NH_3 + CH_3C{\equiv}C{:}^-$

Hexane or liquid ammonia could be used; ethanol is too acidic and would lead to $CH_3CH_2O^-$ (ethoxide ion) instead of the desired alkynide ion.

(c)

Hexane or ethanol could be used; liquid ammonia is too strong a base and would lead to $^+NH_4$ instead of the desired anilinium ion.

3.46 (a,b)

The uncharged structure on the left is the more important resonance form.

(c) Since DMF does not bind with (solvate) anions, their electron density remains high and their size small, both of which make nucleophiles more reactive.

3.47 (a)

(b)

(c)

3.48 The most acidic hydrogen atoms in formamide are bonded to the nitrogen atom. They are acidic due to the electron-withdrawing effect of the carbonyl group and the fact that the resulting conjugate base can be stabilized by resonance delocalization of the negative charge into the carbonyl group. The electrostatic potential map shows deep blue color near the hydrogen atoms bonded to the nitrogen atom, consistent with their relative acidity.

QUIZ

3.1 Which of the following is the strongest acid?

(a) $CH_3CH_2CO_2H$ (b) CH_3CH_3 (c) CH_3CH_2OH (d) $CH_2{=}CH_2$

3.2 Which of the following is the strongest base?

(a) CH_3ONa (b) $NaNH_2$ (c) CH_3CH_2Li (d) $NaOH$ (e) CH_3CO_2Na

3.3 Dissolving $NaNH_2$ in water will give:

(a) A solution containing solvated Na^+ and $^-NH_2$ ions.

(b) A solution containing solvated Na^+ ions, HO^- ions, and NH_3.

(c) NH_3 and metallic Na.

(d) Solvated Na^+ ions and hydrogen gas.

(e) None of the above.

3.4 Which base is strong enough to convert $(CH_3)_3COH$ into $(CH_3)_3CONa$ in a reaction that goes to completion?

(a) $NaNH_2$ (b) CH_3CH_2Na (c) $NaOH$ (d) CH_3CO_2Na

(e) More than one of the above.

3.5 Which would be the strongest acid?

(a) $CH_3CH_2CH_2CO_2H$ (b) $CH_3CH_2CHFCO_2H$ (c) $CH_3CHFCH_2CO_2H$

(d) $CH_2FCH_2CH_2CO_2H$ (e) $CH_3CH_2CH_2CH_2OH$

3.6 Which would be the weakest base?

(a) CH_3CO_2Na (b) CF_3CO_2Na (c) CHF_2CO_2Na (d) CH_2FCO_2Na

3.7 What acid-base reaction (if any) would occur when NaF is dissolved in H_2SO_4?

3.8 The pK_a of $CH_3\overset{+}{N}H_3$ equals 10.6; the pK_a of $(CH_3)_2\overset{+}{N}H_2$ equals 10.7. Which is the stronger base, CH_3NH_2 or $(CH_3)_2NH$?

3.9 Supply the missing reagents.

(a)

$$CH_3CH_2C\equiv CH + \boxed{} \xrightarrow{\text{hexane}} CH_3CH_2C\equiv C{:}^- \; Li^+ + CH_3CH_3$$

(b)

$$\boxed{}$$

$$\downarrow$$

$$CH_3CH_2C\equiv CD + LiOD$$

3.10 Supply the missing intermediates and reagents.

(a)

$$CH_3Br + 2\,Li \longrightarrow \boxed{} + LiBr$$

(b)

$$\boxed{}$$

(c)

$$\underset{\displaystyle CH_3\!\overset{\textstyle |}{}}{CH_3CHCH_2OT} + LiOT \xleftarrow{\;T_2O\;} \boxed{}$$

4 NOMENCLATURE AND CONFORMATIONS OF ALKANES AND CYCLOALKANES

SOLUTIONS TO PROBLEMS

4.1

or
$CH_3(CH_2)_5CH_3$
Heptane

or
$(CH_3)_2CHCH_2CH_2CH_2CH_3$
2-Methylhexane

or
$CH_3CH_2CH(CH_3)CH_2CH_2CH_3$
3-Methylhexane

or
$(CH_3)_3CCH_2CH_2CH_3$
2,2-Dimethylpentane

or
$(CH_3CH_2)_2C(CH_3)_2$
3,3-Dimethylpentane

or
$(CH_3)_2CHCH(CH_3)CH_2CH_3$
2,3-Dimethylpentane

or
$(CH_3)_2CHCH_2CH(CH_3)_2$
2,4-Dimethylpentane

or
$(CH_3CH_2)_3CH$
3-Ethylpentane

or
$(CH_3)_3CCH(CH_3)_2$
2,2,3-Trimethylbutane

4.2   (d); it represents 3-methylpentane

4.3   CH$_3$CH$_2$CH$_2$CH$_2$CH$_2$CH$_3$ hexane

4.4   (a,b)

2-Methylheptane

3-Methylheptane

4-Methylheptane

2,2-Dimethylhexane   2,3-Dimethylhexane   2,4-Dimethylhexane

2,5-Dimethylhexane   3,3-Dimethylhexane   3,4-Dimethylhexane

3-Ethylhexane

2,2,3-Trimethylpentane   2,2,4-Trimethylpentane

2,3,3-Trimethylpentane   2,3,4-Trimethylpentane

3-Ethyl-2-methylpentane   3-Ethyl-3-methylpentane

2,2,3,3-Tetramethylbutane

4.5 (a)

Pentyl   1-Methylbutyl   1-Ethylpropyl

2-Methylbutyl   3-Methylbutyl   1,2-Dimethylpropyl

1,1-Dimethylpropyl

(b) See the answer to Review Problem 4.1 for the formulas and names of C_7H_{16} isomers.

4.6 (a)

Cl
1-Chlorobutane

Cl
1-Chloro-2-methylpropane

Cl
2-Chlorobutane

Cl
2-Chloro-2-methylpropane

(b)

Br
1-Bromopentane

Br
1-Bromo-3-methylbutane

Br
2-Bromopentane

Br
1-Bromo-2-methylbutane

Br
3-Bromopentane

Br
2-Bromo-3-methylbutane

Br
1-Bromo-2,2-dimethylpropane

Br
2-Bromo-2-methylbutane

4.7 (a)

1-Butanol

2-Methyl-1-propanol

2-Butanol

2-Methyl-2-propanol

(b)

1-Pentanol

3-Methyl-1-butanol

2-Pentanol

2-Methyl-1-butanol

3-Pentanol

3-Methyl-2-butanol

2,2-Dimethyl-1-propanol

2-Methyl-2-butanol

4.8 (a) 1,1-Dimethylethylcyclopentane or
tert-butylcyclopentane

(b) 1-Methyl-2-(2-methylpropyl)cyclohexane or
1-isobutyl-2-methylcyclohexane

(c) Butylcyclohexane

(d) 1-Chloro-2,4-dimethylcyclohexane

(e) 2-Chlorocyclopentanol

(f) 3-(1,1-Dimethylethyl)cyclohexanol or 3-*tert*-butylcyclohexanol

4.9 (a) 2-Chlorobicyclo[1.1.0]butane

(b) Bicyclo[3.2.1]octane

(c) Bicyclo[2.1.1]hexane

(d) 9-Chlorobicyclo[3.3.1]nonane

(e) 2-Methylbicyclo[2.2.2]octane

(f)

Bicyclo[3.1.0]hexane or

bicyclo[2.1.1]hexane

4.10 (a) *trans*-3-Heptene

(b) 2,5-Dimethyl-2-octene

(c) 4-Ethyl-2-methyl-l-hexene

(d) 3,5-Dimethylcyclohexene

(e) 4-Methyl-4-penten-2-ol

(f) 2-Chloro-3-methylcyclohex-3-en-1-ol

4.11 (a) (b) (c)

(d) (e) (f)

(g) (h) (i)

(j)

4.12

1-Hexyne 2-Hexyne 3-Hexyne 3,3-Dimethyl-1-butyne

3-Methyl-1-pentyne 4-Methyl-1-pentyne 4-Methyl-2-pentyne

4.13

4.14 $\log K_{eq} = \dfrac{\Delta G^\circ}{-2.303RT} = \dfrac{-7600\ \text{J}}{(-2.303)(8.314\ \text{J K}^{-1})(298\ \text{K})} = 1.32$

$K_{eq} = 21.38$

Let e = amount of equatorial form

and a = amount of axial form

then, $K_{eq} = \dfrac{e}{a} = 21.38$

$e = 21.38a$

$\% e = \dfrac{21.38a}{a + 21.38a} \times 100 = 95.5\%$

4.15 (a)

(c) No

4.16 (a-d)

More stable because
larger group is equatorial
and so preferred at equilibrium

Less stable because
larger group is axial

4.17

all equatorial
(more stable)

all axial
(less stable)

4.18 (a,b)

Less stable because the large
tert-butyl group is axial
(more potential energy)

More stable because the large
tert-butyl group is equatorial
(less potential energy)

4.19

$$\xrightarrow[\text{pressure}]{\substack{H_2 \\ \text{Pd, Pt or Ni}}}$$

$$\xleftarrow[\text{pressure}]{\substack{H_2 \\ \text{Pd, Pt or Ni}}}$$

$$\xrightarrow[\text{pressure}]{\substack{H_2 \\ \text{Pd, Pt or Ni}}}$$

$$\xrightarrow[\text{pressure}]{\substack{H_2 \\ \text{Pd, Pt or Ni}}}$$

4.20 (a) C_6H_{14} = formula of alkane
$\underline{C_6H_{12}}$ = formula of 2-hexene
    H_2 = difference = 1 pair of hydrogen atoms
Index of hydrogen deficiency = 1

   (b) C_6H_{14} = formula of alkane
$\underline{C_6H_{12}}$ = formula of methylcyclopentane
    H_2 = difference = 1 pair of hydrogen atoms
Index of hydrogen deficiency = 1

   (c) No, all isomers of C_6H_{12}, for example, have the same index of hydrogen deficiency.

   (d) No

(e) C_6H_{14} = formula of alkane

$\underline{C_6H_{10}}$ = formula of 2-hexyne

H_4 = difference = 2 pairs of hydrogen atoms

Index of hydrogen deficiency = 2

(f) $C_{10}H_{22}$ (alkane)

$\underline{C_{10}H_{16}}$ (compound)

H_6 = difference = 3 pairs of hydrogen atoms

Index of hydrogen deficiency = 3

The structural possibilities are thus

3 double bonds

1 double bond and 1 triple bond

2 double bonds and 1 ring

1 double bond and 2 rings

3 rings

1 triple bond and 1 ring

4.21 (a) $C_{15}H_{32}$ = formula of alkane

$\underline{C_{15}H_{24}}$ = formula of zingiberene

H_8 = difference = 4 pairs of hydrogen atoms

Index of hydrogen deficiency = 4

(b) Since 1 mol of zingiberene absorbs 3 mol of hydrogen, one molecule of zingiberene must contain three double bonds. (We are told that molecules of zingiberene do not contain any triple bonds.)

(c) If a molecule of zingiberene has three double bonds and an index of hydrogen deficiency equal to 4, it must have one ring. (The structural formula for zingiberene can be found in Review Problem 23.2.)

4.22   molecular formula $C_{10}H_{16}O_2$

$C_{10}H_{22}$ = formula for alkane

$\underline{C_{10}H_{16}}$ = formula for unsaturated ester (ignoring oxygens)

H_6 = difference = 3 pairs of hydrogen atoms

IHD = 3

molecular formula $C_{10}H_{18}O_2$

$C_{10}H_{22}$ = formula for alkane

$\underline{C_{10}H_{18}}$ = formula for hydrogenation product (ignoring oxygens)

H_4 = difference = 2 pairs of hydrogen atoms

IHD = 2

Nomenclature and Isomerism

4.23   (a)

(b)

(c)

(d)

(e)

(f)

(g)

(h)

(i)

(j)

(k)

(l)

(m)   or

(n)

(o)

4.24   (a) 5-Ethyl-7-isopropyl-2,3-dimethyldecane

(b) 2,2,5-Trimethylhexane

(c) 4-Bromo-6-chloro-3-methyloctane

(d) 2-Methyl-1-butanol

(e) 2-Bromobicyclo[3.3.1]nonane

(f) *trans*-1-Bromo-3-fluorocyclopentane

(g) 5,6-Dimethyl-2-heptane

(h) 7-Chlorobicyclo[2.2.1]heptane

4.25 The two secondary carbon atoms in *sec*-butyl alcohol are equivalent; however, there are three five-carbon alcohols (pentyl alcohols) that contain a secondary carbon atom.

4.26 (a)

H$_3$C—C—C—CH$_3$

2,2,3,3-Tetramethylbutane

(b) Cyclohexane

(c) 1,1-Dimethylcyclobutane

(d) Bicyclo[2.2.2]octane

4.27

CH$_3$—C—CH$_2$CHCH$_3$ or   2,2,4-trimethylpentane

CH$_3$CH$_2$—C—CHCH$_3$ or   2,3,3-trimethylpentane

CH$_3$—C—CHCH$_2$CH$_3$ or   2,2,3-trimethylpentane

4.28

Cyclopentane   Methylcyclobutane   *cis*-1,2-Dimethylcyclopropane

trans-1,2-Dimethylcyclopropane   1,1-Dimethylcyclopropane   Ethylcyclopropane

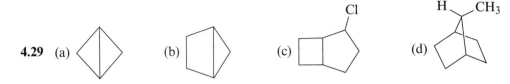

4.29 (a)   (b)   (c)   (d)

4.30 S – A + 1 = N
For cubane, S = 12 and A = 8. Thus 12 – 8 + 1 = N; N = 5 rings in cubane.

4.31 (a)

(b)

4.32 A homologous series is one in which each member of the series differs from the one preceding it by a constant amount, usually a CH_2 group. A homologous series of alkyl halides would be the following:

$$CH_3X$$
$$CH_3CH_2X$$
$$CH_3(CH_2)_2X$$
$$CH_3(CH_2)_3X$$
$$CH_3(CH_2)_4X$$
etc.

Hydrogenation

4.33

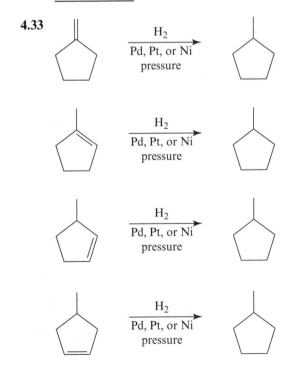

$$\xrightarrow[\substack{\text{Pd, Pt, or Ni} \\ \text{pressure}}]{\text{H}_2}$$

$$\xrightarrow[\substack{\text{Pd, Pt, or Ni} \\ \text{pressure}}]{\text{H}_2}$$

$$\xrightarrow[\substack{\text{Pd, Pt, or Ni} \\ \text{pressure}}]{\text{H}_2}$$

$$\xrightarrow[\substack{\text{Pd, Pt, or Ni} \\ \text{pressure}}]{\text{H}_2}$$

4.34 (a) Each of the desired alkenes must have the same carbon skeleton as 2-methylbutane,

$$C-C-C-C$$
with C above the second carbon; they are therefore

$$+ \ H_2 \quad \xrightarrow[\substack{C_2H_5OH \\ \text{pressure}}]{Pd, \ Pt, \ or \ Ni}$$

(b)

4.35

2,3-Dimethylbutane

From two alkenes

$$\xrightarrow[\substack{Pd, \ Pt, \ or \ Ni \\ \text{pressure}}]{H_2}$$

Conformations and Stability

4.36

Least stable   Most stable

4.37

This conformation is *less stable* because 1,3-diaxial interactions with the large *tert*-butyl group cause considerable repulsion.

This conformation is *more stable* because 1,3-diaxial interactions with the smaller methyl group are less repulsive.

4.38

4.39 (a) Pentane would boil higher because its chain is unbranched. Chain-branching lowers the boiling point.

(b) Heptane would boil higher because it has the larger molecular weight and would, because of its larger surface area, have larger van der Waals attractions.

(c) 2-Chloropropane because it is more polar and because it has a larger molecular weight.

(d) 1-Propanol would boil higher because its molecules would be associated with each other through hydrogen-bond formation.

(e) Propanone (CH_3COCH_3) would boil higher because its molecules are more polar.

4.40 C_4H_6

Bicyclo[1.1.0]butane   1-Butyne

The IR stretch at ~2250 cm^{-1} for the alkyne $C{\equiv}C$ bond distinguishes these two compounds.

4.41 *trans*-1,2-Dimethylcyclopropane would be more stable because there is less crowding between its methyl groups.

Less stable   More stable

4.42 For 1,2-disubstituted cyclobutanes, the trans isomer is *e,e*; the cis isomer is *a,e*, and so the less stable of the two. For 1,3-disubstituted cyclobutanes, the cis isomer is *e,e* and more stable than the *a,e* trans isomer.

Trans   Cis   Trans   Cis

4.43 (a)

More stable conformation because both alkyl groups are equatorial

(b)

More stable because larger group is equatorial

(c)

$(CH_3)_3C$ ⬡ CH_3 ⇌ $C(CH_3)_3$ ⬡ CH_3

More stable conformation because
both alkyl groups are equatorial

(d)

CH_3 $(CH_3)_3C$ ⬡ ⇌ $C(CH_3)_3$ ⬡ CH_3

More stable because larger group
is equatorial

4.44 Certainly it is expected that the alkyl groups would prefer the equatorial disposition in a case such as this, and indeed this is true in the case of *all-trans*-1,2,3,4,5,6-hexaethylcyclohexane, which does have all ethyl groups equatorial.

Apparently, the torsional and steric effects resulting from equatorial isopropyl groups destabilize the all-equatorial conformation to a greater degree than axial isopropyl groups destabilize the all-axial conformation. The fully axial structure assigned on the basis of X-ray studies is also supported by calculations.

4.45 If the cyclobutane ring were planar, the C—Br bond moments would exactly cancel in the trans isomer. The fact that *trans*-1,3-dibromocyclobutane has a dipole moment shows the ring is not planar.

Planar form   Bent form
$\mu = 0$ D $\mu \neq 0$ D

Synthesis

4.46 (a) H_2; Pd, Pt, or Ni catalyst, pressure

(b) H_2; Pd, Pt, or Ni catalyst, pressure

(c) Predicted IR absorption frequencies for reactants in parts (a) and (b) are the following:

(1) *trans*-5-Methyl-2-hexene has an absorption at approximately 964 cm^{-1}, characteristic of C—H bending in a trans-substituted alkene.

(2) The alkene double bond of the reactant is predicted to have a stretching absorption between 1580 and 1680 cm^{-1}.

Challenge Problems

4.47 If *trans*-1,3-di-*tert*-butylcyclohexane were to adopt a chair conformation, one *tert*-butyl group would have to be axial. It is, therefore, more energetically favorable for the molecule to adopt a twist boat conformation.

4.48 (a) More rules are needed (see Chapter 7) to indicate relative stereochemistry for these 1-bromo-2-chloro-1-fluoroethenes.

(b) Bicyclo[4.4.0]decane (or decalin)

(c) Bicyclo[4.4.0]dec-2-ene (or Δ^1-octalin)

(d) Bicyclo[4.4.0]dec-1-ene (or $\Delta^{1(8a)}$-octalin)
NOTE: The common name decalin comes from decahydronaphthalene, the derivative of naphthalene that has had all of its five double bonds converted to single bonds by addition of 10 atoms of hydrogen. Octalin similarly comes from octahydronaphthalene and contains one surviving double bond. When using these common names derived from naphthalene, their skeletons are usually numbered like that of naphthalene. When, as in case (d), a double bond does not lie between the indicated carbon and the next higher numbered carbon, its location is specified as shown.

Also, the symbol Δ is one that has been used with common names to indicate the presence of a double bond at the position specified by the accompanying superscript number(s).

4.49 The cyclohexane ring in *trans*-1-*tert*-butyl-3-methylcyclohexane is in a chair conformation. The ring in *trans*-1,3-di-*tert*-butylcyclohexane is in a twist-boat conformation. In *trans*-1-*tert*-butyl-3-methylcyclohexane the *tert*-butyl group can be equatorial if the methyl group is axial. The energy cost of having the methyl group axial is apparently less than that for adopting the twist-boat ring conformation. In *trans*-1,3-di-*tert*-butylcyclohexane the potential energy cost of having one *tert*-butyl group equatorial and the other axial is apparently greater than having the ring adopt a twist-boat conformation so that both can be approximately equatorial.

4.50 All of the nitrogen-containing five-membered rings of Vitamin B$_{12}$ contain at least one sp^2-hybridized atom (in some cases there are more). These atoms, because of the trigonal planar geometry associated with them, impose some degree of planarity on the nitrogen-containing five-membered rings in B$_{12}$. Furthermore, 13 atoms in sequence around the cobalt are sp^2-hybridized, a fact that adds substantial resonance stabilization to this part of the ring system. The four five-membered nitrogen-containing rings that surround the cobalt in roughly a plane and whose nitrogens lend their unshared pairs to the cobalt comprise what is called a corrin ring. The corrin ring may look familiar to you, and for good reason, because it is related to the porphyrin ring of heme. (Additional question: What geometry is associated with the cobalt atom and its six bound ligands? Answer: octahedral, or square bipyramidal.)

4.51 The form of a benzene ring occurs 16 times in buckminsterfullerene. The other eight facets in the 24 faces of a "buckyball" are five-membered rings. Every carbon of buckminster-fullerene is approximately sp^2-hybridized.

QUIZ

4.1 Consider the properties of the following compounds:

NAME	FORMULA	BOILING POINT (°C)	MOLECULAR WEIGHT
Ethane	CH_3CH_3	−88.2	30
Fluoromethane	CH_3F	−78.6	34
Methanol	CH_3OH	+64.7	32

Select the answer that explains why methanol boils so much higher than ethane or fluoromethane, even though they all have nearly equal molecular weights.

(a) Ion-ion forces between molecules.

(b) Weak dipole-dipole forces between molecules.

(c) Hydrogen bonding between molecules.

(d) van der Waals forces between molecules.

(e) Covalent bonding between molecules.

4.2 Select the correct name of the compound whose structure is

(a) 2,5-Diethyl-6-methyloctane

(b) 4,7-Diethyl-3-methyloctane

(c) 4-Ethyl-3,7-dimethylnonane

(d) 6-Ethyl-3,7-dimethylnonane

(e) More than one of the above

4.3 Select the correct name of the compound whose structure is $CH_3\overset{\overset{\displaystyle CH_3}{|}}{C}HCH_2Cl$.

(a) Butyl chloride

(b) Isobutyl chloride

(c) *sec*-Butyl chloride

(d) *tert*-Butyl chloride

(e) More than one of the above

4.4 The structure shown in Problem 4.2 has:

(a) 1°, 2°, and 3° carbon atoms

(b) 1° and 2° carbon atoms only

(c) 1° and 3° carbon atoms only

(d) 2° and 3° carbon atoms only

(e) 1°, 3°, and 4° carbon atoms only

4.5 How many isomers are possible for C_3H_7Br?

(a) 1   (b) 2   (c) 3   (d) 4   (e) 5

4.6 Which isomer of 1,3-dimethylcyclohexane is more stable?

(a) cis   (b) trans   (c) Both are equally stable   (d) Impossible to tell

4.7 Which is the lowest energy conformation of *trans*-1,4-dimethylcyclohexane?

(a)

(b)

(c)

(d)

(e) More than one of the above

4.8 Supply the missing structures

(a)

ring flip

(b)

2-Bromobicyclo[2.2.1]heptane

(c) Newman projection for a gauche form of 1,2-dibromoethane

4.9 Supply the missing reagents in the box:

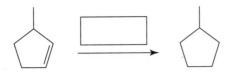

4.10 The most stable conformation of *trans*-1-isopropyl-3-methylcyclohexane:

5 STEREOCHEMISTRY: CHIRAL MOLECULES

SOLUTIONS TO PROBLEMS

5.1 (a) Achiral   (c) Chiral   (e) Chiral   (g) Chiral

      (b) Achiral   (d) Chiral   (f) Chiral   (h) Achiral

5.2 (a) Yes   (b) No   (c) No

5.3 (a) They are the same molecule.   (b) They are enantiomers.

5.4 (a), (b), (e), (g), and (i) do not have chirality centers.

(c)

$$
\begin{array}{c}
CH_3 \\
| \\
H{-}C{-}Cl \\
| \\
CH_2 \\
| \\
CH_3 \quad (1)
\end{array}
\qquad
\begin{array}{c}
CH_3 \\
| \\
Cl{-}C{-}H \\
| \\
CH_2 \\
| \\
CH_3 \quad (2)
\end{array}
$$

(d)

$$
\begin{array}{c}
CH_3 \\
| \\
H{-}C{-}CH_2OH \\
| \\
CH_2 \\
| \\
CH_3 \quad (1)
\end{array}
\qquad
\begin{array}{c}
CH_3 \\
| \\
HOCH_2{-}C{-}H \\
| \\
CH_2 \\
| \\
CH_3 \quad (2)
\end{array}
$$

(f)

$$
\begin{array}{c}
CH_3 \\
| \\
H{-}C{-}Br \\
| \\
CH_2 \\
| \\
CH_2 \\
| \\
CH_3 \quad (1)
\end{array}
\qquad
\begin{array}{c}
CH_3 \\
| \\
Br{-}C{-}H \\
| \\
CH_2 \\
| \\
CH_2 \\
| \\
CH_3 \quad (2)
\end{array}
$$

(h)

CH_3
|
CH_2
|
H—C—CH_3
|
CH_2
|
CH_2
|
CH_3 **(1)**

CH_3
|
CH_2
|
H_3C—C—H
|
CH_2
|
CH_2
|
CH_3 **(2)**

(j)

CH_3
|
H—C—CH_2Cl
|
CH_2
|
CH_3 **(1)**

CH_3
|
$ClCH_2$—C—H
|
CH_2
|
CH_3 **(2)**

5.5 (a)

Limonene

(R)

(S)

(b)

Thalidomide

(R)

(S)

5.6 (a)

(d)

(b)

(e)

(c)

(f)

5.7 The following items possess a plane of symmetry, and are, therefore, achiral.

(a) A screwdriver

(b) A baseball bat (ignoring any writing on it)

(h) A hammer

5.8 In each instance below, the plane defined by the page is a plane of symmetry.

(a)

(b)

(e)

(g)

(i)

5.9

(S)   (R)

5.10   (c) (1) is (S)
          (2) is (R)

      (d) (1) is (S)
          (2) is (R)

      (f) (1) is (S)
          (2) is (R)

      (h) (1) is (S)
          (2) is (R)

      (j) (1) is (S)
          (2) is (R)

5.11   (a) $-Cl$ > $-SH$ > $-OH$ > $-H$

      (b) $-CH_2Br$ > $-CH_2Cl$ > $-CH_2OH$ > $-CH_3$

      (c) $-OH$ > $-CHO$ > $-CH_3$ > $-H$

      (d) $-C(CH_3)_3$ > $-CH=CH_2$ > $-CH(CH_3)_2$ > $-H$

      (e) $-OCH_3$ > $-N(CH_3)_2$ > $-CH_3$ > $-H$

      (f) $-OPO_3H_2$ > $-OH$ > $-CHO$ > $-H$

5.12   (a) (S)   (b) (R)   (c) (S)   (d) (R)

5.13   (a) Enantiomers

      (b) Two molecules of the same compound

      (c) Enantiomers

5.14

(S)-(+)-Carvone   (R)-(−)-Carvone

5.15   (a)   Enant. excess $= \dfrac{\text{observed specific rotation}}{\text{specific rotation of pure enantiomer}} \times 100$

         $= \dfrac{+1.15}{+5.76} \times 100$

         $= 20.0\%$

      (b) Since the (R) enantiomer (see Section 5.8C) is +, the (R) enantiomer is present in excess.

5.16 (a)

(S)-Methyldopa

(b)

(S)-Penicillamine

(c)

(S)-Ibuprofen

5.17

(R) configuration

5.18

5.19 (a) Diastereomers.

(b) Diastereomers in each instance.

(c) No, diastereomers have different melting points.

(d) No, diastereomers have different boiling points.

(e) No, diastereomers have different vapor pressures.

STUDY AID

An Approach to the Classification of Isomers

We can classify isomers by asking and answering a series of questions:

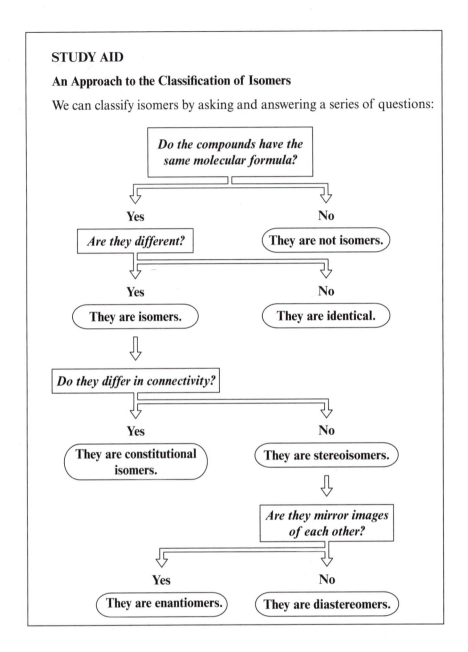

5.20 (a) It would be optically active.

(b) It would be optically active.

(c) No, because it is a meso compound.

(d) No, because it would be a racemic form.

5.21 (a) Represents **A**

(b) Represents **C**

(c) Represents **B**

5.22 (a)

(1)   (2)

Enantiomers

(3)

Meso compound

(b)

(1)   (2)

Enantiomers

(3)

Meso compound

(c)

(1)   (2)

Enantiomers

(3)

Meso compound

(d)

(1)   (2)

Enantiomers

(3)   (4)

Enantiomers

(e)

(1)   (2)

Enantiomers

(3)   (4)

Enantiomers

(f)

(1)   (2)

Enantiomers

(3)

Meso compound

5.23 **B** is (2*S*,3*S*)-2,3-Dibromobutane
    C is (2*R*,3*S*)-2,3-Dibromobutane

5.24 (a) **(1)** is (2*S*,3*S*)-2,3-Dichlorobutane
        (2) is (2*R*,3*R*)-2,3-Dichlorobutane
        (3) is (2*R*,3*S*)-2,3-Dichlorobutane

    (b) **(1)** is (2*S*,4*S*)-2,4-Pentanediol
        (2) is (2*R*,4*R*)-2,4-Pentanediol
        (3) is (2*R*,4*S*)-2,4-Pentanediol

    (c) **(1)** is (2*R*,3*R*)-1,4-Dichloro-2,3-difluorobutane
        (2) is (2*S*,3*S*)-1,4-Dichloro-2,3-difluorobutane
        (3) is (2*R*,3*S*)-1,4-Dichloro-2,3-difluorobutane

    (d) **(1)** is (2*S*,4*S*)-4-Chloro-2-pentanol
        (2) is (2*R*,4*R*)-4-Chloro-2-pentanol
        (3) is (2*S*,4*R*)-4-Chloro-2-pentanol
        (4) is (2*R*,4*S*)-4-Chloro-2-pentanol

    (e) **(1)** is (2*S*,3*S*)-2-Bromo-3-fluorobutane
        (2) is (2*R*,3*R*)-2-Bromo-3-fluorobutane
        (3) is (2*S*,3*R*)-2-Bromo-3-fluorobutane
        (4) is (2*R*,3*S*)-2-Bromo-3-fluorobutane

    (f) **(1)** is (2*R*,3*R*)-Butanedioic acid
        (2) is (2*S*,3*S*)-Butanedioic acid
        (3) is (2*R*,3*S*)-Butanedioic acid

5.25

Chloramphenicol

5.26 (a) **A**   C1, *R*; C2, *R*
            B   C1, *S*; C2, *S*

(b,c)

optically inactive, a meso compound

5.27 (a) No   (b) Yes   (c) No   (d) No   (e) Diastereomers

(f) Diastereomers

5.28

Meso compound   Enantiomers

5.29 (a)

(1R,2R)

Enantiomers
(both trans)

(1S,2S)

(1R,2S)

Enantiomers
(both cis)

(1S,2R)

(b)

(1S,3R)

Enantiomers
(both cis)

(1R,3S)

(1S,3S)

Enantiomers
(both trans)

(1R,3R)

(c)

Achiral (trans)

Achiral (cis)

5.30 See Problem 5.29. The molecules in (c) are achiral, so they have no (R, S) designation.

5.31

(S)-(−)-Glyceraldehyde   (S)-(+)-Glyceric acid   (S)-(−)-Isoserine

(S)-(−)-Isoserine   (R)-(+)-3-Bromo-   (S)-(+)-Lactic acid
                         2-hydroxypropanoic
                         acid

5.32 (a)

(R)-Glyceraldehyde   (S)-Glyceraldehyde

(b)

(+)-Tartaric acid   (−)-Tartaric acid

(R)-Lactic acid   (S)-Lactic acid

Problems

Chirality and Stereoisomerism

5.33 (a), (b), (f), and (g) only

5.34 (a) Seven.

(b) (R)- and (S)-3-Methylhexane and (R)- and (S)-2,3-dimethylpentane.

5.35

5.36

(R) configuration

5.37 (a)

(b) Two, indicated by asterisks in (a)

(c) Four

(d) Because a trans arrangement of the one carbon bridge is structurally impossible. Such a molecule would have too much strain.

5.38 (a) **A** is (2*R*,3*S*)-2,3-dichlorobutane; **B** is (2*S*,3*S*)-2,3-dichlorobutane; **C** is (2*R*,3*R*)-2,3-dichlorobutane.

(b) **A**

5.39 (a)

(other answers are possible)

(c) [structure] and [structure] (other answers are possible)

(d) [structure] and [structure]

(e) [structure] and [structure]

(other answers are possible)

5.40 (a) Same: (*S*)

(b) Enantiomers: left (*S*); right (*R*)

(c) Diastereomers: left (1*S*, 2*S*); right (1*R*, 2*S*)

(d) Same: (1*S*, 2*S*)

(e) Diastereomers: left (1*S*, 2*S*); right (1*S*, 2*R*)

(f) Constitutional isomers: both achiral

(g) Diastereomers: left, cis (4*S*); right, trans (4*R*)

(h) Enantiomers: left (1*S*, 3*S*); right (1*R*, 3*R*)

(i) Same: no chirality centers

(j) Different conformers of the same molecule (interconvertable by a ring flip): (1R, 2S)

(k) Diastereomers: left (1R, 2S); right (1R, 2R)

(l) Same: (1R, 2S)

(m) Diastereomers: no chirality center in either

(n) Constitutional isomers: no chirality center in either

(o) Diastereomers: no chirality center in either

(p) Same: no chirality center

(q) Same: no chirality center

5.41 All of these molecules are expected to be planar. Their stereochemistry is identical to that of the corresponding chloroethenes. (a) can exist as cis and trans isomers. Only one compound exists in the case of (b) and (c).

5.42 (a) diastereomers   (c) enantiomers

(b) enantiomers   (d) same compound

5.43

D (racemic)

E (achiral)

5.44

(or enantiomer)   (achiral)
F   G

5.45

(or enantiomer)   (achiral)
H   I

or

(or enantiomer)   (achiral)

5.46

Aspartame

5.47 (a)

(1)   CH₃   (2)   CH₃   (3)   CH₃   (4)   CH₃

(b) (3) and (4) are chiral and are enantiomers of each other.

(c) Three fractions: a fraction containing (1), a fraction containing (2), and a fraction containing (3) and (4) [because, being enantiomers, (3) and (4) would have the same vapor pressure].

(d) None

5.48

Diastereomer
(one center changed)

Enantiomer
(mirror image)
all centers changed

5.49 (a)

(b) No, they are not superposable.

(c) No, and they are, therefore, enantiomers of each other.

(d)

(e) No, they are not superposable.

(f) Yes, and they are, therefore, just different conformations of the same molecule.

5.50 (a)

(b) Yes, and, therefore, *trans*-1,4-diethylcyclohexane is achiral.

(c) No, they are different orientations of the same molecule.

(d) Yes, *cis*-1,4-diethylcyclohexane is a stereoisomer (a diastereomer) of *trans*-1, 4-diethylcyclohexane.

cis-1,4-Diethylcyclohexane

(e) No, it, too, is superposable on its mirror image. (Notice, too, that the plane of the page constitutes a plane of symmetry for both *cis*-1,4-diethylcyclohexane and for *trans*-1, 4-diethylcyclohexane as we have drawn them.)

5.51 *trans*-1,3-Diethylcyclohexane can exist in the following enantiomeric forms.

trans-1,3-Diethylcyclohexane enantiomers

cis-1,3-Diethylcyclohexane consists of achiral molecules because they have a plane of symmetry. [The plane of the page (below) is a plane of symmetry.]

cis-1,3-Diethylcyclohexane (meso)

Challenge Problems

5.52 (a) Since it is optically active and not resolvable, it must be the meso form:

CO₂H

H—C—OH

H—C—OH

CO₂H

(meso)

(b)

CO₂H

H—C—OH

HO—C—H

CO₂H

(R, R)

CO₂H

HO—C—H

H—C—OH

CO₂H

(S, S)

(c) No   (d) A racemic mixture

5.53 (a) $[\alpha]_D = \dfrac{-30}{(0.10\,\text{g/mL})(1.0\,\text{dm})} = -300$

(b) $[\alpha]_D = \dfrac{+165}{(0.05\,\text{g/mL})(1.0\,\text{dm})} = +3300$

The two rotation values can be explained by recognizing that this is a powerfully optically active substance and that the first reading, assumed to be −30, was really +330. Making this change the $[\alpha]_D$ becomes +3300 in both cases.

(c) No, the apparent 0 rotation could actually be + or −360 (or an integral multiple of these values).

5.54 Yes, it could be a *meso* form or an enantiomer whose chirality centers, by rare coincidence, happen to cancel each other's activities.

5.55 A compound $C_3H_6O_2$ has an index of hydrogen deficiency of 1. Thus, it could possess a carbon-carbon double bond, a carbon-oxygen double bond, or a ring.

The IR spectral data rule out a carbonyl group but indicate the presence of an —OH group.

No stable structure having molecular formula $C_3H_6O_2$ with a $C{=}C$ bond can exist in stereoisomeric forms but 1,2-cyclopropanediol can exist in three stereoisomeric forms.

Only ethylene oxide (oxirane) derivatives are possible for Y.

QUIZ

5.1 Describe the relationship between the two structures shown.

(a) Enantiomers   (b) Diastereomers   (c) Constitutional isomers

(d) Conformations   (e) Two molecules of the same compound

5.2 Which of the following molecule(s) possess(es) a plane of symmetry?

(a) H—C—Cl with F up, Br down   (b) F—C—Cl with F up, Br down   (c) cyclohexane with H, Cl, Br, H

(d) More than one of these   (e) None of these

5.3 Give the (R, S) designation of the structure shown:

(a) (R)   (b) (S)   (c) Neither, because this molecule has no chirality center.

(d) Impossible to tell

5.4 Select the words that best describe the following structure:

(a) Chiral   (b) Meso form   (c) Achiral   (d) Has a plane of symmetry

(e) More than one of these

5.5 Select the words that best describe what happens to the optical rotation of the alkene shown when it is hydrogenated to the alkane according to the following equation:

(a) Increases (b) Changes to zero (c) Changes sign

(d) Stays the same (e) Impossible to predict

5.6 There are two compounds with the formula C_7H_{16} that are capable of existing as enantiomers. Write three-dimensional formulas for the (S) isomer of each.

5.7 Compound A is optically active and is the (S) isomer.

5.8 Compound B is a hydrocarbon with the minimum number of carbon atoms necessary for it to possess a chirality center and, as well, alternative stereochemistries about a double bond.

B

5.9   Which is untrue about the following structure?

(a) It is the most stable of the possible conformations.

(b) $\mu = 0\,D$

(c) It is identical to its mirror image.

(d) It is optically active.

(e) (R,S) designations cannot be applied.

5.10

$$
\begin{array}{c}
CH_2OH \\
| \\
H-C-OH \\
| \\
HO-C-H \\
| \\
H-C-OH \\
| \\
CH_3
\end{array}
$$

is a Fischer projection of one of _____ stereoisomers.

(a) 2   (b) 3   (c) 4   (d) 7   (e) 8

6 NUCLEOPHILIC REACTIONS: PROPERTIES AND SUBSTITUTION REACTIONS OF ALKYL HALIDES

SOLUTIONS TO PROBLEMS

6.1 (a) *cis*-1-Bromo-2-methylcyclohexane

(b) *cis*-1-Bromo-3-methylcyclohexane

(c) 2,3,4-Trimethylheptane

6.2 (a) 3°   (b) alkenyl (vinylic)   (c) 2°   (d) aryl   (e) 1°

6.3 (a) $CH_3-\ddot{\underset{..}{I}}:$ + $CH_3CH_2-\ddot{\underset{..}{O}}:^-$ ⟶ $CH_3-\ddot{\underset{..}{O}}-CH_2CH_3$ + $:\ddot{\underset{..}{I}}:^-$

Substrate   Nucleophile   Leaving group

(b) $:\ddot{\underset{..}{I}}:^-$ + $CH_3CH_2-\ddot{\underset{..}{Br}}:$ ⟶ $CH_3CH_2-\ddot{\underset{..}{I}}:$ + $:\ddot{\underset{..}{Br}}:^-$

Nucleophile   Substrate   Leaving group

(c) $2\ CH_3\ddot{\underset{..}{O}}H$ + $(CH_3)_3C-\ddot{\underset{..}{C}}l:$ ⟶ $(CH_3)_3C-\ddot{\underset{..}{O}}-CH_3$ + $:\ddot{\underset{..}{C}}l:^-$ + $CH_3\overset{+}{\underset{..}{O}}H_2$

Nucleophile   Substrate   Leaving group

(d) $\wedge\!\!\!\wedge\ddot{\underset{..}{Br}}:$ + $^-:C\equiv N:$ ⟶ $\wedge\!\!\!\wedge C\equiv N:$ + $:\ddot{\underset{..}{Br}}:^-$

Substrate   Nucleophile   Leaving group

(e) ⟨benzene⟩$CH_2\ddot{\underset{..}{Br}}:$ + $2\ \ddot{N}H_3$ ⟶ ⟨benzene⟩$CH_2\ddot{N}H_2$ + $:\ddot{\underset{..}{Br}}:^-$ + $^+NH_4$

Substrate   Nucleophile   Leaving group

6.4

6.5

6.6 (a) We know that when a secondary alkyl halide reacts with hydroxide ion by substitution, the reaction occurs with *inversion of configuration* because the reaction is S_N2.
If we know that the configuration of $(-)$-2-butanol (from Section 5.8C) is that shown
here, then we can conclude that $(+)$-2-chlorobutane has the opposite configuration.

(b) Again the reaction is S_N2. Because we now know the configuration of $(+)$-2-
chlorobutane to be (S) [cf., part (a)], we can conclude that the configuration of
$(-)$-2-iodobutane is (R).

$(+)$-2-Iodobutane has the (S) configuration.

6.7 (b) < (c) < (a) in order of increasing stability

6.8 (a, b)

6.9

$(CH_3)_3C$ ⟶ CH₃ / OCH₃   and $(CH_3)_3C$ ⟶ OCH₃ / CH₃

6.10 (c) is most likely to react by an S_N1 mechanism because it is a tertiary alkyl halide, whereas (a) is primary and (b) is secondary.

6.11 (a) Being primary halides, the reactions are most likely to be S_N2, with the nucleophile in each instance being a molecule of the solvent (i.e., a molecule of ethanol).

  (b) Steric hindrance is provided by the substituent or substituents on the carbon β to the carbon bearing the leaving group. With each addition of a methyl group at the β carbon (below), the number of pathways open to the attacking nucleophile becomes fewer.

6.12 $NC^- > CH_3O^- > CH_3CO_2^- > CH_3CO_2H > CH_3OH$
Order of decreasing nucleophilicity in methanol

6.13 $CH_3S^- > CH_3O^- > CH_3CO_2^- > CH_3SH > CH_3OH$

Order of decreasing nucleophilicity in methanol

6.14 **Protic solvents** are those that have an H bonded to an oxygen or nitrogen (or to another strongly electronegative atom). Therefore, the protic solvents are formic acid, $\overset{\displaystyle O}{\overset{\displaystyle \|}{HC}}OH$; formamide, $\overset{\displaystyle O}{\overset{\displaystyle \|}{HC}}NH_2$; ammonia, NH_3; and ethylene glycol, $HOCH_2CH_2OH$. In other words, choices (a), (d), (f), and (h).

Aprotic solvents lack an H bonded to a strongly electronegative element. Aprotic solvents in this list are acetone, $CH_3\overset{\displaystyle O}{\overset{\displaystyle \|}{C}}CH_3$; acetonitrile, $CH_3C\equiv N$; sulfur dioxide, SO_2; and trimethylamine, $N(CH_3)_3$. In other words, choices (b), (c), (e), and (g).

6.15 The reaction is an S_N2 reaction. In the polar aprotic solvent (DMF), the nucleophile (CN^-) will be relatively unencumbered by solvent molecules, and, therefore, it will be more reactive than in ethanol. As a result, the reaction will occur faster in N,N-dimethylformamide.

6.16 (a) CH_3O^-

(b) H_2S

(c) $(CH_3)_3P$

6.17 (a) Increasing the percentage of water in the mixture increases the polarity of the solvent. (Water is more polar than methanol.) Increasing the polarity of the solvent increases the rate of the solvolysis because separated charges develop in the transition state. The more polar the solvent, the more the transition state is stabilized (Section 6.13D).

(b) In an S_N2 reaction of this type, the charge becomes dispersed in the transition state:

$$I^- + CH_3CH_2-Cl \longrightarrow \left[\overset{CH_3}{\underset{H\ H}{\overset{\delta-}{I}----\overset{|}{C}----\overset{\delta-}{Cl}}} \right]^{\ddagger} \longrightarrow ICH_2CH_3 + Cl^-$$

Reactants   Transition state
Charge is concentrated   *Charge is dispersed*

Increasing the polarity of the solvent increases the stabilization of the reactant I^- more than the stabilization of the transition state, and thereby increases the free energy of activation, thus decreasing the rate of reaction.

6.18 $CH_3OSO_2CF_3 > CH_3I > CH_3Br > CH_3Cl > CH_3F > {}^{14}CH_3OH$

(Most reactive)   (Least reactive)

6.19 (a) $CH_3CH_2O^-$ Na^+ + [structure: C with CH_3, CH_2CH_3, H (S), bonded to Br] $\xrightarrow[S_N2]{\text{inversion}}$ CH_3CH_2O-C [with CH_3, CH_2CH_3, (R) H] + Na^+ Br^-

(b) $CH_3\overset{O}{\overset{\|}{C}}O^-$ Na^+ + [structure: C with CH_3, CH_2CH_3, H (S), bonded to Br] $\xrightarrow[S_N2]{\text{inversion}}$ $CH_3\overset{O}{\overset{\|}{C}}O-C$ [with CH_3, CH_2CH_3, (R) H] + Na^+ Br^-

(c) HS^- Na^+ + [structure: C with CH_3, CH_2CH_3, H (S), bonded to Br] $\xrightarrow[S_N2]{\text{inversion}}$ $HS-C$ [with CH_3, CH_2CH_3, (R) H] + Na^+ Br^-

(d) CH_3S^- Na^+ + [structure: C with CH_3, CH_2CH_3, H (S), bonded to Br] $\xrightarrow[S_N2]{\text{inversion}}$ CH_3S-C [with CH_3, CH_2CH_3, (R) H] + Na^+ Br^-

Relative Rates of Nucleophilic Substitution

6.20 (a) 1-Bromopropane would react more rapidly because, being a primary halide, it is less hindered.

(b) 1-Iodobutane, because iodide ion is a better leaving group than chloride ion.

(c) 1-Chlorobutane, because the carbon bearing the leaving group is less hindered than in 1-chloro-2-methylpropane.

(d) 1-Chloro-3-methylbutane, because the carbon bearing the leaving group is less hindered than in 1-chloro-2-methylbutane.

(e) 1-Chlorohexane, because it is a primary halide. Phenyl halides are unreactive in S_N2 reactions.

6.21 (a) Reaction (1), because ethoxide ion is a stronger nucleophile than ethanol.

(b) Reaction (2), because the ethyl sulfide ion is a stronger nucleophile than the ethoxide ion in a protic solvent. (Because sulfur is larger than oxygen, the ethyl sulfide ion is less solvated and it is more polarizable.)

(c) Reaction (2), because triphenylphosphine [$(C_6H_5)_3P$] is a stronger nucleophile than triphenylamine. (Phosphorus atoms are larger than nitrogen atoms.)

(d) Reaction (2), because in an S_N2 reaction the rate depends on the concentration of the substrate and the nucleophile. In reaction (2) the concentration of the nucleophile is twice that of the reaction (1).

6.22 (a) Reaction (2), because bromide ion is a better leaving group than chloride ion.

(b) Reaction (1), because water is a more polar solvent than methanol, and S_N1 reactions take place faster in more polar solvents.

(c) Reaction (2), because the concentration of the substrate is twice that of reaction (1). The major reaction would be E2. (However, the problem asks us to consider that small portion of the overall reaction that proceeds by an S_N1 pathway.)

(d) Considering only S_N1 reactions, as the problem specifies, both reactions would take place at the same rate because S_N1 reactions are independent of the concentration of the nucleophile. The predominant process in this pair of reactions would be E2, however.

(e) Reaction (1), because the substrate is a tertiary halide. Phenyl halides are unreactive in S_N1 reactions.

Synthesis

6.23 (a) $\quad$ Br + NaOH $\longrightarrow$ OH + NaBr

(b) $\quad$ Br + NaI $\longrightarrow$ I + NaBr

(c) $\quad$ Br + ONa $\longrightarrow$ O + NaBr

(d) $\quad$ Br + CH_3SNa $\longrightarrow$ S–CH_3 + NaBr

(e) $\quad$ Br + $\overset{O}{\underset{}{\parallel}}$ ONa $\longrightarrow$ O + NaBr

(f) $\quad$ Br + NaN_3 $\longrightarrow$ N_3 + NaBr

(g) $\quad$ Br + $:N(CH_3)_3$ $\longrightarrow$ $\overset{+}{N}(CH_3)_3$ Br^-

(h) $\quad$ Br + NaCN $\longrightarrow$ $\equiv N$ + NaBr

(i) $\quad$ Br + NaSH $\longrightarrow$ SH + NaBr

6.24 Possible methods are given here.

(a) $CH_3Cl \xrightarrow[\substack{CH_3OH \\ S_N2}]{I^-} CH_3I$ $\qquad$ (b) $\quad$ Cl $\xrightarrow[\substack{CH_3OH \\ S_N2}]{I^-}$ I

(c) $CH_3Cl \xrightarrow[\substack{CH_3OH/H_2O \\ S_N2}]{HO^-} CH_3OH$

(d) $\wedge\!\!\!\diagdown Cl \xrightarrow[\substack{CH_3OH/H_2O \\ S_N2}]{HO^-} \wedge\!\!\!\diagdown OH$

(e) $CH_3Cl \xrightarrow[\substack{CH_3OH \\ S_N2}]{HS^-} CH_3SH$

(f) $\wedge\!\!\!\diagdown Cl \xrightarrow[\substack{CH_3OH \\ S_N2}]{HS^-} \wedge\!\!\!\diagdown SH$

(g) $CH_3I \xrightarrow[DMF]{^-CN} CH_3CN$

(h) $\wedge\!\!\!\diagdown Br \xrightarrow[DMF]{^-CN} \wedge\!\!\!\diagdown CN$

(i) $CH_3OH \xrightarrow[(-H_2)]{NaH} CH_3ONa \xrightarrow[CH_3OH]{CH_3I} CH_3OCH_3$

(j) $\wedge\!\!\!\diagdown OH \xrightarrow[(-H_2)]{NaH} \wedge\!\!\!\diagdown ONa \xrightarrow{CH_3I} \wedge\!\!\!\diagdown OMe$

6.25 (a) The reaction will not take place because the leaving group would have to be a methyl anion, a very powerful base, and a very poor leaving group.

(b) The reaction will not take place because the leaving group would have to be a hydride ion, a very powerful base, and a very poor leaving group.

(c) The reaction will not take place because the leaving group would have to be a carbanion, a very powerful base, and a very poor leaving group.

(d) The reaction will not take place because the leaving group would have to be a CH_3O^- ion, a strong base, and a very poor leaving group.

(e) The reaction will not take place because the first reaction that would take place would be an acid-base reaction that would convert the ammonia to an ammonium ion. An ammonium ion, because it lacks an electron pair, is not nucleophilic.

$$NH_3 + CH_3\overset{+}{O}H_2 \rightleftharpoons \overset{+}{N}H_4 + CH_3OH$$

6.26 (a)

(b)

(c)

(d)

(e)

(f)

(g) $Na^+ HO^-$ +

(R)-2-Bromopentane   (S)-2-Pentanol

(h) Na$^+$I$^-$

(S)-2-Chloro-4-methylpentane   (R)-2-Iodo-4-methylpentane

(i)

(j) Na$^+$ $^-$:C≡N: +

(S)-2-Bromobutane

(k)

General S$_N$1 and S$_N$2

6.27 (a), (b), and (c) are all S$_N$2 reactions and, therefore, proceed with inversion of configuration. The products are

(a)   (b)   (c)

(d) is an S$_N$1 reaction. The carbocation that forms can react with either nucleophile (H$_2$O or CH$_3$OH) from either the top or bottom side of the molecule. Four substitution products (below) would be obtained. (Considerable elimination by an E1 path would also occur.)

6.28 (a) S$_N$2 because the substrate is a 1° halide.

(b) Rate = k [CH$_3$CH$_2$Cl][I$^-$]

   = 5 × 10^{-5} L mol^{-1}s^{-1} × 0.1 mol L^{-1} × 0.1 mol L^{-1}

   Rate = 5 × 10^{-7} mol L^{-1}s^{-1}

(c) 1×10^{-6} mol L^{-1}s^{-1}

(d) 1×10^{-6} mol L^{-1}s^{-1}

(e) 2×10^{-6} mol L^{-1}s^{-1}

6.29 (a) $CH_3\overset{-}{N}H$ because it is the stronger base.

(b) CH_3O^- because it is the stronger base.

(c) CH_3SH because sulfur atoms are larger and more polarizable than oxygen atoms.

(d) $(C_6H_5)_3P$ because phosphorus atoms are larger and more polarizable than nitrogen atoms.

(e) H_2O because it is the stronger base.

(f) NH_3 because it is the stronger base.

(g) HS^- because it is the stronger base.

(h) HO^- because it is the stronger base.

6.30 (a)

(b)

6.31

6.32 Iodide ion is a good nucleophile and a good leaving group; it can rapidly convert an alkyl chloride or alkyl bromide into an alkyl iodide, and the alkyl iodide can then react rapidly with another nucleophile. With methyl bromide in water, for example, the following reaction can take place:

6.33 1-Bromobicyclo[2.2.1]heptane is unreactive in an S_N2 reaction because it is a tertiary halide and its ring structure makes the backside of the carbon bearing the leaving group completely inaccessible to attack by a nucleophile.

1-Bromobicyclo[2.2.1]heptane is unreactive in an S_N1 reaction because the ring structure makes it impossible for the carbocation that must be formed to assume the required trigonal planar geometry around the positively charged carbon. Any carbocation formed from 1-bromobicyclo[2.2.1]heptane would have a trigonal pyramidal arrangement of the $-CH_2-$ groups attached to the positively charged carbon (make a model). Such a structure does not allow stabilization of the carbocation by overlap of sp^3 orbitals from the alkyl groups (see Fig. 6.7).

6.34 The cyanide ion has two nucleophilic atoms; it is what is called an ambident nucleophile.

$$^{-}:C\equiv N:$$

It can react with a substrate using either atom, although the carbon atom is more nucleophilic.

$$Br-CH_2CH_3 + {}^{-}:C\equiv N: \longrightarrow CH_3CH_2-C\equiv N:$$

$$^{-}:C\equiv N: + CH_3CH_2-Br \longrightarrow CH_3CH_2-\ddot{N}=C:$$

6.35 (a)

I, F, H, H (cyclopentene structure)

(Formation of this product depends on the fact that bromide ion is a much better leaving group than fluoride ion.)

(b)

Cl, I (substituted alkyl chain)

(Formation of this product depends on the greater reactivity of 1° substrates in S_N2 reactions.)

(c)

S, S (ring)

(Here two S_N2 reactions produce a cyclic molecule.)

(d) Cl⌒⌒OH + NaH $\xrightarrow[\text{Et}_2\text{O}]{-H_2}$ Na$^+$ $:\ddot{O}^-$ ... Cl ... $\longrightarrow$ (ring with O)

(e) ⎯≡ + NaNH$_2$ $\xrightarrow{\text{liq. NH}_3}$ ⎯≡: $^-$ Na$^+$ $\xrightarrow[(-\text{NaI})]{CH_3-I}$ ⎯≡⎯

6.36   The rate-determining step in the S_N1 reaction of *tert*-butyl bromide is the following:

$$(CH_3)_3C-Br \xrightleftharpoons[\times]{slow} (CH_3)_3C^+ + Br^-$$

$$\xrightarrow{H_2O} (CH_3)_3C\overset{+}{O}H_2$$

$(CH_3)_3C^+$ is so unstable that it reacts almost immediately with one of the surrounding water molecules, and, for all practical purposes, no reverse reaction with Br^- takes place. Adding a common ion (Br^- from NaBr), therefore, has no effect on the rate.

Because the $(C_6H_5)_2CH^+$ cation is more stable, a reversible first step occurs and adding a common ion (Br^-) slows the overall reaction by increasing the rate at which $(C_6H_5)_2CH^+$ is converted back to $(C_6H_5)_2CHBr$.

$$(C_6H_5)_2CHBr \rightleftharpoons (C_6H_5)_2\overset{+}{C}H + Br^-$$

$$\xrightarrow{H_2O} (C_6H_5)_2CH\overset{+}{O}H_2$$

6.37   Two different mechanisms are involved. $(CH_3)_3CBr$ reacts by an S_N1 mechanism, and apparently this reaction takes place faster. The other three alkyl halides react by an S_N2 mechanism, and their reactions are slower because the nucleophile (H_2O) is weak. The reaction rates of CH_3Br, CH_3CH_2Br, and $(CH_3)_2CHBr$ are affected by the steric hindrance, and thus their order of reactivity is $CH_3Br > CH_3CH_2Br > (CH_3)_2CHBr$.

6.38   The nitrite ion is an *ambident nucleophile*; that is, it is an ion with two nucleophilic sites. The equivalent oxygen atoms (because of resonance) and the nitrogen atom are nucleophilic.

Nucleophilic site

$$\overset{..}{\underset{..}{O}}=\overset{..}{N}-\overset{..}{\underset{..}{O}}:^-$$

Nucleophilic site

6.39   (a) The transition state has the form:

$$\overset{\delta^+}{Nu}----\overset{}{R}-----\overset{\delta^-}{L}$$

in which charges are developing. The more polar the solvent, the better it can solvate the transition state, thus lowering the free energy of activation and increasing the reaction rate.

(b) The transition state has the form:

$$\overset{\delta^+}{R}--------\overset{\delta^+}{L}$$

in which the charge is becoming dispersed. A polar solvent is less able to solvate this transition state than it is to solvate the reactant. The free energy of activation, therefore, will become somewhat larger as the solvent polarity increases, and the rate will be slower.

6.40 (a)

(b) HO⟍⟋⟍Cl   +   some alkene

6.41 (a) In an S_N1 reaction the carbocation intermediate reacts rapidly with any nucleophile it encounters in a Lewis acid-Lewis base reaction. In the case of the S_N2 reaction, the leaving group departs only when "pushed out" by the attacking nucleophile and some nucleophiles are better than others.

(b) ^-CN is a much better nucleophile than ethanol and hence the nitrile is formed in the

S_N2 reaction of ⟍⟋⟍Cl. In the case of ⤶Cl, the *tert*-butyl cation reacts chiefly with the nucleophile present in higher concentration, here the ethanol solvent.

Challenge Problems

6.42 (a) The entropy term is slightly favorable. (The enthalpy term is highly unfavorable.)

(b)
$$\Delta G° = \Delta H° - T\Delta S°$$
$$= 26.6 \text{ kJ mol}^{-1} - (298)(0.00481 \text{ kJ mol}^{-1})$$
$$= 25.2 \text{ kJ mol}^{-1}$$

The hydrolysis process will not occur to any significant extent.

(c)
$$\log K_{eq} = \frac{-\Delta G°}{2.303RT}$$
$$= \frac{-25.2 \text{ kJ mol}^{-1}}{(2.303)(0.008314 \text{ kJ mol}^{-1}\text{K}^{-1})(298 \text{ K})}$$
$$= -4.4165$$
$$K_{eq} = 10^{-4.4165} = 3.85 \times 10^{-5}$$

(d) The equilibrium is very much more favorable in aqueous solution because solvation of the products (ethanol, hydronium ions, and chloride ions) takes place and thereby stabilizes them.

6.43 The mechanism for the reaction, in which a low concentration of HO^- is used in the presence of Ag_2O, involves the participation of the carboxylate group. In step 1 (see following reaction) an oxygen of the carboxylate group attacks the chirality center from the back side and displaces bromide ion. (Silver ion aids in this process in much the same way that protonation assists the ionization of an alcohol.) The configuration of the chirality center inverts in step 1, and a cyclic ester called an α-lactone forms.

The highly strained three-membered ring of the α-lactone opens when it is attacked by a water molecule in step 2. *This step also takes place with an inversion of configuration.*

The net result of two inversions (in steps 1 and 2) is an overall *retention of configuration*.

6.44 (a) and (b)

(c) The reaction takes place with retention of configuration.

(d)

6.45 (a)

CH$_3$O—CH(H)(CH$_3$)—CH$_2$—Cl →[N$_3^-$] CH$_3$O—CH(H)(CH$_3$)—CH$_2$—N$_3$

A → **B**

(b) No change of configuration occurs, just a change in the relative priority of a group at the chirality center.

6.46 Comparison of the molecular formulas of starting material and product indicates a loss of HCl. The absence of IR bands in the 1620–1680 cm^{-1} region rules out the presence of the alkene function.

A nucleophilic substitution agrees with the evidence:

(–H$_2$O)

6.47 Regarding the S$_N$2 reaction, there is extreme steric hindrance for attack by the nucleophile from the back side with respect to the leaving group due to atoms on the other side of the rigid ring structure, as the following model shows.

For the S$_N$1 reaction, formation of a carbocation would require that the bridgehead carbon approach trigonal planar geometry, which would lead to a carbocation of extremely high energy due to the geometric constraints of the bicyclic ring.

6.48 The lobe of the LUMO that would accept electron density from the nucleophile is buried within the bicyclic ring structure of 1-bromobicyclo[2.2.1]heptane (the large blue lobe), effectively making it inaccessible for approach by the nucleophile.

6.49 (a) The LUMO in an S$_N$1 reaction is the orbital that includes the vacant p orbital in our simplified molecular orbital diagrams of carbocations. (b) The large lobes above and below the trigonal planar carbon atom of the isopropyl group are the ones that would interact with a nucleophile. These are the lobes associated with stylized p orbitals we draw in simplified diagrams of carbocations. (c) The HOMO for this carbocation shows the contribution of electron density from a nonbonding electron pair of the ether oxygen that is adjacent to the carbocation, This is evident by the lobes that extend over these two atoms and encompass the bond between them. In effect, this orbital model represents the resonance hybrid we can draw where a nonbonding electron pair from oxygen is shifted to the bonding region between the carbon and oxygen.

(b)

A nucleophile could contribute electron density to either lobe of the carbocation p-orbital indicated by these arrows.

(c)

The HOMO of this carbocation shows contribution of electron density from the ether oxygen to the adjacent carbon.
This is evident by the lobes that encompass both atoms and extend over the bond between them.
These lobes are indicated by the arrows.

QUIZ

6.1 Which of the following reactions would give the best yield?

(a) CH_3ONa + (isopropyl bromide) $\longrightarrow$ CH_3—O—(isopropyl)

(b) (isopropyl)—ONa + CH_3Br $\longrightarrow$ CH_3—O—(isopropyl)

(c) CH_3OH + (isopropyl bromide) $\xrightarrow{heat}$ CH_3—O—(isopropyl)

6.2 A kinetic study yielded the following reaction rate data:

Experiment Number	Initial Concentrations		Initial Rate of Disappearance of R—Br and Formation of R—OH
	[HO⁻]	[R—Br]	
1	0.50	0.50	1.00
2	0.50	0.25	0.50
3	0.25	0.25	0.25

Which of the following statements best describe this reaction?

(a) The reaction is second order.

(b) The reaction is first order.

(c) The reaction is S_N1.

(d) Increasing the concentration of HO⁻ has no effect on the rate.

(e) More than one of the above.

6.3 There are four compounds with the formula C_4H_9Br. List them in order of decreasing reactivity in an S_N2 reaction.

☐ > ☐ > ☐ > ☐

6.4 Which S_N2 reaction will occur most rapidly. (Assume the concentrations and temperatures are all the same.)

(a) CH_3O^- + ⌒F → ⌒O—CH_3 + F^-

(b) CH_3O^- + ⌒I → ⌒O—CH_3 + I^-

(c) CH_3O^- + ⌒Cl → ⌒O—CH_3 + Cl^-

(d) CH_3O^- + ⌒Br → ⌒O—CH_3 + Br^-

6.5 Provide three-dimensional structures for the missing boxed structures and formulas for missing reagents.

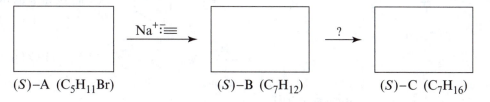

☐ $\xrightarrow{Na^+ \colon\!\!\equiv}$ ☐ $\xrightarrow{?}$ ☐

(S)–A ($C_5H_{11}Br$)   (S)–B (C_7H_{12})   (S)–C (C_7H_{16})

7 ALKENES AND ALKYNES I: PROPERTIES AND SYNTHESIS. ELIMINATION REACTIONS OF ALKYL HALIDES

SOLUTIONS TO PROBLEMS

7.1 (a) (*E*)-1-Bromo-1-chloro-1-pentene or (*E*)-1-Bromo-1-chloropent-1-ene

(b) (*E*)-2-Bromo-1-chloro-1-iodo-1-butene or (*E*)-2-Bromo-1-chloro-1-iodobut-1-ene

(c) (*Z*)-3,5-Dimethyl-2-hexene or (*Z*)-3,5-Dimethylhex-2-ene

(d) (*Z*)-1-Chloro-1-iodo-2-methyl-1-butene or (*Z*)-1-Chloro-1-iodo-2-methylbut-1-ene

(e) (*Z*,4*S*)-3,4-Dimethyl-2-hexene or (*Z*,4*S*)-3,4-Dimethylhex-2-ene

(f) (*Z*,3*S*)-1-Bromo-2-chloro-3-methyl-1-hexene or
(*Z*,3*S*)-1-Bromo-2-chloro-3-methylhex-1-ene

7.2

< < Order of increasing stability

7.3 (a), (b)

2-Methyl-1-butene
(disubstituted)

$\xrightarrow[\text{pressure}]{\text{H}_2 \quad \text{Pt}}$

$\Delta H° = -119 \text{ kJ mol}^{-1}$

3-Methyl-1-butene
(monosubstituted)

$\xrightarrow[\text{pressure}]{\text{H}_2 \quad \text{Pt}}$

$\Delta H° = -127 \text{ kJ mol}^{-1}$

2-Methyl-2-butene
(trisubstituted)

$\xrightarrow[\text{pressure}]{\text{H}_2 \quad \text{Pt}}$

$\Delta H° = -113 \text{ kJ mol}^{-1}$

(c) Yes, because hydrogenation converts each alkene into the same product.

(d)

(trisubstituted)   (disubstituted)   (monosubstituted)

Notice that this predicted order of stability is confirmed by the heats of hydrogenation. 2-Methyl-2-butene evolves the least heat; therefore, it is the most stable. 3-Methyl-1-butene evolves the most heat; therefore, it is the least stable.

(e)

1-Pentene   *cis*-2-Pentene   *trans*-2-Pentene

(f) Order of stability: *trans*-2-pentene > *cis*-2-pentene > 1-pentene

7.4 (a) 2,3-Dimethyl-2-butene would be the more stable because the double bond is tetrasubstituted. 2-Methyl-2-pentene has a trisubstituted double bond.

(b) *trans*-3-Hexene would be the more stable because alkenes with trans double bonds are more stable than those with cis double bonds.

(c) *cis*-3-Hexene would be more stable because its double bond is disubstituted. The double bond of 1-hexene is monosubstituted.

(d) 2-Methyl-2-pentene would be the more stable because its double bond is trisubstituted. The double bond of *trans*-2-hexene is disubstituted.

7.5   *Four stereoisomers.* However only *two fractions* are found as enantiomer pairs are indistinguishible by distillation.

enantiomers   enantiomers

7.6

7.7 (a)

(trisubstituted,
more stable)

Major product

+

(monosubstituted,
less stable)

Minor product

(b)

(tetrasubstituted,
more stable)

Major product

+

(disubstituted,
less stable)

Minor product

7.8 *t*-BuOK in *t*-BuOH

7.9 An anti coplanar transition state allows the molecule to assume the more stable staggered conformation,

whereas a syn coplanar transition state requires the molecule to assume the less stable eclipsed conformation.

7.10 *cis*-1-Bromo-4-*tert*-butylcyclohexane can assume an anti coplanar transition state in which the bulky *tert*-butyl group is equatorial.

The conformation (above), because it is relatively stable, is assumed by most of the molecules present, and, therefore, the reaction is rapid.

On the other hand, for *trans*-1-bromo-4-*tert*-butylcyclohexane to assume an anti copla-nar transition state, the molecule must assume a conformation in which the large *tert*-butyl group is axial:

Such a conformation is of high energy; therefore, very few molecules assume this confor-mation. The reaction, consequently, is very slow.

7.11 (a) Anti coplanar elimination can occur in two ways with the cis isomer.

cis-1-Bromo-2-methylcyclohexane

(b) Anti coplanar elimination can occur in only one way with the trans isomer.

trans-1-Bromo-2-methylcyclohexane

7.12 The better yield will be obtained by using the secondary halide, 1-bromo-1-phenylethane, because the desired reaction is E2. Using the primary halide will result in substantial S_N2 reaction as well, producing the alcohol as well as the desired alkene.

7.13 Isobutyl bromide is more sterically hindered than ethyl bromide because of the methyl groups on the β carbon atom.

Isobutyl bromide   Ethyl bromide

This steric hindrance causes isobutyl bromide to react more slowly in S_N2 reactions and to give relatively more elimination (by an E2 path) when a strong base is used.

7.14 (a) (1) CH_3—CH—$\ddot{O}H$ + H—$\ddot{O}$—S—$\ddot{O}$—H $\rightleftharpoons$
  | ‖
  CH_3 $\ddot{O}$

CH_3—CH—$\overset{+}{\underset{H}{O}}$—H + $^-\!:\!\ddot{O}$—S—$\ddot{O}$—H
  |
  CH_3

(2) CH_3—CH—$\overset{+}{\underset{H}{O}}$—H $\rightleftharpoons$ CH_3—$\overset{+}{C}H$ + H_2O
  | |
  CH_3 CH_3

(3) CH_3—$\overset{+}{C}H$—CH_2—H + $^-\!:\!\ddot{O}SO_3H$ $\rightleftharpoons$ CH_3—CH=CH_2 + $HOSO_3H$

(b) By donating a proton to the —OH group of the alcohol in step (1), the acid allows the loss of a relatively stable, weakly basic, leaving group (H_2O) in step (2). In the absence of an acid, the leaving group would have to be the strongly basic HO^- ion, and such steps almost never occur.

7.15

1°   <   2°   <   3°

Order of increasing case of dehydration

7.16 (1) $CH_3\overset{CH_3}{\underset{CH_3}{\overset{|}{\underset{|}{C}}}}CH_2\ddot{O}H$ + H—A $\rightleftharpoons$ $CH_3\overset{CH_3}{\underset{CH_3}{\overset{|}{\underset{|}{C}}}}CH_2\overset{+}{O}H_2$ + :A$^-$

(2) $CH_3\overset{CH_3}{\underset{CH_3}{\overset{|}{\underset{|}{C}}}}CH_2$—$\overset{+}{\ddot{O}}H_2$ $\rightleftharpoons$ $CH_3\overset{CH_3}{\underset{CH_3}{\overset{|}{\underset{|}{C}}}}\overset{+}{C}H_2$ + H_2O

1° Carbocation

(3) $CH_3\overset{CH_3}{\underset{CH_3}{\overset{|}{\underset{|}{C}}}}\overset{+}{C}H_2$ $\longrightarrow$ $\left[CH_3\overset{CH_3}{\overset{|}{C}}\text{—}CH_2 \atop {}_+CH_3\right]^{\ddagger}$ $\longrightarrow$ $CH_3\overset{CH_3}{\underset{+}{\overset{|}{C}}}$—$CH_2$—$CH_3$

1° Carbocation   Transition state   3° Carbocation

[Steps (2) and (3), ionization and rearrangement, may occur simultaneously.]

(4) $CH_3-\overset{\overset{CH_3}{|}}{\underset{+}{C}}-\underset{\underset{:A^-}{\overset{|}{H}}}{CH}-CH_3 \longrightarrow \overset{CH_3}{\underset{CH_3}{}}C=C\overset{H}{\underset{CH_3}{}} + HA$

2-Methyl-2-butene

7.17 $CH_3CH_2CHCH_2-\ddot{O}H + H-\overset{+}{\ddot{O}}-H \underset{(+H_2O)}{\overset{(-H_2O)}{\rightleftharpoons}} CH_3CH_2CHCH_2-\overset{+}{O}H_2 \underset{(+H_2O)}{\overset{(-H_2O)}{\rightleftharpoons}}$
$\qquad\qquad |\qquad\qquad\quad |$
$\qquad\qquad CH_3\qquad\qquad\quad H\qquad\qquad\qquad\qquad CH_3$

2-Methyl-1-butanol

$CH_3CH_2-\overset{\overset{H}{|}}{\underset{\underset{CH_3}{|}}{C}}-\overset{+}{CH_2} \xrightarrow[\text{shift}^*]{\text{1,2-hydride}} CH_3-CH-\overset{+}{C}-CH_3 \underset{\qquad}{\overset{:\ddot{O}H_2}{\rightleftharpoons}}$
$\qquad\qquad\qquad\qquad\qquad\qquad\qquad\qquad\quad |$
$\text{1° Cation}\qquad\qquad\qquad\qquad CH_3$
$\qquad\qquad\qquad\qquad\qquad\qquad\text{3° Cation}$

$CH_3CH=\underset{\underset{CH_3}{|}}{C}-CH_3 + H_3O^+$

2-Methyl-2-butene

$CH_3CHCH_2CH_2-\ddot{O}H + H-\overset{+}{\ddot{O}}-H \underset{(+H_2O)}{\overset{(-H_2O)}{\rightleftharpoons}} CH_3CHCH_2CH_2-\overset{+}{O}H_2 \underset{(+H_2O)}{\overset{(-H_2O)}{\rightleftharpoons}}$
$\quad |\qquad\qquad\qquad\qquad |$
$\quad CH_3\qquad\qquad\qquad H\qquad\qquad\quad CH_3$

3-Methyl-1-butanol

$CH_3CH-\overset{\overset{H}{|}}{CH}-\overset{+}{CH_2} \xrightarrow[\text{shift}^*]{\text{1,2-hydride}} CH_3-\overset{\overset{H}{|}}{\underset{\underset{CH_3}{|}}{C}}-\overset{+}{CH}-CH_3 \underset{\qquad}{\overset{:\ddot{O}H_2}{\rightleftharpoons}}$
$\quad |$
$\quad CH_3$

$CH_3C=CH-CH_3 + H_3O^+$
$\quad |$
$\quad CH_3$

2-Methyl-2-butene

* The hydride shift may occur simultaneously with the preceding step.

7.18

7.19 (a) $CH_3CH{=}CH_2 + NaNH_2 \longrightarrow$ No reaction

(b) $CH_3C{\equiv}C{-}H + Na^+\ ^-{:}\ddot{N}H_2 \longrightarrow CH_3C{\equiv}C{:}^-Na^+ + {:}NH_3$

Stronger   Stronger   Weaker   Weaker
acid   base   base   acid

(c) $CH_3CH_2CH_3 + NaNH_2 \longrightarrow$ No reaction

(d) $CH_3C{\equiv}C{:}^- + H{-}\ddot{O}CH_2CH_3 \longrightarrow CH_3C{\equiv}CH + {}^-{:}\ddot{O}CH_2CH_3$

Stronger   Stronger   Weaker   Weaker
base   acid   acid   base

(e) $CH_3C{\equiv}C{:}^- + H{-}\overset{+}{N}H_3 \longrightarrow CH_3C{\equiv}CH + {:}NH_3$

Stronger   Stronger acid   Weaker   Weaker
base   acid   base

7.20

7.21 (a) $\underset{\underset{\displaystyle CH_3CCH_3}{\overset{\displaystyle \|}{}}}{\overset{\displaystyle O}{}} \xrightarrow[0\,°C]{PCl_5} CH_3CCl_2CH_3 \xrightarrow[\substack{\text{mineral oil, heat} \\ (2)\ ^+NH_4}]{(1)\ 3\ NaNH_2} CH_3C\equiv CH$

(b) $CH_3CH_2CHBr_2 \xrightarrow[\substack{\text{mineral oil, heat} \\ (2)\ ^+NH_4}]{(1)\ 3\ NaNH_2} CH_3C\equiv CH$

(c) $CH_3CHBrCH_2Br \xrightarrow{\text{[same as (b)]}} CH_3C\equiv CH$

(d) $CH_3CH=CH_2 \xrightarrow{Br_2} \underset{\underset{\displaystyle Br}{|}}{CH_3CHCH_2Br} \xrightarrow{\text{[same as (b)]}} CH_3C\equiv CH$

7.22 $\underset{\underset{\displaystyle CH_3}{|}}{\overset{\overset{\displaystyle CH_3}{|}}{CH_3-C-C\equiv C-H}} + Na^+\ ^-:NH_2 \xrightarrow[(-NH_3)]{} \underset{\underset{\displaystyle CH_3}{|}}{\overset{\overset{\displaystyle CH_3}{|}}{CH_3-C-C\equiv C:^-}}\ Na^+$

(Starting the synthesis with 1-propyne and attempting to alkylate with a *tert*-butyl substrate would not work because elimination would occur instead of substitution.)

$\downarrow CH_3-I$

$\underset{\underset{\displaystyle CH_3}{|}}{\overset{\overset{\displaystyle CH_3}{|}}{CH_3-C-C\equiv C-CH_3}}$

7.23

Compound A

7.24 $\xrightarrow[\text{(2) } NH_4Cl]{\text{(1) Li, } C_2H_5NH_2,\ -78\,°C}$

(*E*)-2-Nonene

7.25   Route 1

$$HC \equiv CCH_2 \overset{\underset{|}{CH_3}}{C}HCH_3 \xrightarrow[(-NH_3)]{NaNH_2} Na^{+-}{:}C \equiv CCH_2 \overset{\underset{|}{CH_3}}{C}HCH_3 \xrightarrow[(-NaBr)]{CH_3-Br}$$

$$CH_3-C \equiv CCH_2 \overset{\underset{|}{CH_3}}{C}HCH_3 \xrightarrow[\substack{Pd, Pt, \text{ or } Ni \\ pressure}]{H_2} CH_3CH_2CH_2CH_2 \overset{\underset{|}{CH_3}}{C}HCH_3$$

Route 2

$$HC \equiv CH \xrightarrow[(-NH_3)]{NaNH_2} HC \equiv C{:}^- Na^+ \xrightarrow[(-NaBr)]{Br-CH_2CH_2 \overset{\underset{|}{CH_3}}{C}HCH_3}$$

$$HC \equiv C-CH_2CH_2 \overset{\underset{|}{CH_3}}{C}HCH_3 \xrightarrow[\substack{Pd, Pt, \text{ or } Ni \\ pressure}]{H_2} CH_3CH_2CH_2CH_2 \overset{\underset{|}{CH_3}}{C}HCH_3$$

Route 3

$$HC \equiv C \overset{\underset{|}{CH_3}}{C}HCH_3 \xrightarrow[(-NH_3)]{NaNH_2} Na^{+-}{:}C \equiv C \overset{\underset{|}{CH_3}}{C}HCH_3 \xrightarrow[(-NaBr)]{CH_3CH_2Br}$$

$$CH_3CH_2-C \equiv C \overset{\underset{|}{CH_3}}{C}HCH_3 \xrightarrow[Pd, Pt, \text{ or } Ni]{H_2} CH_3CH_2CH_2CH_2 \overset{\underset{|}{CH_3}}{C}HCH_3$$

7.26 (a) Undecane

Etc. (using the other alkyne and alkyl halide homologue pairs)

$$HC\equiv C\!:^-$$
$$+$$
$$X(CH_2)_8CH_3$$

$$CH_3C\equiv C\!:^-$$
$$+$$
$$X(CH_2)_7CH_3$$

$$CH_3CH_2C\equiv C\!:^-$$
$$+$$
$$X(CH_2)_6CH_3$$

$$CH_3CH_2CH_2C\equiv C\!:^-$$
$$+$$
$$X(CH_2)_5CH_3$$

2-Methylheptadecane

(after hydrogenation of the alkyne from one of the possible retrosynthetic disconnections)

X + $^-$ (or homologous pairs)

$\equiv^-$ + X

(Note that $\underset{}{\diagup}\!X$ + $^-\equiv$ is not a good choice because the alkyl halide is branched at the carbon adjacent to the one which bears the halogen. Neither would $\underset{X}{\diagup}$ + $^-\equiv$ work because the alkyl halide is secondary. Both of these routes would lead to elimination instead of substitution.)

(b) For any pair of reactants above that is a feasible retrosynthetic disconnection, the steps for the synthesis would be

$$RC\equiv C-H \xrightarrow[(-NH_3)]{NaNH_2} R-C\equiv C\!:^- \xrightarrow[\substack{(R'\ is \\ primary\ and \\ unbranched\ at\ the \\ second\ carbon)}]{R'-X} R-C\equiv C-R'$$

(a terminal alkyne; R = alkyl, H)

(an alkynide anion)

$\Big\downarrow \substack{H_2 \\ Pd,\ Pt,\ or\ Ni \\ pressure}$

$$R-CH_2CH_2-R'$$

7.27 (a) We designate the position of the double bond by using the *lower* number of the two numbers of the doubly bonded carbon atoms, and the chain is numbered from the end nearer the double bond. The correct name is *trans*-2-pentene.

$\overset{5}{}\diagdown\!\!\overset{3}{\diagup}\!\!\diagdown\!\!\overset{1}{\diagup}$ **not** $\overset{1}{}\diagdown\!\!\overset{3}{\diagup}\!\!\diagdown\!\!\overset{5}{\diagup}$

(b) We must choose the longest chain for the base name. The correct name is 2-methylpropene.

(c) We use the lower number of the two doubly bonded carbon atoms to designate the position of the double bond. The correct name is 1-methylcyclohexene.

(d) We must number the ring starting with the double bond in the direction that gives the substituent the lower number. The correct name is 3-methylcyclobutene.

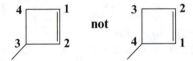

(e) We number in the way that gives the double bond *and the substituent* the lower number. The correct name is (*Z*)-2-chloro-2-butene or (*Z*)-2-chlorobut-2-ene.

(f) We number the ring starting with the double bond so as to give the substituents the lower numbers. The correct name is 3,4-dichlorocyclohexene.

7.28 (a) (b) (c)

(d) (e) (f)

(g) (h) (i)

(j) (k) (l)

7.29 (a)

(2Z,4R)-4-Bromo-2-hexene
or
(2Z,4R)-4-Bromohex-2-ene

(2Z,4S)-4-Bromo-2-hexene
or
(2Z,4S)-4-Bromohex-2-ene

(2E,4R)-4-Bromo-2-hexene
or
(2E,4R)-4-Bromohex-2-ene

(2E,4S)-4-Bromo-2-hexene
or
(2E,4S)-4-Bromohex-2-ene

(b)

(3R,4Z)-3-Chloro-1,4-hexadiene
or
(3R,4Z)-3-Chlorohexa-1,4-diene

(3S,4Z)-3-Chloro-1,4-hexadiene
or
(3S,4Z)-3-Chlorohexa-1,4-diene

(3R,4E)-3-Chloro-1,4-hexadiene
or
(3R,4E)-3-Chlorohexa-1,4-diene

(3S,4E)-3-Chloro-1,4-hexadiene
or
(3S,4E)-3-Chlorohexa-1,4-diene

(c)

(2E,4R)-2,4-Dichloro-2-pentene
or
(2E,4R)-2,4-Dichloropent-2-ene

(2Z,4R)-2,4-Dichloro-2-pentene
or
(2Z,4R)-2,4-Dichloropent-2-ene

(2E,4S)-2,4-Dichloro-2-pentene
or
(2E,4S)-2,4-Dichloropent-2-ene

(2Z,4S)-2,4-Dichloro-2-pentene
or
(2Z,4S)-2,4-Dichloropent-2-ene

(d)

(3R,4Z)-5-Bromo-3-chloro-4-
hexen-1-yne
or
(3R,4Z)-5-Bromo-3-chlorohex-
4-en-1-yne

(3S,4Z)-5-Bromo-3-chloro-4-
hexen-1-yne
or
(3S,4Z)-5-Bromo-3-chlorohex-
4-en-1-yne

(3R,4E)-5-Bromo-3-chloro-4-
hexen-1-yne
or
(3R,4E)-5-Bromo-3-chlorohex-
4-en-1-yne

(3S,4E)-5-Bromo-3-chloro-4-
hexen-1-yne
or
(3S,4E)-5-Bromo-3-chlorohex-
4-en-1-yne

An IUPAC* rule covers those cases in which a double bond and a triple bond occur in the same molecule:

> Numbers as low as possible are given to double and triple bonds as a set, even though this may at times give "-yne" a lower number than "-ene." If a choice remains, preference for low locants is given to the double bonds.

*International Union of Pure and Applied Chemistry, www.iupac.org.

7.30 (a) (E)-3,5-Dimethyl-2-hexene or (E)-3,5-dimethylhex-2-ene

(b) 4-Chloro-3-methylcyclopentene

(c) 6-Methyl-3-heptyne or 6-methylhept-3-yne

(d) 1-sec-Butyl-2-methylcyclohexene or 1-methyl-2-(1-methylpropyl)cyclohexene

(e) (4Z,3R)-3-Chloro-4-hepten-1-yne or (4Z,3R)-3-chlorohept-4-en-1-yne

(f) 2-Pentyl-1-heptene or 2-pentylhept-1-ene

7.31 1-Pentanol > 1-pentyne > 1-pentene > pentane
(See Section 3.8 for the explanation.)

Mechanisms of Dehydrohalogenation, Dehydration, and Nucleophilic Substitution

7.32

7.33 Dehydration of *trans*-2-methylcyclohexanol proceeds through the formation of a carbocation (through an E1 reaction of the protonated alcohol) and leads preferentially to the more stable alkene. 1-Methylcyclohexene (below) is more stable than 3-methylcyclohexene (the minor product of the dehydration) because its double bond is more highly substituted.

(major)
Trisubstituted
double bond

(minor)
Disubstituted
double bond

Dehydrohalogenation of *trans*-1-bromo-2-methylcyclohexane is an E2 reaction and must proceed through an anti coplanar transition state. Such a transition state is possible only for the elimination leading to 3-methylcyclohexene (cf. Review Problem 7.11).

3-Methylcyclohexene

7.34 (a)

major   minor

(b)

only product

(c)

major   minor
(+ stereoisomer)

(d)

major   minor

(e)

major   minor

(f)

only product

7.35 (a)

major   minor

(d)

major   minor
(+ stereoisomer)

(b)

only product

(e)

major   minor

(c)

major   minor
(+ stereoisomer)

7.36 *tert*-Butyl alcohol and *tert*-butyl methyl ether are formed via an S_N1 mechanism. The rate of the reaction is independent of the concentration of methoxide ion (from sodium methoxide). This, however, is only one reaction that causes *tert*-butyl bromide to disappear. A competing reaction that also causes *tert*-butyl bromide to disappear is an E2 reaction in which methoxide ion reacts with *tert*-butyl bromide. This reaction is dependent on the concentration of methoxide ion; therefore, increasing the methoxide ion concentration causes an increase in the rate of disappearance of *tert*-butyl bromide.

7.37 (a) The major product would be (by an S_N2 mechanism) because the substrate is primary and the nucleophile-base is not hindered. Some would be produced by an E2 mechanism.

(b) The major product would be (by an E2 mechanism), even though the substrate is primary, because the base is a hindered strong base. Some would be produced by an S_N2 mechanism.

(c) For all practical purposes, (by an E2 mechanism) would be the only product because the substrate is tertiary and the base is strong.

(d) Same answer as (c) above.

(e) (formed by an S_N2 mechanism) would, for all practical purposes, be the only product. Iodide ion is a very weak base and a good nucleophile.

(f) Because the substrate is tertiary and the base weak, an S_N1 reaction (solvolysis) will occur, accompanied by elimination (E1). At 25°C, the S_N1 reaction would predominate.

(g) [also (Z)] (by an E2 mechanism) would be the major product because the substrate is secondary and the base/nucleophile is a strong base. Some of the ether

would be formed by an S_N2 pathway.

(h) The major product would be (by an S_N2 mechanism) because the acetate ion is a weak base. Some and might be formed by an E2 pathway.

(i) [also (Z)] and (by E2) would be major products, and

[(S) isomer] (by S_N2) would be the minor product.

(j) (by S_N1) would be the major product. [also (Z)],

[also (Z)], and (by E1) would be minor products.

(k) (by S_N2) would be the only product.

(l) The reaction will not take place by an S_N2 mechanism because the substrate is a tertiary halide, and is, therefore, not susceptible to S_N2 attack because of the steric hindrance. (A very small amount of S_N1 reaction may take place, but the main reaction will be E2 to produce an alkene.)

7.38 $\underset{\substack{| \\ OH \quad 3°}}{\overset{\substack{CH_3 \\ |}}{CH_3CCH_2CH_3}}$ > $\underset{\substack{| \\ OH \quad 2°}}{\overset{\substack{CH_3 \\ |}}{CH_3CHCHCH_3}}$ > $\underset{1°}{CH_3CH_2CH_2CH_2CH_2OH}$

7.39 (a) A hydride shift occurs.

hydride shift
(may be concerted
with departure of
the leaving group)

major
product

(b) A methanide shift occurs.

major product

(c) A methanide shift occurs.

major product

(d) The required anti coplanar transition state leads only to (Z) alkene:

(Z) only

7.40 Only the deuterium atom can assume the anti coplanar orientation necessary for an E2 reaction to occur.

7.41 The alkene cannot be formed because the double bond in the product is too highly strained. Recall that the atoms at each carbon of a double bond prefer to be in the same plane.

7.42 (a) You should use a strong base, such as RO⁻, at a higher temperature to bring about an E2 reaction.

(b) Here we want an S_N1 reaction. We use ethanol as the solvent *and as the nucleophile*, and we carry out the reaction at a low temperature so that elimination will be minimized.

7.43 Reaction (2) would give the better yield because the desired reaction is an S_N2 reaction, and the substrate is a methyl halide. Use of reaction (1) would, because the substrate is a secondary halide, result in considerable elimination by an E2 pathway.

7.44 (a)

(b)

(c)

(d)

(e)

Synthesis of Alkenes and Alkynes

7.45 (a) ![propyl chloride] $\xrightarrow[\text{$t$-BuOH}]{\text{$t$-BuOK}}$![propene]

(b) ![2-chloropropane] $\xrightarrow[\text{EtOH}]{\text{EtONa}}$![propene]

(c) ![1-propanol] OH $\xrightarrow{\text{HA, heat}}$![propene]

(d) ![2-propanol] OH $\xrightarrow{\text{HA, heat}}$![propene]

(e) ![1,2-dibromopropane] Br, Br $\xrightarrow[\text{(2) HA}]{\text{(1) NaNH}_2 \text{ (3 equiv.)}}$![propyne] $\xrightarrow[\text{Ni}_2\text{B (P-2)}]{\text{H}_2 \text{ (1 equiv.)}}$![propene]

(f) ![propyne] $\xrightarrow[\text{Ni}_2\text{B (P-2)}]{\text{H}_2}$![propene]

7.46 (a) ![bromocyclopentane] Br $\xrightarrow[\substack{\text{EtOH} \\ \text{heat}}]{\text{EtONa}}$![cyclopentene]

(b) ![cyclopentanol] OH $\xrightarrow{\text{HA, heat}}$![cyclopentene]

7.47 (a) ![2-bromopentane] Br $\xrightarrow[\text{$t$-BuOH}]{\text{$t$-BuOK}}$![pentene] major + ![pentene] minor [+(Z)]

(b) ![1-bromo-3-methylbutane] Br $\xrightarrow[\text{EtOH}]{\text{EtONa}}$![3-methyl-1-butene] only product

(c)

EtONa / EtOH →   only product

(d)

EtONa / EtOH →   only product

(e)

EtONa / EtOH →   +

major   minor

(f)

EtONa / EtOH →

7.48   (a)   ≡≡ $\xrightarrow[\text{liq. NH}_3]{\text{NaNH}_2}$ ≡≡:⁻ Na⁺ $\xrightarrow[(-\text{NaI})]{\text{CH}_3\!-\!\text{I}}$ ≡≡——

(b)   ≡≡ $\xrightarrow[\text{liq. NH}_3]{\text{NaNH}_2}$ ≡≡:⁻ Na⁺ $\xrightarrow[(-\text{NaBr})]{\text{Br}}$ ≡≡——

(c)   ——≡≡ $\xrightarrow[\text{liq. NH}_3]{\text{NaNH}_2}$ ——≡≡:⁻ Na⁺ $\xrightarrow[(-\text{NaI})]{\text{CH}_3\!-\!\text{I}}$ ——≡≡——
[from (a)]

(d)   ——≡≡—— $\xrightarrow[\text{Ni}_2\text{B (P-2)}]{\text{H}_2}$
[from (c)]

(e)   ——≡≡—— $\xrightarrow[\text{(2)}\quad\text{NH}_4\text{Cl}]{\text{(1) Li,}\quad\text{NH}_2}$
[from (c)]

(f)   ≡≡:⁻ Na⁺ $\xrightarrow[(-\text{NaBr})]{\text{Br}}$ ≡≡——
[from (a)]

(g)   ≡≡ $\xrightarrow[\text{liq. NH}_3]{\text{NaNH}_2}$ ≡≡:⁻ Na⁺ $\xrightarrow[(-\text{NaI})]{\text{CH}_3\!-\!\text{I}}$ ≡≡——
[from (f)]

(h) [from (g)] $\xrightarrow[\text{Ni}_2\text{B (P-2)}]{\text{H}_2}$

(i) [from (g)] $\xrightarrow[\text{(2)} \quad \text{NH}_4\text{Cl}]{\text{(1) Li,} \quad \text{NH}_2}$

(j) [from (a)] $\equiv\!\!\!\!\text{:}^-\ \text{Na}^+ \quad \xrightarrow[\text{(}-\text{NaBr)}]{\text{Br}}$ ... $\xrightarrow[\text{liq. NH}_3]{\text{NaNH}_2}$ ... $\equiv\!\!\!\!\text{:}^-\ \text{Na}^+$ ... Br

(k) [from (j)] $\equiv\!\!\!\!\text{:}^-\ \text{Na}^+ \xrightarrow{\text{D}_2\text{O}}$ ... D

(l) [from (c)] $\xrightarrow[\text{Ni}_2\text{B (P-2)}]{\text{D}_2}$ ... D   D

7.49 (a) Three

(b) One

Index of Hydrogen Deficiency

7.50 (a) Caryophyllene has the same molecular formula as zingiberene (Review Problem 4.21); thus it, too, has an index of hydrogen deficiency equal to 4. That 1 mol of caryophyllene absorbs 2 mol of hydrogen on catalytic hydrogenation indicates the presence of two double bonds per molecule.

(b) Caryophyllene molecules must also have two rings. (See Review Problem 23.2 for the structure of caryophyllene.)

7.51 (a) $C_{30}H_{62}$ = formula of alkane

$C_{30}H_{50}$ = formula of squalene

H_{12} = difference = 6 pairs of hydrogen atoms

Index of hydrogen deficiency = 6

(b) Molecules of squalene contain six double bonds.

(c) Squalene molecules contain no rings. (See Review Problem 23.2 for the structural formula of squalene.)

Structure Elucidation

7.52   That **I** and **J** rotate plane-polarized light in the same direction tells us that **I** and **J** are not enantiomers of each other. Thus, the following are possible structures for **I, J,** and **K**. (The enantiomers of **I, J,** and **K** would form another set of structures, and other answers are possible as well.)

I
Optically active

J
Optically active

K
Optically active

7.53   The following are possible structures:

L

M

(other answers are possible as well)

N
Optically inactive
but resolvable

Synthesizing the Material

7.54 (a)

(b)

(c)

(d)

7.55 (a)

(b)

(c)

7.56 (a)

A (X = halogen)

(b)

B

7.57

Challenge Problems

7.58 The IR evidence indicates that C possesses both an alkene function and a hydroxyl group. An E2 reaction on this substrate produces enantiomeric unsaturated alcohols.

C (racemic)

(a)   +   (b)

(R)   (S)

7.59

E
Optically active (the enantiomeric form is an equally valid answer)

F
Optically inactive and nonresolvable

G
Optically active (the
enantiomeric form is an
equally valid answer)

H
Optically inactive and
nonresolvable

7.60 (a) We are given (Section 7.3A) the following heats of hydrogenation:

$$\textit{cis}\text{-2-Butene} + H_2 \xrightarrow{\text{Pt}} \text{butane} \quad \Delta H° = -120 \text{ kJ mol}^{-1}$$

$$\textit{trans}\text{-2-Butene} + H_2 \xrightarrow{\text{Pt}} \text{butane} \quad \Delta H° = -115 \text{ kJ mol}^{-1}$$

Thus, for

$$\textit{cis}\text{-2-Butene} \longrightarrow \textit{trans}\text{-2-butene} \quad \Delta H° = -5.0 \text{ kJ mol}^{-1}$$

(b) Converting *cis*-2-butene into *trans*-2-butene involves breaking the π bond. Therefore, we would expect the energy of activation to be at least as large as the π-bond strength, that is, at least 264 kJ mol^{-1}.

(c)

```
Free Energy →

          ΔG‡ ≥  264 kJ mol⁻¹
cis-2-Butene
ΔG° = −5.0 kJ mol⁻¹   trans-2-Butene

      Reaction coordinate →
```

7.61 (a) With either the (1*R*,2*R*)- or the (1*S*,2*S*)-1,2-dibromo-1,2-diphenylethane, only one conformation will allow an anti coplanar arrangement of the H- and Br-. In either case, the elimination leads only to (*Z*)-1-bromo-1,2-diphenylethene:

(1*R*,2*R*)-1,2-Dibromo-1,2-diphenylethane
(anti coplanar orientation of H- and -Br)

(*Z*)-1-Bromo-1,2-diphenylethene

(1*S*,2*S*)-1,2-Dibromo-1,2-diphenylethane
(anti coplanar orientation of H- and -Br)

(*Z*)-1-Bromo-1,2-diphenylethene

(b) With (1*R*,2*S*)-1,2-dibromo-1,2-diphenylethane, only one conformation will allow an anti coplanar arrangement of the H- and Br-. In either case, the elimination leads only to (*E*)-1-bromo-1,2-diphenylethene:

(1*R*,2*S*)-1,2-Dibromo-1,2-diphenylethane   (*E*)-1-Bromo-1,2-diphenylethene
(anti coplanar orientation of H and Br)

(c) With (1*R*,2*S*)-1,2-dibromo-1,2-diphenylethane, only one conformation will allow an anti coplanar arrangement of both bromine atoms. In this case, the elimination leads only to (*E*)-1,2-diphenylethene:

(1*R*,2*S*)-1,2-Dibromo-1,2-diphenylethane   (*E*)-1,2-Diphenylethene
(anti coplanar orientation of both -Br atoms)

7.62 (a)

(b) No, tetrasubstituted double bonds usually show no C=C stretching absorption in their infrared spectra.

7.63

7.64 (a) Three

(b) Six

QUIZ

7.1 Which conditions/reagents would you employ to obtain the best yields in the following reaction?

    (a) H_2O/heat   (c) *t*-BuOK in *t*-BuOH, heat

    (b) EtONa in EtOH   (d) Reaction cannot occur as shown

7.2 Which of the following names is incorrect?

    (a) 1-Butene   (b) *trans*-2-Butene   (c) (*Z*)-2-Chloro-2-pentene

    (d) 1,1-Dimethylcyclopentene   (e) Cyclohexene

7.3 Select the major product of the reaction

7.4 Supply the missing reagents.

7.5 Arrange the following alkenes in order of decreasing stability. 1-Pentene, *cis*-2-pentene, *trans*-2-pentene, 2-methyl-2-butene

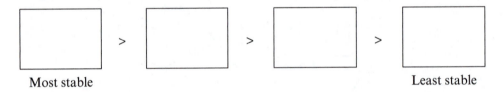

Most stable   Least stable

7.6 Complete the following synthesis.

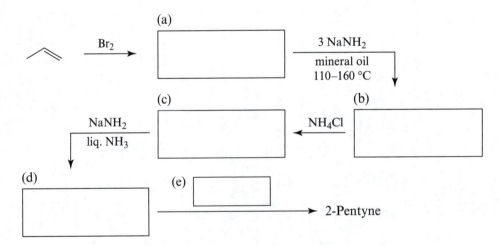

7.7 Supply the missing reactants, reagents, intermediates, or products.

8

ALKENES AND ALKYNES II: ADDITION REACTIONS

SOLUTIONS TO PROBLEMS

8.1   CH₃CHCH₂I
              |
              Br
2-Bromo-1-iodopropane

8.2 (a)

(b)

(c)

8.3 (a)

2° carbocation

1,2-hydride shift

+ 3° carbocation

Cl⁻

Cl

2-Chloro-2-methylbutane
(from rearranged carbocation)

Cl⁻

Cl

2-Chloro-3-methylbutane
(from unrearranged carbocation)

Unrearranged
2° carbocation

(b)

3-Chloro-2,2-dimethylbutane
(from unrearranged carbocation)

1,2-methanide
shift

2-Chloro-2,3-dimethylbutane
(from rearranged carbocation)

8.4 (a)

(from dilute
H_2SO_4)

$+ H_3O^+$

(b) Use a high concentration of water because we want the carbocation produced to react with water. And use a strong acid whose conjugate base is a very weak nucleophile. (For this reason we would not use HI, HBr, or HCl.) An excellent method, therefore, is to use dilute sulfuric acid.

(c) Use a low concentration of water (i.e., use concentrated H_2SO_4) and use a higher temperature to encourage elimination. Distill cyclohexene from reaction mixture as it is formed, so as to draw the equilibrium toward product.

(d) 1-Methylcyclohexanol would be the product because a 3° carbocation would be formed as the intermediate.

$+ H_3O^+$

8.5

(i.e. $CH_3CH_2OSO_3H$)

8.6

+ methanide migration

8.7 The order reflects the relative ease with which these alkenes accept a proton and form a carbocation. $(CH_3)_2C=CH_2$ reacts faster because it leads to a tertiary cation,

$$(CH_3)_2C=CH_2 \xrightarrow{H_3O^+} CH_3-\overset{CH_3}{\underset{CH_3}{\overset{+}{C}}} \quad 3° \text{ Carbocation}$$

$CH_3CH=CH_2$ leads to a secondary cation,

$$CH_3CH=CH_2 \xrightarrow{H_3O^+} CH_3-\overset{H}{\underset{CH_3}{\overset{+}{C}}} \quad 2° \text{ Carbocation}$$

and $CH_2=CH_2$ reacts the slowest because it leads to a primary carbocation.

$$CH_2=CH_2 \xrightarrow{H_3O^+} CH_3-\overset{H}{\underset{H}{\overset{+}{C}}} \quad 1° \text{ Carbocation}$$

Recall that formation of the carbocation is the rate-determining step in acid-catalyzed hydration and that the order of stabilities of carbocations is the following:

$$3° > 2° > 1° > {}^+CH_3$$

8.8

8.9 (a)

$$\text{(isobutylene)} \xrightarrow[\text{(2) NaBH}_4,\ \text{HO}^-]{\text{(1) Hg(OAc)}_2/\text{THF}-\text{H}_2\text{O}} \text{(}t\text{-butanol with OH)}$$

(b)

$$\xrightarrow[\text{(2) NaBH}_4,\ \text{HO}^-]{\text{(1) Hg(OAc)}_2/\text{THF}-\text{H}_2\text{O}} \text{(isopropanol, OH)}$$

(c)

$$\xrightarrow[\text{(2) NaBH}_4,\ \text{HO}^-]{\text{(1) Hg(OAc)}_2/\text{THF}-\text{H}_2\text{O}} \text{(product with OH)}$$

(can also be used.)

8.10 (a)

$$\text{C}=\text{C} + {}^+\text{HgOCCF}_3 \longrightarrow \overset{\delta^+}{-}\text{C}-\text{C}- \ (\text{HgOCCF}_3, \delta^+) \xrightarrow{\text{R}-\ddot{\text{O}}-\text{H}} -\text{C}-\text{C}-\text{O} \ (\text{HgOCCF}_3)$$

$$\xrightarrow{-\text{HA}} \overset{\text{RO}}{-}\text{C}-\text{C}-\text{O} \ (\text{HgOCCF}_3)$$

(b)

$$\text{CH}_3-\overset{\text{CH}_3}{\underset{}{\text{C}}}=\text{CH}_2 \xrightarrow[\substack{\text{THF}-\text{CH}_3\text{OH} \\ \text{solvomercuration}}]{\text{Hg(OCCF}_3)_2} \text{CH}_3-\overset{\text{CH}_3}{\underset{\text{OCH}_3}{\text{C}}}-\text{CH}_2\text{HgOCCF}_3 \xrightarrow[\text{demercuration}]{\text{NaBH}_4/\text{HO}^-}$$

$$\overset{\text{CH}_3}{\underset{\text{OCH}_3}{\text{CH}_3\text{CCH}_3}} + \text{Hg} + \text{CF}_3\text{COO}^-$$

(c) The electron-withdrawing fluorine atoms in mercuric trifluoroacetate enhance the electrophilicity of the cation. Experiments have demonstrated that for the preparation of tertiary alcohols in satisfactory yields, the trifluoroacetate must be used rather than the acetate.

8.11 (a)

$$3 \ \diagdown\!\!\diagup\!\!\diagdown \xrightarrow{\text{BH}_3\text{:THF}} (\text{B with three propyl groups})$$

(b) 3 $\quad$ $\xrightarrow{\text{BH}_3\text{:THF}}$

(c) 3 $\quad$ $\xrightarrow{\text{BH}_3\text{:THF}}$
(or cis isomer)

(d) 3 $\quad$ $\xrightarrow[\substack{\text{syn addition} \\ \text{anti Markovnikov}}]{\text{BH}_3\text{:THF}}$ $\quad$ + enantiomer

8.12 $\quad$ 2 CH$_3$C=CHCH$_3$ $\xrightarrow{\text{BH}_3\text{:THF}}$ $\left(\begin{array}{c} \text{CH}_3 \\ | \\ \text{CH}_3\text{CH}-\text{CH}- \\ | \\ \text{CH}_3 \end{array}\right)_2$BH

with CH$_3$ above C

Disiamylborane

8.13 (a) $\quad$ $\xrightarrow[\text{(2) H}_2\text{O}_2,\ \text{HO}^-]{\text{(1) BH}_3\text{:THF}}$ $\quad$ OH

(b) $\quad$ $\xrightarrow[\text{(2) H}_2\text{O}_2,\ \text{HO}^-]{\text{(1) BH}_3\text{:THF}}$ $\quad$ OH

(c) $\quad$ $\xrightarrow[\text{(2) H}_2\text{O}_2,\ \text{HO}^-]{\text{(1) BH}_3\text{:THF}}$ $\quad$ OH
[or (Z) isomer]

(d) $\quad$ $\xrightarrow[\text{(2) H}_2\text{O}_2,\ \text{HO}^-]{\text{(1) BH}_3\text{:THF}}$ $\quad$ OH

(e) $\quad$ CH$_3$ $\quad$ $\xrightarrow[\text{(2) H}_2\text{O}_2,\ \text{HO}^-]{\text{(1) BH}_3\text{:THF}}$ $\quad$ CH$_3$ $\quad$ OH $\quad$ + enantiomer

(f)

$$\xrightarrow[\text{(2) } H_2O_2, HO^-]{\text{(1) } BH_3:THF}$$

8.14 (a) $3 \ CH_3CHCH{=}CH_2 \xrightarrow{BH_3:THF} \left(CH_3CHCH_2CH_2 \right)_3{-}B \xrightarrow{CH_3CO_2D}$

with CH_3 below, and CH_3 below in the product

$3 \ CH_3CHCH_2CH_2D$

$\quad\quad\; CH_3$

(b) $3 \ CH_3C{=}CHCH_3 \xrightarrow{BH_3:THF} CH_3CH{-}CH{-}CH_3 \xrightarrow{CH_3CO_2D}$

$\quad\quad\; CH_3 \quad\quad\quad\quad\quad\quad\quad | \quad\quad | $ with B above and CH_3 below

$3 \ CH_3CHCHDCH_3$

$\quad\quad\; CH_3$

(c) 3 $\xrightarrow{BH_3:THF}$ (+ enantiomer) $\xrightarrow{CH_3CO_2D}$

3 (+ enantiomer)

(d) 3 $\xrightarrow{BD_3:THF}$ (+ enantiomer) $\xrightarrow{CH_3CO_2T}$

3 (+ enantiomer)

8.15

Bromonium ion

from (a)   from (b)   Racemic mixture

Because paths (a) and (b) occur at equal rates, these enantiomers are formed at equal rates and in equal amounts.

8.16

$Nu = H_2O$

or Br^-

or Cl^-

8.17 (a)

t-BuOK / CHCl$_3$

+ enantiomer

(b)

t-BuOK / CHBr$_3$

(c)

$CH_2I_2/Zn(Cu)$ / Et$_2$O

8.18

CHBr$_3$ / t-BuOK

8.19

8.20 (a)

(b)

(racemic)

(c)

(racemic)

8.21 (a)

(b)

(c)

8.22 (a)

(b)

[or (*E*) isomer]

(c)

8.23 Ordinary alkenes *are* more reactive toward electrophilic reagents than alkynes. But the alkenes obtained from the addition of an electrophilic reagent to an alkyne have at least one electronegative atom (Cl, Br, etc.) attached to a carbon atom of the double bond.

or

These alkenes are less reactive than alkynes toward electrophilic addition because the electronegative group makes the double bond "electron poor."

8.24 The molecular formula and the formation of octane from **A** and **B** indicate that both compounds are unbranched octynes. Since **A** yields only on ozonolysis, **A** must be the symmetrical octyne . The IR absorption for **B** shows the presence of a terminal triple bond. Hence **B** is .

Since **C** (C_8H_{12}) gives HO on ozonolysis, **C** must be cyclooctyne. This is supported by the molecular formula of **C** and the fact that it is converted to **D**, C_8H_{16} (cyclooctane), on catalytic hydrogenation.

8.25 By converting the 3-hexyne to *cis*-3-hexene using H_2/Ni_2B (P-2).

Then, addition of bromine to *cis*-3-hexene will yield (3*R*,4*R*), and (3*S*,4*S*)-3,4-dibromohexane as a racemic form.

Alkenes and Alkynes Reaction Tool Kit

8.26 The answers to all parts except (b), (j), (l), and (n) are formed as racemic mixtures.

(a)   (b)   (c)   (d)

(e)   (f)   (g)   (h)

(i)   (j)   (k)

(l) $+ CO_2$ (m)   (n)

8.27 The answers to (g), (h), (k), and (n) are formed as racemic mixtures.

(a)   (b)   (c)   (d)

(e)   (f)   (g)   (h)

(i)   (j)   (k)

(l)   (m)   (n)

8.28 (a)   (b)

enantiomers

(c)

enantiomers

(d) [structure] + [structure] (e) 2 [structure]

enantiomers

8.29 (a) [structure] (b) [structure] (c) [structure]

(d) [structure] (e) [structure] (f) [structure]

(g) [structure] and [structure] [An E2 reaction would take place

when [structure] Na⁺ is treated with [structure]]

8.30 (a) [structure] (b) Br Br [structure] (c) [structure]

(d) [structure] (e) [structure] (f) [structure]

(g) [structure] (h) [structure] (i) [structure]

(j) [structure]
(2 molar
equivalents)

(k) [structure]
(2 molar
equivalents)

(l) No reaction

8.31 (a)

(b)

(c)

(d)

8.32 (a)

(b) Then as in (a)

(c)

(d)

(e)

8.33 (a)

(b)

(c)

(d)

8.34 (a)

+ enantiomer

(b)

+ enantiomer

(c)

+ enantiomer

8.35

8.36 The rate-determining step in each reaction is the formation of a carbocation when the alkene accepts a proton from HI. When 2-methylpropene reacts, it forms a 3° carbocation (the most stable); therefore, it reacts fastest. When ethene reacts, it forms a 1° carbocation (the least stable); therefore, it reacts the slowest.

3° cation

2° cation

1° cation

8.37

Loss of H$_2$O
and 1,2-hydride
shift

8.38

(−H$_2$O)

2° carbocation

1,2-
Methanide
shift

3° carbocation

8.39 (a)

(1) OsO$_4$
(2) NaHSO$_3$, H$_2$O

HO   OH

H   H
syn addition

+ enantiomer

(b)

(1) OsO$_4$
(2) NaHSO$_3$, H$_2$O

HO   OH

H
syn addition

+ enantiomer

(c)

Br$_2$

Br   Br

H   H
anti addition

+ enantiomer

(d)

Br$_2$

Br   Br

H   H

H
anti addition

+ enantiomer

8.40 (a) $(2S, 3R)$- [the enantiomer is $(2R, 3S)$-]

(b) $(2S, 3S)$- [the enantiomer is $(2R, 3R)$-]

(c) $(2S, 3R)$- [the enantiomer is $(2R, 3S)$-]

(d) $(2S, 3S)$- [the enantiomer is $(2R, 3R)$-]

8.41 Because of the electron-withdrawing nature of chlorine, the electron density at the double bond is greatly reduced and attack by the electrophilic bromine does not occur.

8.42

I (major)   +   II (minor)

Though II is the product predicted by application of the Zaitsev rule, it actually is less stable than I due to crowding about the double bond. Hence I is the major product (by about a 4:1 ratio).

8.43

8.44

8.45

Enantiomers of each also formed.

8.46 (a) $C_{10}H_{22}$ (saturated alkane)

$\underline{C_{10}H_{16}}$ (formula of myrcene)

$H_6 = 3$ pairs of hydrogen atoms

Index of hydrogen deficiency (IHD) = 3

(b) Myrcene contains no rings because complete hydrogenation gives $C_{10}H_{22}$, which corresponds to an alkane.

(c) That myrcene absorbs three molar equivalents of H_2 on hydrogenation indicates that it contains three double bonds.

(d)

(e) Three structures are possible; however, only one gives 2,6-dimethyloctane on complete hydrogenation. Myrcene is therefore

8.47 (a)

2,6,10-Trimethyldodecane

(b) Four

(c)

8.48

Limonene

8.49 The hydrogenation experiment discloses the carbon skeleton of the pheromone.

$$C_{13}H_{24}O \xrightarrow[\text{Pt}]{2\ H_2}$$

OH $C_{13}H_{28}O$

Codling moth
pheromone

3-Ethyl-7-methyl-1-decanol

The ozonolysis experiment allows us to locate the position of the double bonds.

$$\xrightarrow[\text{(2) Me}_2\text{S}]{\text{(1) O}_3}$$

Codling moth pheromone

General Problems

8.50 *Retrosynthetic analysis*

$$\xRightarrow[\text{nikov addition}]{\text{Markov-}}$$ + HBr

$$\xRightarrow[\text{nikov addition}]{\text{Markov-}}$$ + HBr $\Longrightarrow$ HC≡CH + Br

Synthesis

8.51 (a) 1-Pentyne has IR absorption at about 3300 cm^{-1} due to the C-H stretching of its terminal triple bond. Pentane does not absorb in that region.

   (b) 1-Pentene absorbs in the 1620–1680 cm^{-1} region due to the alkene function. Pentane does not exhibit absorption in that region.

   (c) See parts (a) and (b).

   (d) 1-Bromopentane shows C-Br absorption in the 515–690 cm^{-1} region while pentane does not.

   (e) For 1-pentyne, see (a). The interior triple bond of 2-pentyne gives relatively weak absorption in the 2100–2260 cm^{-1} region.

   (f) For 1-pentene, see (b). 1-Pentanol has a broad absorption band in the 3200–3550 cm^{-1} region.

   (g) See (a) and (f).

   (h) 1-Bromo-2-pentene has double bond absorption in the 1620–1680 cm^{-1} region which 1-bromopentane lacks.

   (i) 2-Penten-1-ol has double bond absorption in the 1620–1680 cm^{-1} region not found in 1-pentanol.

8.52   The index of hydrogen deficiency of **A**, **B**, and **C** is two.

$$\frac{\begin{array}{c} C_6H_{14} \\ C_6H_{10} \end{array}}{H_4} = \text{2 pairs of hydrogen atoms}$$

This result suggests the presence of a triple bond, two double bonds, a double bond and a ring, or two rings.

$$\textbf{A, B, C} \xrightarrow{\text{Br}_2} \text{all decolorize a solution of bromine}$$

$$\downarrow \begin{array}{l} \text{cold} \\ \text{concd} \\ \text{H}_2\text{SO}_4 \end{array}$$

all soluble

The fact that **A**, **B**, and **C** all decolorize Br$_2$ and dissolve in concd. H$_2$SO$_4$ suggests they all have a carbon-carbon multiple bond.

$$\text{A, B} \xrightarrow[\text{Pt}]{\text{excess } H_2} \diagdown\diagup\diagdown\diagup \quad ; \text{C} \xrightarrow[\text{Pt}]{\text{excess } H_2} C_6H_{12}$$

$$(C_6H_{10}) \qquad\qquad (C_6H_{14}) \quad (C_6H_{10})$$

A must be a terminal alkyne, because of IR absorption at about 3300 cm^{-1}.

Since **A** gives hexane on catalytic hydrogenation, **A** must be 1-hexyne.

$$\diagdown\diagup\diagdown\diagup\diagdown \xleftarrow[\text{Pt}]{2H_2} \diagdown\diagup\diagdown\!\!\!\equiv \xrightarrow[\text{(2) } H_3O^+]{\text{(1) KMnO}_4,\ HO^-,\ \text{heat}}$$

$$\mathbf{A}$$

$$\diagdown\diagup\diagdown\diagup\!\!\overset{O}{\underset{OH}{C}} + CO_2\uparrow$$

This is confirmed by the oxidation experiment

$$\mathbf{B} \xrightarrow[\text{(2) } H_3O^+]{\text{(1) KMnO}_4,\ HO^-,\ \text{heat}} 2 \diagup\!\!\overset{O}{\underset{OH}{C}}$$

Hydrogenation of **B** to hexane shows that its chain is unbranched, and the oxidation experiment shows that **B** is 3-hexyne, $\diagdown\!\!\!\equiv\!\!\!\diagup$.

Hydrogenation of **C** indicates a ring is present.

Oxidation of **C** shows that it is cyclohexene.

$$\bigcirc\!\!\diagup \underset{\mathbf{C}}{} \xrightarrow[\text{(2) } H_3O^+]{\text{(1) KMnO}_4,\ HO^-,\ \text{heat}} HO\overset{O}{\underset{}{\diagdown}}\diagup\diagdown\diagup\overset{OH}{\underset{O}{\diagup}}$$

8.53 (a) Four

(b) $CH_3(CH_2)_5\diagdown\underset{\overset{|}{HO}\,H}{C}\diagup\overset{CH_2}{\diagdown}\underset{H}{\overset{}{C}}\!\!=\!\!\underset{H}{\overset{}{C}}\diagdown(CH_2)_7CO_2H$ + enantiomer

$CH_3(CH_2)_5\diagdown\underset{\overset{|}{HO}\,H}{C}\diagup\overset{CH_2}{\diagdown}\underset{H}{\overset{}{C}}\!\!=\!\!\underset{(CH_2)_7CO_2H}{\overset{H}{C}}$ + enantiomer

8.54 Hydroxylations by OsO$_4$ are syn hydroxylations (cf. Section 8.16). Thus, maleic acid must be the cis-dicarboxylic acid:

$$\underset{H}{\overset{H}{\diagdown}}\!\!\overset{CO_2H}{\underset{\parallel}{C}}\atop\underset{H}{\overset{}{C}}\!\!\diagdown CO_2H \xrightarrow[\text{syn hydroxylation}]{\substack{\text{(1) OsO}_4,\ \text{pyr} \\ \text{(2) NaHSO}_3/H_2O}} \underset{\text{meso-Tartaric acid}}{\overset{HO_2C}{\underset{H}{\diagdown}}\!\!\overset{}{\underset{}{C}}\!\!\diagup OH}$$

Maleic acid

Fumaric acid must be the trans-dicarboxylic acid:

8.55 (a) The addition of bromine is an anti addition. Thus, fumaric acid yields a meso compound.

A meso compound

(b) Maleic acid adds bromine to yield a racemic form.

8.56

Each carbocation can combine with chloride ion to form a racemic alkyl chloride.

8.57

(+) −A
(C$_7$H$_{11}$Br)

$\xrightarrow[\text{ROOR}]{\text{HBr}}$ B   +   C

Opt.active   Opt.inactive   *t*-BuOK

both C$_7$H$_{12}$Br$_2$

(±) −A

D $\xrightarrow[\text{(2) Me}_2\text{S}]{\text{(1) O}_3}$ 2 H—CHO + (cyclopentane-1,3-dione)

(C$_7$H$_{10}$)

(+) A

(+) A
(optically active)

(a meso compound) C

B $\xrightarrow[\text{1 molar eq.}]{t\text{-BuOK}}$ (+) A = (+) A

C $\xrightarrow[\text{1 molar eq.}]{t\text{-BuOK}}$ (+) A + (−) A

A $\xrightarrow[\text{1 molar eq.}]{t\text{-BuOK}}$ $\xrightarrow[\text{(2) Me}_2\text{S}]{\text{(1) O}_3}$ + 2 H—CHO

8.58

D
Optically active
(the other enantiomer is
an equally valid answer)

$\xrightarrow[\text{Pt}]{\text{H}_2}$

E
Optically inactive
(nonresolvable)

Synthesize the Material

8.59 (a)

(±)

(b)

(c)

(d)

8.60 (a)

Products are the same due to symmetry

and   (±)-**A**

(b)

(±)-**B**   and   (±)-**B**

(c)

1) *t*-BuOK, *t*-BuOH

2) H₂, Pd/C

and

C

8.61 (a)

2) Br₂ (1 equiv)

1) NaNH₂, MeI

A

(b)

3) Me₂S
2) O₃

or

1) KOH, EtOH

B

or

Note: multiple answers for B are possible as long as R is an alkyl chain

(c)

3) KMnO₄, HO⁻, heat

or

2) H₂, Lindlar's catalyst or Ni₂B (P-2)

2) Li, liq. NH₃

1) NaNH₂ (2 equiv.), Δ

C

Challenge Problems

8.62

8.63

QUIZ

8.1 A hydrocarbon whose molecular formula is C_7H_{12}, on catalytic hydrogenation (excess H_2/Pt), yields C_7H_{16}. The original hydrocarbon adds bromine and also exhibits an IR absorption band at 3300 cm^{-1}. Which of the following is a plausible choice of structure for the original hydrocarbon?

(a)   (b)   (c)   (d)   (e)

8.2 Select the major product of the dehydration of the alcohol,

(a)   (b)   (c)   (d)   (e)

8.3 Give the major product of the reaction of *cis*-2-pentene with bromine.

(a)	(b)	(c)	(d)	(e) A racemic
CH₃	CH₃	CH₃	CH₃	mixture of
H——Br	Br——H	H——Br	Br——H	(c) and (d)
H——Br	Br——H	Br——H	H——Br	
CH₂CH₃	CH₂CH₃	CH₂CH₃	CH₂CH₃	

8.4 The compound shown here is best prepared by which sequence of reactions?

(a) + NaNH$_2$ $\longrightarrow$ then Br $\longrightarrow$ product

(b) + NaNH$_2$ $\longrightarrow$ then $\longrightarrow$ product

(c) + H$_2$ $\xrightarrow{\text{Pt}}$ product

(d) $\xrightarrow[\text{C}_2\text{H}_5\text{OH}]{\text{NaOC}_2\text{H}_5}$ product

8.5 A compound whose formula is C$_6$H$_{10}$ (Compound **A**) reacts with H$_2$/Pt in excess to give a product C$_6$H$_{12}$, which does not decolorize Br$_2$. Compound **A** does not show IR absorption in the 3200–3400 cm^{-1} region.

Ozonolysis of **A** gives 1 mol of and 1 mol of =O. Give the structure of **A**.

(a)

(b)

(d)

(c)

(e)

8.6 Compound **B** (C$_5$H$_{10}$) does not dissolve in cold, concentrated H$_2$SO$_4$. What is **B**?

(a)

(b)

(c)

(d)

8.7 Which reaction sequence converts cyclohexene to *cis*-1,2-cyclohexanediol? That is,

(a) H$_2$O$_2$

(b) (1) O$_3$ (2) Me$_2$S

(c) (1) OsO$_4$ (2) NaHSO$_3$/H$_2$O

(d) (1) (2) H$_3$O$^+$/H$_2$O

(e) More than one of these

8.8 Which of the following sequences leads to the best synthesis of the compound $\diagdown\!\!\diagup\!\!\equiv$?
(Assume that the quantities of reagents are sufficient to carry out the desired reaction.)

9 NUCLEAR MAGNETIC RESONANCE AND MASS SPECTROMETRY: TOOLS FOR STRUCTURE DETERMINATION

SOLUTIONS TO PROBLEMS

9.1

$$CH_3\overset{\overset{\displaystyle O}{\|}}{C}\underset{(b)}{\diagdown}\underset{(c)\ \ (a)}{OCH_2CH_3}$$

(a) δ 0.8–1.10
(b) δ 2.1–2.6
(c) δ 3.3–3.9

9.2 The presence of two signals in the ^1H NMR spectrum signifies two unique proton environments in the molecule. The higher frequency (larger chemical shift) signal, (a), is consistent with protons and a chlorine on the same carbon. Its triplet nature indicates two hydrogens on the adjacent carbon.

The lower frequency (smaller chemical shift) quintet (b) indicates four adjacent and equivalent hydrogens, which requires two hydrogens on each of two carbons.

Integration data indicate a ratio of (approximately) 2:1, which is actually 4:2 in this case.

Of the isomeric structures with the formula $C_3H_6Cl_2$, only $\underset{(a)}{Cl}\diagup\overset{(b)}{\diagup}\diagdown\underset{(a)}{Cl}$ fits the evidence.

9.3 (a) We see below that if we replace a methyl hydrogen by Cl, then rotation of the methyl group or turning the whole molecule end-for-end gives structures that represent the same compound. This means that all of the methyl hydrogens are equivalent.

replace H with Cl

turn

rotate CH_2Cl group

All of these are the same compound.

Replacing a ring hydrogen by Cl gives another compound, but replacing each ring hydrogen in turn gives the same compound. This shows that all four ring hydrogens are equivalent and that 1,4-dimethylbenzene has only two sets of chemical shift equivalent protons.

All of these are the same compound.

(b) As in (a) we replace a methyl hydrogen by Cl yielding the compound shown. Essentially free rotation of the $-CH_2Cl$ group means that all orientations of that group give the same compound.

Substitution into the other methyl group produces the same compound, as seen when we flip and rotate to form that structure.

Thus the methyl hydrogens are chemical shift equivalent.

Ring substitution produces two different compounds.

Again, flipping and rotating members of each pair demonstrates their equivalence. There are then, three sets of chemical shift equivalent protons—one for the methyl hydrogens and one each for the two different ring hydrogens.

(c)

on replacement of a methyl hydrogen by Cl.

Again, different orientations (conformations) do not result in different compounds. Substitution in the other methyl group yields the same compound.

Replacement of ring hydrogens produces three compounds:

Four sets of chemical shift equivalent hydrogens are present in this isomer: one for the methyl hydrogens and three for the chemically non-equivalent ring hydrogens.

9.4 (a) One

(b) Two

(c) Two

(d) One

(e) Two

(f) Two

9.5 (a)

Diastereomers

(b) Six

(a)
CH_3
$(b)H—C—OH(c)$
$(d)H—C—H \ (e)$
CH_3
(f)

(c) Six signals

9.6 (a) Two

(b) Three

(c) Four

(d) Two

(e) Four

Br $H(c)$
$(a)H—C—C—(d)$
H Br
(b)

(f) Two

$(a) \quad (a)$
$(b) \quad (b)$

(g) Three

(c)
$(b) \quad (b)$
$(a) \quad (a)$

(h) Four

$(c)H \quad H(d)$
$(b) \quad (b)$
$(a) \quad (a)$

(i) Six

$(b) \quad (d)$
$(a) \quad (c) \quad H(e)$
$H(f)$

(j) Five

$Cl \quad H(c)$
$(a)H—C—C—(d)$
$H \quad OH(e)$
(b)

(k) Three

(b)
(a)
(c)

(l) Six

(b)
$(c) \ H \ (a) \quad (b)$
$H \quad H$
$(d)H \quad (a)$
$(e)H \quad H \ (d)$
$H \quad H$
$(f) \quad (c)$

(m) Five

(e)
$(c) \quad (d)$
(b)
(c)
(a)
(b)

(n) Three

(a)
(c)
(b)
O

(o) Three

$O \quad (c)$
$(a) \quad (b) \quad (a)$
$(a) \quad (b)$
$(c) \quad O$

(p) Four

(a)
$(b) \quad (d)$
$(c) \quad (c)$
$(c) \quad (b)$
(d)
(a)

9.7 The determining factors here are the number of chlorine atoms attached to the carbon atoms bearing protons and the deshielding that results from chlorine's electronegativity. In 1,1,2-trichloroethane the proton that gives rise to the triplet is on a carbon atom that bears two chlorines, and the signal from this proton is at higher frequency. In 1,1,2,3,3-pentachloropropane, the proton that gives rise to the triplet is on a carbon atom that bears only one chlorine; the signal from this proton is at lower frequency.

9.8

The signal from the three equivalent protons designated (*a*) should be split into a doublet by the proton *(b)*. This doublet, because of the electronegativity of the two attached chlorines, should occur at higher frequency ($\delta \sim$ 5–6). It should have a relative area (integration value) three times that of (b).

The signal for the proton designated (*b*) should be split into a quartet by the three equivalent protons (*a*). The quartet should occur at lower frequency ($\delta \sim$ 1–2).

9.9 **A** C_3H_7I is

(a) doublet δ 1.9

(b) septet δ 4.35

B $C_2H_4Cl_2$ is

(a) doublet δ 2.08

(b) quartet δ 5.9

9.10 (a)

$\phi = 180°$ $J_{ax,ax} = 8\text{-}10$ Hz

(b)

$\phi = 60°$ $J_{ax,eq} = 2\text{-}3$ Hz

9.11   A chair conformation with both the bromine and the chlorine in equatorial positions is consistent with $J_{1,2} = 7.8$ Hz, since the hydrogens would be axial.

$$J_{ax,ax} = 7.8 \text{ Hz}$$

trans-1-Bromo-2-chlorocyclohexane

9.12   NMR coupling constants would be distinctive for the protons indicated. Compound **A** would have an equatorial-axial proton NMR coupling constant of 2-3 Hz between the indicated protons, whereas **B** would have an axial-axial coupling constant of 8-10 Hz.

A	**B**
$J_{a,b} = 2\text{-}3$ Hz	$J_{a,b} = 8\text{-}10$ Hz

9.13   A single unsplit signal, because the environment of the proton rapidly changes, reversibly, from axial to equatorial.

9.14   **A** is 1-bromo-3-methylbutane. The following are the signal assignments:

(a) δ 23
(b) δ 27
(c) δ 32
(d) δ 42

B is 2-bromo-2-methylbutane. The following are the signal assignments:

(a) δ 11
(b) δ 33
(c) δ 40
(d) δ 68

C is 1-bromopentane. The following are the signal assignments.

(a) δ 14
(b) δ 23
(c) δ 30
(d) δ 33
(e) δ 34

9.15 A peak at $M^+ - 15$ involves the loss of a methyl radical and the formation of a 1° or 2° carbocation.

or

A peak at $M^+ - 29$ arises from the loss of an ethyl radical and the formation of a 2° carbocation.

Since a secondary carbocation is more stable than a methyl carbocation, and since there are two cleavages that form secondary carbocations by loss of an ethyl radical, the peak at $M^+ - 29$ is more intense.

9.16 After loss of a pi bonding electron through electron impact ionization, both peaks arise from allylic fragmentations:

Allyl radical m/z 57

m/z 41
Allyl cation

9.17 The spectrum given in Fig. 9.29 is that of butyl isopropyl ether. The main clues are the peaks at m/z 101 and m/z 73 due to the following fragmentations.

m/z 101

m/z 73

Butyl propyl ether (Fig. 9.30) has no peak at m/z 101 but has a peak at m/z 87 instead.

m/z 87

9.18 (a) First calculate the expected masses of the compound shown for $C_6H_4 + {}^{79}Br + {}^{79}Br$ (M) = 234 m/z, $C_6H_4 + {}^{81}Br + {}^{79}Br$ (M+2) = 236 m/z, and $C_6H_4 + {}^{81}Br + {}^{81}Br$ (M+4) = 238 m/z.

The relative ratio of ${}^{79}Br$ and ${}^{81}Br$ is 1:1. Therefore if there are two bromines in the molecule we must simplify the expression of (1:1)(1:1) = 1:2:1.

The ratio of the peaks will be M (234), M+2 (236), M+4 (238) = (1:2:1)

(b) First calculate the expected masses of the compound shown for $C_6H_4 + {}^{35}Cl + {}^{35}Cl$ (M) = 146 m/z, $C_6H_4 + {}^{35}Cl + {}^{37}Cl$ (M+2) = 148 m/z, and $C_6H_4 + {}^{37}Cl + {}^{37}Cl$ (M+4) = 150 m/z.

The relative ratio of ${}^{35}Cl$ and ${}^{37}Cl$ is 3:1. Therefore if there are two chlorines in the molecule we must simplify the expression of (3:1)(3:1) = 9:6:1.

The ratio of the peaks will be M (146), M+2 (148), M+4 (150) = (9:6:1).

9.19 (1) We hypothesize that the highest m/z peaks with significant intensity, 78 and 80, relate to a molecular ion with isotopic contributions, i.e, that 78 and 80 are M^+ and M+2 peaks, respectively.

(2) Given that the ratio of m/z 78 is about three times the intensity of m/z 80, we hypothesize that chlorine is the likely isotope, i.e., that ${}^{35}Cl$ and ${}^{37}Cl$ are part of the M^+ and M+2 peaks, respectively.

(3) The base peak with a m/z of 43 indicates a $C_3H_7{}^+$ fragment.

(4) The m/z peak at 43 plus either 35 or 37 for the isotopes of chlorine accounts for the M^+ and M+2 at 78 and 80 and suggests the formula C_3H_7Cl.

(5) Based on the ^1H NMR, with a small septet (1H) and a large doublet (2x CH_3's) means that the data indicate the compound is 2-chloropropane.

Cl

2-Chloropropane

NMR Spectroscopy

9.20 Each number represents another signal, with the highest number indicating the total number of proton NMR signals for the molecule. Repeated use of the same number indicates positions bearing homotopic hydrogens.

(a)

(b)

(c)

(d)

(e)

(f)

(g)

(h)

(i)

* Two numbers at the same carbon indicates diastereotopic protons.

9.21 Each number represents another signal, with the highest number indicating the total number of carbon NMR signals for the molecule. Repeated use of the same number indicates positions bearing homotopic carbons.

(a)

(b)

(c)

(d)

(e)

(f)

(g)

(h)

(i)

9.22

Chemical Shift	Splitting	Integration	Structure
(a) 0.9	t	3H	CH$_3$ adjacent to CH$_2$
(b) 1.2	s	6H	Two equivalent CH$_3$ groups adjacent to no hydrogens
(c) 1.3	q	2H	CH$_2$ adjacent only to CH$_3$
(d) 2.0	s	1H	OH

(d) HO (b) (b) (c) (a)

9.23   Compound **G** is 2-bromobutane. Assignments are as follows:

(a) Br (b) (d) (c)

(a) triplet δ 1.05
(b) multiplet δ 1.82
(c) doublet δ 1.7
(d) multiplet δ 4.1

Compound **H** is 2,3-dibromopropene. Assignments are as follows:

(b) H (a) Br
H Br
(c)

(a) δ 4.2
(b) δ 6.05
(c) δ 5.64

Without an evaluation of the alkene coupling constants, it would be impossible to specify whether the alkene has (E), (Z), or geminal substitution. We do know from the integral values and number of signals that the alkene is disubstituted. For the analysis of the alkene substitution pattern these typical alkene coupling constants are useful: H—C=C—H (trans), $J = 12–18$ Hz; H—C=C—H (cis), $J = 6–12$ Hz; C=CH$_2$ (geminal), $J = 0–3$ Hz. Since the alkene signal for compound **H** shows little or no coupling, the data support the designation of **H** as 2,3-dibromopropene.

9.24 Compound **O** is 1,4-cyclohexadiene and **P** is cyclohexane.

(a) δ 26.0   CH₂ (DEPT)
(b) δ 124.5   CH (DEPT)

9.25 The molecular formula of **Q** (C₇H₈) indicates an index of hydrogen deficiency (Section 4.17) of four. The hydrogenation experiment suggests that **Q** contains two double bonds (or one triple bond). Compound **Q**, therefore, must contain two rings.

Bicyclo[2.2.1]hepta-2,5-diene. The following reasoning shows one way to arrive at this conclusion: there is only one signal (δ 144) in the region for a doubly bonded carbon. This fact indicates that the doubly bonded carbon atoms are all equivalent. That the signal at δ 144 is a CH group in the DEPT spectra indicates that each of the doubly bonded carbon atoms bears one hydrogen atom. Information from the DEPT spectra tells us that the signal at δ 85 is a —CH₂— group and the signal at δ 50 is a —C—H group.

The molecular formula tells us that the compound must contain two —C—H groups, and since only one signal occurs in the ¹³C spectrum, these —C—H groups must be equivalent. Putting this all together, we get the following:

(a) δ 50   CH (DEPT)
(b) δ 85   CH₂ (DEPT)
(c) δ 144   CH (DEPT)

9.26 (a) ¹H NMR

Compound A = δ 0.9 (t, 3 H), δ 2.0 (s, 3 H), δ 4.1 (q, 2 H)
Compound B = δ 0.9 (t, 3 H), δ 2.0 (q, 2 H), δ 4.1 (s, 3 H)

(b)   ¹H NMR

¹³C NMR

(c) ^{13}C NMR

9.27 That **S** decolorizes bromine indicates that it is unsaturated. The molecular formula of **S** allows us to calculate an index of hydrogen deficiency equal to 1. Therefore, we can conclude that **S** has one double bond.

The ^{13}C spectrum shows the doubly bonded carbon atoms at δ 130 and δ 135. In the DEPT spectra, one of these signals (δ 130) is a carbon that bears no hydrogen atoms; the other (δ 135) is a carbon that bears one hydrogen atom. We can now arrive at the following partial structure.

The three most upfield signals (δ 19, δ 28, and δ 31) all arise from methyl groups. The signal at δ 32 is a carbon atom with no hydrogen atoms. Putting these facts together allows us to arrive at the following structure.

(a) δ 19	*(d)* δ 32
(b) δ 28	*(e)* δ 130
(c) δ 31	*(f)* δ 135

Although the structure just given is the actual compound, other reasonable structures suggested by these data are

and

Mass Spectrometry

9.28 The compound is butanal. The peak at *m/z* 44 arises from a McLafferty rearrangement.

The peak at *m/z* 29 arises from a fragmentation producing an acylium ion.

H—≡O⁺   +
m/z 29

9.29

m/z: 144 (100.0%), 145 (10.0%)

m/z: 126 (100.0%), 127 (9.9%)   + H_2O

m/z: 111 (100.0%), 112 (8.8%)

m/z: 87 (100.0%), 88 (5.6%)

9.30

m/z: 100 (100.0%), 101 (6.7%)

m/z: 58 (100.0%), 59 (3.4%)   McLafferty rearrangement

m/z: 85 (100.0%), 86 (5.5%)

m/z: 43 (100.0%), 44 (2.2%)

9.31 (a)

m/z: 114 (100.0%), 115 (7.8%) → m/z: 71 (100.0%), 72 (4.4%) $+ \cdot C_3H_7$

m/z: 71 (100.0%), 72 (4.4%) $+ \cdot C_3H_7$

(b)

m/z: 86 (100.0%), 87 (5.6%) m/z: 68 (100.0%), 69 (5.5%) $+ H_2O$

(c)

m/z: 142 (100.0%), 143 (10.0%) m/z: 58 (100.0%), 59 (3.4%)

9.32 First calculate the expected masses of the compound shown for $C_6H_4 + {}^{79}Br + {}^{35}Cl$ (M) $= 190$ m/z, $C_6H_4 + {}^{37}Cl + {}^{79}Br$ (M+2) $= 192$ m/z or $C_6H_4 + {}^{35}Cl + {}^{81}Br$ (M+2) $= 192$ m/z, and $C_6H_4 + {}^{37}Cl + {}^{81}Br$ (M+4) $= 194$ m/z.

The relative ratio of ${}^{79}Br$ and ${}^{81}Br$ is 1:1 and ${}^{35}Cl$ and ${}^{37}Cl$ is 3:1. Therefore if you have 1 bromine and 1 chlorine in the molecule we must simplify the expression of (3:1)(1:1) = 3:4:1.

The ratio of the peaks will be M (190), M+2 (192), M+4 (194) = (3:4:1) or (77%:100%: 24%).

9.33 Ethyl bromide will have a significant M+2 for the ${}^{81}Br$ isotope at 110. The ratio of the peak at m/z 108 and m/z 110 will be approximately 1:1. The spectrum will also show peaks at M+1 and M+3 for the presence of a ${}^{13}C$ isotope (2.2 %)

Methoxybenzene will have a small peak, M+1, for the ${}^{13}C$ isotope (7.8 %)

9.34 The ion, $CH_2 = \overset{+}{N}H_2$, produced by the following fragmentation.

m/z 30

Integrated Structure Elucidation

9.35 (a)

(a) Singlet, δ 1.28 (9H)
(b) Singlet, δ 1.35 (1H)

(b)

(a) Doublet, δ 1.71 (6H)
(b) Septet, δ 4.32 (1H)

(c)

(a) Triplet, δ 1.05 (3H)   C=O, near 1720 cm^{-1} (s)
(b) Singlet, δ 2.13 (3H)
(c) Quartet, δ 2.47 (2H)

(d)

(a) Singlet, δ 2.43 (1H)   O—H, 3200–3550 cm^{-1} (br)
(b) Singlet, δ 4.58 (2H)
(c) Multiplet, δ 7.28 (5H)

(e)

(a) Doublet, δ 1.04 (6H)
(b) Multiplet, δ 1.95 (1H)
(c) Doublet, δ 3.35 (2H)

(f)

(a) Singlet, δ 2.20 (3H)   C=O, near 1720 cm^{-1} (s)
(b) Singlet, δ 5.08 (1H)
(c) Multiplet, δ 7.25 (10H)

(g)

(a) Triplet, δ 1.08 (3H)   C=O (acid), 1715 cm^{-1} (s)
(b) Multiplet, δ 2.07 (2H)   O—H, 2500–3500 cm^{-1} (br)
(c) Triplet, δ 4.23 (1H)
(d) Singlet, δ 10.97 (1H)

(h)

(a) Triplet, δ 1.25 (3H)
(b) Quartet, δ 2.68 (2H)
(c) Multiplet, δ 7.23 (5H)

(i) (a) ⌇⌇⌇ O ⌇⌇ (b) (c) OH (d)

(a) Triplet, δ 1.27 (3H)   C=O (acid), 1715 cm⁻¹ (s)
(b) Quartet, δ 3.66 (2H)   O—H, 2500–3550 cm⁻¹ (br)
(c) Singlet, δ 4.13 (2H)
(d) Singlet, δ 10.95 (1H)

(j) NO₂ (a) (b) (a)

(a) Doublet, δ 1.55 (6H)
(b) Septet, δ 4.67 (1H)

(k) CH₃ (a) O (b) O (a) CH₃ (b)

(a) Singlet, δ 3.25 (6H)
(b) Singlet, δ 3.45 (4H)

(l) O (c) (b) (a)

(a) Doublet, δ 1.10 (6H)   C=O, near 1720 cm⁻¹ (s)
(b) Singlet, δ 2.10 (3H)
(c) Septet, δ 2.50 (1H)

(m) (c) (b) (a) Br

(a) Doublet, δ 2.00 (3H)
(b) Quartet, δ 5.15 (1H)
(c) Multiplet, δ 7.35 (5H)

9.36   Compound **E** is phenylacetylene, $C_6H_5C{\equiv}CH$. We can make the following assignments in the IR spectrum:

The IR spectrum of compound **E** (Problem 9.36). Compound **F** is $C_6H_5CBr_2CHBr_2$.

9.37   Compound **J** is *cis*-1,2-dichloroethene.

We can make the following IR assignments:

3125 cm^{-1}, alkene C—H stretching
1625 cm^{-1}, C=C stretching
 695 cm^{-1}, out-of-plane bending of cis C—H bonds

The ^{1}H NMR spectrum indicates the hydrogens are equivalent.

9.38   (a) Compound **K** is,

(*a*) Singlet, δ 2.15   C=O, near 1720 cm^{-1} (s)
(*b*) Quartet, δ 4.25
(*c*) Doublet, δ 1.35
(*d*) Singlet, δ 3.75

(b) When the compound is dissolved in D_2O, the —OH proton (*d*) is replaced by a deuteron, and thus the ^{1}H NMR absorption peak disappears.

9.39   The IR absorption band at 1745 cm^{-1} indicates the presence of a $\diagdown\!\!\!C\!\!=\!\!O$ group in a five-membered ring, and the signal at δ 220 can be assigned to the carbon of the carbonyl group.

There are only two other signals in the ^{13}C spectrum; the DEPT spectra suggest two equivalent sets of two —CH$_2$— groups each. Putting these facts together, we arrive at cyclopentanone as the structure for **T**.

(*a*) δ 23
(*b*) δ 38
(*c*) δ 220

9.40   (1) Molecular ion = 96 *m/z*

Potential molecular formula = C_7H_{14} or C_6H_8O

(2) The presence of a strong C=O absorption at 1685 cm^{-1} in the IR and the integrals of the ^{1}H NMR spectra totaling 8, and the appearance of 6 unique carbons in the ^{13}C spectra lead to C_6H_8O as the correct molecular formula.

(3) Degree of unsaturation for $C_6H_8O = 3$

(4) ^1H NMR

Letter	ppm	Splitting	Integration	Conclusion
(a)	1.9	s	3H	CH_3 (allylic)
(b)	2.2	q	2H	CH_2 (allylic)
(c)	3.0	t	2H	CH_2 (adjacent to ketone and CH_2)
(d)	6.5	t	1H	CH (adjacent to CH_2)

(5) ^{13}C NMR

1. $\delta 207$ – (ketone)
2. $\delta 145$ – (CH – alkene)
3. $\delta 139$ – (C – alkene)
4. $\delta 37$ – (CH_2 alpha to ketone)
5. $\delta 25$ – (CH_2 allylic)
6. $\delta 16$ – (CH_3 – allylic)

9.41 (1) Molecular ion = 148 m/z
Potential molecular formula = $C_{10}H_{12}O$

(2) IR: 3065 cm^{-1} (CH sp^2 aromatic), 2960 cm^{-1} (CH sp^3), 2760 cm^{-1} (CH aldehyde), 1700 cm^{-1} (C=O conjugate aldehyde), 1600 cm^{-1} (C=C aromatic)

(3) Degree of unsaturation for $C_{10}H_{12}O = 5$

(4) ^1H NMR

Letter	ppm	Splitting	Integration	Conclusion
(a)	1.3	d	6H	Two CH_3 adjacent to CH
(b)	3.1	septet	1H	C-H adjacent to two CH_3
(c)	7.2	d	2H	Aromatic ring disub - para
(d)	7.8	d	2H	Aromatic ring disub - para
(e)	9.8	s	1H	Aldehyde

(5) ^{13}C NMR
  1. $\delta 191$ – CH aldehyde
  2. $\delta 154$ – C aromatic
  3. $\delta 134$ – C aromatic
  4. $\delta 130$ – CH aromatic
  5. $\delta 127$ – CH aromatic
  6. $\delta 36$ – CH aliphatic
  7. $\delta 23$ – CH$_3$ aliphatic

9.42 (1) Molecular ion $= 204$ *m/z*

Potential molecular formula $= C_{15}H_{24}$

(2) IR: 3065 cm^{-1} (C—H sp^2 aromatic), 2960 cm^{-1} (C—H sp^3) 1600 cm^{-1} (C=C aromatic)

(3) Degree of unsaturation for $C_{15}H_{24} = 4$

(4) ^{1}H NMR

Letter	ppm	Splitting	Integration	Conclusion
(a)	1.3	d	6 H	Two CH$_3$ next to CH
(b)	2.8	septet	1 H	CH next to two CH$_3$
(c)	6.9	s	1 H	CH aromatic

(5) ^{13}C NMR
  1. $\delta 148$ – (C aromatic)
  2. $\delta 122$ – (CH aromatic)
  3. $\delta 37$ – (CH aliphatic)
  4. $\delta 23$ – (CH$_3$ aliphatic)

9.43 (1) Molecular Formula = $C_{10}H_{12}O_3$

(2) IR – 3065 cm^{-1} (C—H sp^2 aromatic), 2960 cm^{-1} (C—H sp^3), 1740 cm^{-1} (C=O ester), 1600 cm^{-1} (C=C aromatic)

(3) Degree of unsaturation for $C_{10}H_{12}O_3$ = 5

(4) ^{1}H NMR

Letter	ppm	Splitting	Integration	Conclusion
(a)	3.5	s	2H	CH$_2$
(b)	3.7	s	3H	CH$_3$
(c)	3.8	s	3H	CH$_3$
(d)	6.7	d	2H	CH (aromatic para-sub)
(e)	6.9	d	2H	CH (aromatic para -sub)

The low ppm of the aromatic protons indicates electron donating groups are attached to the ring.

(4) ^{13}C NMR

1. δ171 – (C=O ester)
2. δ160 – (C aromatic)
3. δ130 – (CH aromatic)
4. δ127 – (C aromatic)
5. δ115 – (CH aromatic)
6. δ56 – (CH$_3$ aliphatic)
7. δ52 – (CH$_3$ ester)
8. δ46 – (CH$_2$ benzylic)

Challenge Problems

9.44 The product of the protonation is a relatively stable allylic cation. The six methyl, four methylene, and single methine hydrogens account for the spectral features.

9.45 A McLafferty rearrangement accounts for this outcome, as shown below for butanoic acid. In the case of longer chain carboxylic acids the additional carbons are eliminated in the alkene formed.

m/z 60

9.46 As in the case of IR spectroscopy of alcohols (Section 2.16B), the degree of intermolecular association of the hydroxyl groups is the principal determining factor. If the alcohol is examined in the vapor phase or at very low concentration in a nonpolar solvent such as carbon tetrachloride, the —OH hydrogen absorption occurs at about δ 0.5.

Increasing concentrations of the alcohol up to the neat state (solvent is absent) leads to a progressive shift of the absorption toward the value of δ 5.4.

Hydrogen bonding, , results in a decrease of electron density (de-shielding) at the hydroxyl proton. Thus, the proton peak is shifted downfield. The magnitude of the shift, at some particular temperature, is a function of the alcohol concentration.

9.47 At the lower temperature a DMF molecule is effectively restricted to one conformation due to the resonance contribution of II, which increases the bond order of the C—N bond. At ~ 300 K there is insufficient energy to bring about significant rotation about that bond.

The methyl groups clearly are nonequivalent since each has a unique relationship to the single hydrogen atom. Hence each set of methyl group hydrogens is represented by its own signal (though they are quite close).

At a sufficiently high temperature (> 130°C) there is enough energy available to over-come the barrier to rotation due to the quasi-double bond, $C \cdots N$, and substantially free rotation occurs. The differences between the two sets of methyl group hydrogens are no longer discernable and all six methyl hydrogens are represented by a single signal.

9.48 Formed by fragmentation of the molecular ion, $C_4H_3^+$ possesses the requisite mass. Possible structures are $[CH_2=CH-C\equiv C]^+$ and $[HC=CH-C\equiv CH]^+$, each of which is resonance-stabilized.

9.49

$J_{ab} = 5.3$ Hz

$J_{ac} = 8.2$ Hz

$J_{bc} = 10.7$ Hz

(a)

9 signals

(b)

$J_{ac} = 8.2$ Hz

$J_{ab} = 5.3$ Hz

QUIZ

9.1 Propose a structure that is consistent with each set of the following data.

(a) C_4H_9Br

¹H NMR spectrum
Singlet δ 1.7

(b) $C_4H_7Br_3$

¹H NMR spectrum
Singlet δ 1.95 (3H)
Singlet δ 3.9 (4H)

(c) C_8H_{16}

¹H NMR spectrum
Singlet δ 1.0 (9H)
Singlet δ 1.75 (3H)
Singlet δ 1.9 (2H)
Singlet δ 4.6 (1H)
Singlet δ 4.8 (1H)

IR spectrum
3040, 2950, 1640 cm⁻¹
and other peaks.

(d) $C_9H_{10}O$ ^1H NMR spectrum   IR spectrum
Singlet δ 2.0 (3H)   3100, 3000, 1720,
Singlet δ 3.75 (2H)   740, 700 cm^{-1}
Singlet δ 7.2 (5H)   and other peaks.

(e) $C_5H_7NO_2$ ^1H NMR spectrum   IR spectrum
Triplet δ 1.2 (3H)   2980, 2260, 1750 cm^{-1}
Singlet δ 3.5 (2H)   and other peaks.
Quartet δ 4.2 (2H)   This compound has a nitro group.

9.2 How many ^1H NMR signals would the following compound give?

(a) One   (b) Two   (c) Three   (d) Four   (e) Five

9.3 How many ^1H NMR signals would 1,1-dichlorocyclopropane give?

(a) One   (b) Two   (c) Three   (d) Four   (e) Five

9.4 Which of these C_6H_{14} isomers has the greatest number of ^{13}C NMR signals?

(a) Hexane   (d) 2,2-Dimethylbutane

(b) 2-Methylpentane   (e) 2,3-Dimethylbutane

(c) 3-Methylpentane

9.5 How many ^{13}C NMR signals would be given by the following compound?

(a) 7   (b) 8   (c) 10   (d) 11   (e) 13

9.6 Which of these is a true statement concerning mass spectrometry? (There may be more than one.)

(a) The M^+ peak is always the most prominent (largest m/z).

(b) Only liquids can be analyzed by MS.

(c) Unlike IR and NMR, MS is a destructive method of analysis.

(d) The molecular ion is assigned a value of 100 on the vertical scale.

(e) The initial event in the determination of a mass spectrum in the EI mode is the formation of a radical cation.

9.7 What is the structure of a compound C_5H_{12} which exhibits a prominent MS peak at m/z 57?

10 RADICAL REACTIONS

SOLUTIONS TO PROBLEMS

10.1

$$3° \quad > \quad 2° \quad > \quad 1° \quad > \quad \textbf{Methyl}$$

10.2 The compounds all have different boiling points. They could, therefore, be separated by careful fractional distillation. Or, because the compounds have different vapor pressures, they could easily be separated by gas chromatography. GC/MS (gas chromatography/mass spectrometry) could be used to separate the compounds as well as provide structural information from their mass spectra.

10.3 Their mass spectra would show contributions from the naturally occurring ^{35}Cl and ^{37}Cl isotopes. The natural abundance of ^{35}Cl is approximately 75% and that of ^{37}Cl is approximately 25%. Thus, for CH_3Cl, containing only one chlorine atom, there will be an M^+ peak and an M^++2 peak in roughly a $3:1$ $(0.75:0.25)$ ratio of intensities. For CH_2Cl_2 there will be M^+, M^++2, and M^++4 peaks in roughly a $9:6:1$ ratio, respectively. [The probability of a molecular ion M^+ with both chlorine atoms as ^{35}Cl is $(.75)(.75) = .56$, the probability of an M^++2 ion from one ^{35}Cl and one ^{37}Cl is $2(.75)(.25) = .38$, and the probability of an M^++4 ion peak from both chlorine atoms as ^{37}Cl is $(.25)(.25) = 0.06$; thus, their ratio is $9:6:1$.] For $CHCl_3$ there will be M^+, M^++2, and M^++4, and M^++6 peaks in approximately a $27:27:9:1$ ratio, respectively (based on a calculation by the same method). (This calculation does not take into account the contribution of ^{13}C, ^2H, and other isotopes, but these are much less abundant.)

10.4 The use of a large excess of chlorine allows all of the chlorinated methanes (CH_3Cl, CH_2Cl_2, and $CHCl_3$) to react further with chlorine.

10.5

$$Cl_2 \xrightarrow[\text{or heat}]{hv} 2\,Cl\cdot$$

Step 2a

Step 3a + Cl—Cl ⟶ + Cl·

1,1-dichloroethane

Step 2b Cl· + ⟶ H—Cl +

Step 3b + Cl—Cl ⟶ + Cl·

1,2-dichloroethane

10.6 (a) There is a total of eight hydrogen atoms in propane. There are six equivalent 1° hydrogen atoms, replacement of any one of which leads to propyl chloride, and there are two equivalent 2° hydrogen atoms, replacement of any one of which leads to isopropyl chloride.

If all the hydrogen atoms were equally reactive, we would expect to obtain 75% propyl chloride and 25% isopropyl chloride:

$$\% \text{ Propyl chloride} = {}^6\!/_8 \times 100 = 75\%$$

$$\% \text{ Isopropyl chloride} = {}^2\!/_8 \times 100 = 25\%$$

(b) Reasoning in the same way as in part (a), we would expect 90% isobutyl chloride and 10% *tert*-butyl chloride, if the hydrogen atoms were equally reactive.

$$\% \text{ Isobutyl chloride} = {}^9\!/_{10} \times 100 = 90\%$$

$$\% \textit{ tert-}\text{Butyl chloride} = {}^1\!/_{10} \times 100 = 10\%$$

(c) In the case of propane (see Section 10.5), we actually get more than twice as much isopropyl chloride (55%) than we would expect if the 1° and 2° hydrogen atoms were equally reactive (25%). Clearly, then, 2° hydrogen atoms are more than twice as reactive as 1° hydrogen atoms.

In the case of isobutane, we get almost four times as much *tert*-butyl chloride (37%) as we would get (10%) if the 1° and 3° hydrogen atoms were equally reactive. The order of reactivity of the hydrogens then must be

$$3° \ > \ 2° \ > \ 1°$$

10.7 The hydrogen atoms of these molecules are all equivalent. Replacing any one of them yields the same product.

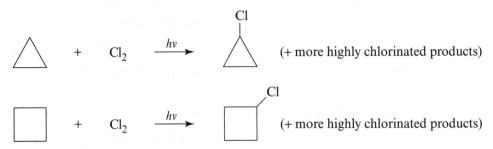

(+ more highly chlorinated products)

(+ more highly chlorinated products)

We can minimize the amounts of more highly chlorinated products formed by using a large excess of the cyclopropane or cyclobutane. (And we can recover the unreacted cyclopropane or cyclobutane after the reaction is over.)

10.8 (a) ⬠   (b)

10.9 (a)

(S)-2-Chloropentane   (2S,4S)-2,4-Dichloro-   (2R,4S)-2,4-Dichloro-
pentane   pentane

(b) They are diastereomers. (They are stereoisomers, but they are not mirror images of each other.)

(c) No, (2R,4S)-2,4-dichloropentane is achiral because it is a meso compound. (It has a plane of symmetry passing through C3.)

(d) No, the achiral meso compound would not be optically active.

(e) Yes, by fractional distillation or by gas chromatography. (Diastereomers have different physical properties. Therefore, the two isomers would have different vapor pressures.)

(f and g)   In addition to the (2S,4S)-2,4-dichloropentane and (2R,4S)-2,4-dichloropentane isomers described previously, we would also get (2S,3S)-2,3-dichloropentane, (2S,3R)-2,3-dichloropentane and the following:

(optically active)   (optically inactive)   (optically active)

10.10 (a) The only fractions that would contain chiral molecules (as enantiomers) would be those containing 1-chloro-2-methylbutane and the two diastereomers of 2-chloro-3-methylbutane. These fractions would not show optical activity, however, because they would contain racemic forms of the enantiomers.

(b) Yes, the fractions containing 1-chloro-2-methylbutane and the two containing the 2-chloro-3-methylbutane diastereomers.

(c) Yes, each fraction from the distillation could be identified on the basis of ^{1}H NMR spectroscopy. The signals related to the carbons where the chlorine atom is bonded would be sufficient to distinguish them. The protons at C1 of 1-chloro-2-methylbutane would be a doublet due to splitting from the single hydrogen at C2. There would be no proton signal for C2 of 2-chloro-2-methylbutane since there are no hydrogens bonded at C2 in this compound; however there would be a strong singlet for the six hydrogens of the geminal methyl groups. The proton signal at C2 of 2-chloro-3-methylbutane would approximately be a quintet, due to combined splitting from the three hydrogens at C1 and the single hydrogen at C3. The protons at C1 of 1-chloro-3-methylbutane would be a triplet due to splitting by the two hydrogens at C2.

10.11 (a) (i)   (ii)

(iii)

(b) (i) No isomers   (ii)   (iii)

(c) (i)   from

(ii)   from

(iii)   from

10.12

10.13

leads to 1-chloro-1-phenylpropane

I

leads to 2-chloro-1-phenylpropane

II

leads to 1-chloro-3-phenylpropane

III

The major product is 1-chloro-1-phenylpropane because **I** is the most stable radical. It is a benzylic radical and therefore is stabilized by resonance.

10.14 (a) *Retrosynthetic Analysis*

Synthesis

(b) *Retrosynthetic Analysis*

Synthesis

(c) *Retrosynthetic Analysis*

Synthesis

[from (a)]

(d) *Retrosynthetic Analysis*

Synthesis

[from (a)]

10.15 Head-to-tail polymerization leads to a more stable radical on the growing polymer chain. In head-to-tail coupling, the radical is 2° (actually 2° benzylic, and as we shall see in Section 15.12A this makes it even more stable). In head-to-head coupling, the radical is 1°.

10.16 (a)

(from initiator)   Monomer

(b)

(from initiator)   Monomer

10.17 In the cationic polymerization of isobutylene (see text), the growing polymer chain has a stable 3° carbocation at the end. In the cationic polymerization of ethene, for example, the intermediates would be much less stable 1° cations.

1° Carbocation

With vinyl chloride and acrylonitrile, the cations at the end of the growing chain would be destabilized by electron-withdrawing groups.

Radical Mechanisms and Properties

10.18 (1) $Br_2 \xrightarrow{hv} 2\,Br^{\cdot}$

(2)

(3)

then 2,3 2,3 etc.

10.19 For formation of , the rate-determining step is:

For formation of Br, the rate-determining step is:

Reaction coordinate

A$^{\bullet}$ is a 3° alkyl radical and more stable than B$^{\bullet}$, a 1° alkyl radical, a difference anticipated by the relative energies of the transition states.

As indicated by the potential energy diagram, the activation energy for the formation of A‡ is less than that for the formation of B‡. The lower energy of A‡ means that a greater fraction of bromine atom-alkane collisions will lead to A$^{\bullet}$ rather than to B$^{\bullet}$.

Note there is a statistical advantage to the formation of B$^{\bullet}$ (6:1) but this is outweighed by the inherently greater reactivity of a 3° hydrogen.

10.20

and are the only monosubstitution products which can be formed from cyclopentane and ethane, respectively. (However, direct iodination of alkanes is not a feasible reaction, practically speaking.)

and would be formed in amounts much larger than the isomeric alternatives due to the highly selective nature of bromine.

Formation of and would be accompanied by considerable amounts of isomeric byproducts in each case.

10.21 *Chain-Initiating Step*

$$Cl_2 \xrightarrow[\text{light}]{\text{heat, } h\nu} 2\,Cl\cdot$$

Chain-Propagating Steps

10.22 (a) Three

I   II   III

Enantiomers as a
racemic form

(b) Only two: one fraction containing I, and another fraction containing the enantiomers II and III as a racemic form. (The enantiomers, having the same boiling points, would distill in the same fraction.)

(c) Both of them.

(d) The fraction containing the enantiomers.

(e) In the ^{1}H spectrum for 1-chlorobutane the signal with the largest chemical shift (highest frequency) would be that for $-CH_2Cl$; it would be a triplet. The corresponding signal for $-CH-$ in either enantiomer of 2-chlorobutane (also with the largest
$\qquad\quad |$
$\qquad\quad Cl$
chemical shift) would be an approximate sextet.

   The DEPT spectra for 1-chlorobutane would identify one CH_3 group and three CH_2 groups; for 2-chlorobutane, two (non-equivalent) CH_3 groups, one CH_2 group, and one CH group would be specified.

(f) Molecular ions from both 1-chlorobutane and the 2-chlorobutane enantiomers would be present (but probably of different intensities). $M^{+\cdot}$ +2 peaks would also be present. Both compounds would likely undergo $C-Cl$ bond cleavage to yield $C_4H_9^+$ cations. The mass spectrum of 1-chlorobutane would probably show loss of a propyl radical by $C-C$ bond cleavage adjacent to the chlorine, resulting in an *m/z* 49 peak for CH_2Cl^+ (and *m/z* 51 from ^{37}Cl). Similar fragmentation in 2-chlorobutane would produce an *m/z* 63 peak for CH_3CHCl^+ (and *m/z* 65).

10.23 (a) Five

(R)-2-
Chlorobutane   A   B   C   D

                                                        E

(b) Five. None of the fractions would be a racemic form.

(c) The fractions containing **A**, **D**, and **E**. The fractions containing **B** and **C** would be optically inactive. (**B** contains no chirality center and **C** is a meso compound.)

10.24 (a) Oxygen-oxygen bonds are especially weak, that is,

$$HO-OH \qquad DH° = 214 \text{ kJ mol}^{-1}$$
$$CH_3CH_2O-OCH_3 \qquad DH° = 184 \text{ kJ mol}^{-1}$$

This means that a peroxide will dissociate into radicals at a relatively low temperature.

$$RO\!-\!OR \xrightarrow{\text{100-200 °C}} 2\,RO\cdot$$

Oxygen-hydrogen single bonds, on the other hand, are very strong. (For HO — H, $DH° = 499$ kJ mol^{-1}.) This means that reactions like the following will be highly exothermic.

$$RO\cdot \;+\; R\!-\!H \longrightarrow RO\!-\!H \;+\; R\cdot$$

(b) *Step 1*

$$\text{(CH}_3)_3C\!-\!O\!-\!O\!-\!C(CH_3)_3 \xrightarrow{\text{heat}} 2\;(CH_3)_3C\!-\!O\cdot$$

Step 2

$$(CH_3)_3C\!-\!O\cdot \;+\; R\!-\!H \longrightarrow (CH_3)_3C\!-\!OH \;+\; R\cdot$$

} Chain Initiation

Step 3 $R\cdot \;+\; Br\!-\!Br \longrightarrow R\!-\!Br \;+\; Br\cdot$

Step 4 $Br\cdot \;+\; R\!-\!H \longrightarrow H\!-\!Br \;+\; R\cdot$

} Chain Propagation

10.25

(3°)   >   (2°)   >   (1°)   ~   (1°)

10.26 Single-barbed arrows show conversion of the enediyne system to a 1,4-benzenoid diradical via the Bergman cycloaromatization. Each alkyne contributes one electron from a pi bond to the new sigma bond. The remaining electrons in each of the pi bonds become the unpaired electrons of the 1,4-benzenoid diradical. The diradical is a highly reactive intermediate that reacts further to abstract hydrogen atoms from the DNA sugar-phosphate backbone. The new radicals formed on the DNA lead to bond fragmentation along the backbone and to double-stranded cleavage of the DNA.

Calicheamicin enediyne
intermediate

1,4-Benzenoid diradical

10.27

$$DH°(\text{kJ mol}^{-1})$$

$CH_2 = CH - H$	465
$(CH_3)_2CH - H$	413
$CH_2 = CHCH_2 - H$	369

(a) It is relatively difficult to effect radical halogenation at H_a because of the large $DH°$ for dissociation of a bond to an alkenyl (vinylic) hydrogen.

(b) Substitution of H_b occurs more readily than H_c because $DH°$ for the generation of an allylic radical is significantly smaller than that for the formation of a simple 2° radical.

10.28 (1)

(2)

(3)

then 2,3 2,3 etc.

Synthesis

10.29 (a) CH_3CH_3 $\xrightarrow[\substack{\text{heat,} \\ \text{light}}]{Br_2}$ ╲╱Br $\xrightarrow[(S_N2)]{NaI}$ ╲╱I

(b) ╲╱Br $\xrightarrow[(S_N2)]{NaOH}$ ╲╱OH $\xrightarrow[(-H_2)]{NaH}$ ╲╱O⁻Na⁺

[from part (a)]

(c) $\xrightarrow[\substack{\text{heat,} \\ \text{light}}]{Br_2}$ $\xrightarrow[\substack{t\text{-BuOH} \\ \text{heat (E2)}}]{t\text{-BuOK}}$

(d)

$$\text{(CH}_3)_2\text{CHCH}_2\text{CH}_3 \xrightarrow[\substack{\text{heat,} \\ \text{light}}]{\text{Br}_2} \text{(2-bromo-2-methylbutane)} \xrightarrow[\substack{\text{EtOH,} \\ \text{heat} \\ \text{(E2)}}]{\text{NaOEt}} \text{(2-methyl-2-butene)}$$

$$\xrightarrow[\substack{\text{ROOR,} \\ \text{heat, light}}]{\text{HBr,}}$$

(by anti-Markovnikov addition)

(e) $\text{CH}_4 \xrightarrow[\substack{\text{heat,} \\ \text{light}}]{\text{Br}_2} \text{CH}_3\text{Br}$

$$\text{H}\!-\!\!\equiv\!\!-\text{H} \xrightarrow[\substack{\text{liq. NH}_3 \\ \text{2) CH}_3\text{Br}}]{\text{1) NaNH}_2} \text{H}\!-\!\!\equiv\!\!- \xrightarrow[\substack{\text{liq. NH}_3 \\ \text{2) CH}_3\text{Br}}]{\text{1) NaNH}_2} \equiv$$

(f) $\text{H}\!-\!\!\equiv\!\!-\text{H} \xrightarrow[\substack{\text{liq. NH}_3 \\ \text{2)} \quad \text{Br} \\ \text{[from part (a)]}}]{\text{1) NaNH}_2} \equiv\!\!-\text{H} \xrightarrow[\substack{\text{or} \\ \text{H}_2, \text{Lindlar's} \\ \text{catalyst}}]{\text{H}_2, \text{Ni}_2\text{B}}$

$$\xrightarrow{\text{HA, H}_2\text{O}} \text{OH}$$

(g) $\underset{\text{[from part (a)]}}{\diagdown\!\!\diagup\!\text{Br}} \; + \; \text{Na}^+\!{}^-\text{N}\!=\!\overset{+}{\text{N}}\!=\!\text{N}^- \xrightarrow{\text{S}_\text{N}2} \diagup\!-\text{N}\!=\!\overset{+}{\text{N}}\!=\!\text{N}^-$

10.30 (a) $\diagdown\!\!\diagup \xrightarrow[hv]{\text{Br}_2} \diagup\!\!\diagdown\!\text{Br} \xrightarrow[\text{EtOH}]{\text{NaOEt}} \diagup\!\!\diagdown \xrightarrow[\text{H}_2\text{SO}_4]{\text{H}_2\text{O}} \diagup\!\!\diagdown\!\text{OH}$

(aqueous solvolysis of 2-bromo-2-methylpropane would also yield the desired alcohol)

(b) $\diagup\!\!\diagdown \xrightarrow[\text{ROOR}]{\text{HBr}} \diagup\!\!\diagdown\!\text{Br}$

[from (a)]

(c) $\diagup\!\!\diagdown \xrightarrow[\substack{(2) \text{ H}_2\text{O}_2, \text{HO}^-}]{(1) \text{ BH}_3 : \text{THF}} \diagup\!\!\diagdown\!\text{OH}$

[from (a)]

(d)

(1)* $\left(\underset{F}{\overset{F}{\underset{F}{\bigg|}}} C - \overset{O}{\overset{\|}{C}} - O^- \right)_2 Hg$ / THF-CH$_3$OH

(2) NaBH$_4$, HO$^-$

→ OCH$_3$

[from (a)]   *[Hg(OAc)$_2$ would also work in step (1)]

(e)

□ $\xrightarrow[hv]{Br_2}$ □—Br $\xrightarrow[EtOH]{NaOEt}$ ◇ $\xrightarrow{Br_2}$ (Br, Br structure)

+ enantiomer

(f)

⬡ $\xrightarrow[hv]{Br_2}$ ⬡—Br $\xrightarrow[EtOH]{NaOEt}$ ◯

$\xrightarrow[H_2O]{Br_2}$ (OH, Br structure) $\xrightarrow[H_2O]{NaOH}$ (epoxide, O)

+ enantiomer

10.31

(a) retro: (structure with O) ⟹ (structure with Br) ⟹ (isobutylene)

synthesis: (isobutylene) $\xrightarrow[hv]{NBS}$ (allylic Br) $\xrightarrow{CH_3ONa}$ (ether product)

(b) retro: Br—CH(OH)—CN ⟹ (CN allyl) ⟹ (Br allyl) ⟹ (propene)

synthesis: ∥ $\xrightarrow[hv]{NBS}$ (allyl Br) $\xrightarrow{NaCN}$ (allyl CN) $\xrightarrow[H_2O]{Br_2}$ Br—CH(OH)—CN

(c)

retro:

synthesis:

NBS
hv

(2)

(1) Hg(OAc)$_2$,
THF-H$_2$O
(2) NaBH$_4$

(1) NaH

Br$_2$ | H$_2$O

Note: H$_2$O with cat. acid would
also hydrate propene

(1) NaNH$_2$ (3 equiv)
(2) NH$_4$Cl

Synthesizing the Material

10.32 (a)

retro:

synthesis:

NBS
hv

NaCN

(b)

retro:

synthesis:

NBS
hv

NaNH$_2$

(c)

retro:

synthesis:

(d)

retro:

synthesis:

(e)

retro:

synthesis:

10.33 (a)

Polystyrene

(b)

(c)

Note: other orders of steps here could work, such as hydrogenation prior to sulfide displacement based on the sequence shown above.

10.34 (a)

Note: bromination at the site indicated is most likely since the position is both allylic and benzylic.

A and **A**

(b)

B

10.35 (a)

Note: in the precursor to the ozonolysis step there could be additional attachments on the alkene. If you had such attachments, make sure your bromide is in the right position for a selective elimination.

(b)

Challenge Problems

10.36 (1)

AIBN

(2)

(3)

(4)

then 3,4 3,4 etc.

10.37 (1)

(from AIBN)

(2) Bu₃Sn· + ⟶ Bu₃SnI +

(3) + Bu₃SnH ⟶ + Bu₃Sn·

then 2,3 2,3 etc.

10.38

Ph· + Bu₃SnH ⟶ PhH + Bu₃Sn·

Bu₃Sn· + ⟶ Bu₃SnBr +

10.39 Besides direct H· abstraction from C5 there would be many H· abstractions from the three methyl groups, leading to:

Any of these radicals could then, besides directly attacking chlorine, intramolecularly abstract H· from C5 (analogous to the "back biting" that explains branching during alkene radical polymerization).

10.40 (1) Ph—C(=O)—O—O—C(=O)—Ph $\xrightarrow{\text{heat}}$ 2 Ph—C(=O)—O·

(2) Ph—C(=O)—O· $\longrightarrow$ Ph· + CO_2

(3) Ph· + [isovaleraldehyde structure]—H(=O) $\longrightarrow$ PhH + [acyl radical structure]·(=O)

(4) [acyl radical]·(=O) $\longrightarrow$ [isobutyl radical]· + CO

(5) [isobutyl radical]· + [aldehyde]—H(=O) $\longrightarrow$ [isobutane] + [acyl radical]·(=O)

then 4,5 4,5 etc.

10.41 HO—C(CH₃)₂—CH₃ + HO· $\longrightarrow$ HO—C(CH₃)₂—CH₂· + HOH

$\xrightarrow{\text{dimerization}}$ HO—C—C—C—C—OH

X

10.42 Unpaired electron density in the methyl radical is localized solely above and below the carbon atom, in the region corresponding to the p orbital of the approximately sp^2-hybridized carbon atom. The ethyl radical shows some unpaired electron density at the adjacent hydrogen atoms, especially the hydrogen atom that in the conformation shown has its H—C sigma bond aligned parallel to the unpaired electron density of the p orbital of the radical. The larger size of the spin density lobe of the hydrogen with its H—C bond parallel to the p orbital of the radical indicates hyperconjugation with the radical. This effect is even more pronounced in the *tert*-butyl radical, where three hydrogen atoms with H—C sigma bonds parallel to the radical p orbital (two hydrogens above the carbon plane and one below in the conformation shown) have larger unpaired electron density volumes than the other hydrogen atoms.

10.43 The sequence of molecular orbitals in O_2 is $\sigma 1s$ (HOMO-7), $\sigma^* 1s$ (HOMO-6), $\sigma 2s$ (HOMO-5), $\sigma^* 2s$ (HOMO-4), $\pi 2p_y$ (HOMO-3), $\pi 2p_z$ (HOMO-2), $\sigma 2p_x$ (HOMO-1), $\pi^* 2p_y$ (HOMO), $\pi^* 2p_z$ (LUMO). Therefore (a) HOMO-3 and HOMO-2 represent bonding pi molecular orbitals, (b) HOMO-1 is a bonding sigma molecular orbital comprised of overlap of the p_x orbitals on each oxygen, and (c) the HOMO and LUMO represent the antibonding pi molecular orbital counterparts to the bonding pi molecular orbitals represented by HOMO-3 and HOMO-2. Note that the s and p orbitals in O_2 are not hybridized. A diagram of the orbitals and their respective energy levels is shown below.

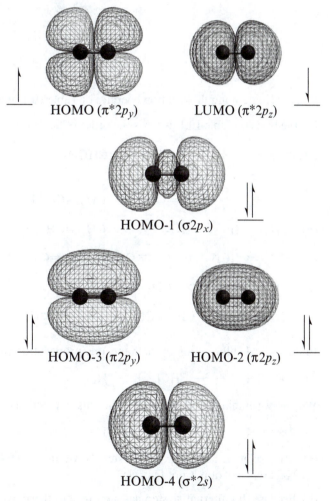

HOMO ($\pi^* 2p_y$)   LUMO ($\pi^* 2p_z$)

HOMO-1 ($\sigma 2p_x$)

HOMO-3 ($\pi 2p_y$)   HOMO-2 ($\pi 2p_z$)

HOMO-4 ($\sigma^* 2s$)

(orbital diagram continues on the next page)

HOMO-5 ($\sigma 2s$)

HOMO-6 ($\sigma^* 1s$)

HOMO-7 ($\sigma 1s$)

QUIZ   Use the single-bond dissociation energies of Table 10.1 (page 205):

10.1 On the basis of Table 10.1, what is the order of decreasing stability of the radicals,

$$HC{\equiv}C\cdot \qquad CH_2{=}\overset{\cdot}{C}H \qquad CH_2{=}CH\overset{\cdot}{C}H_2 \quad ?$$

(a) $HC{\equiv}C\cdot \; > \; CH_2{=}\overset{\cdot}{C}H \; > \; CH_2{=}CH\overset{\cdot}{C}H_2$

(b) $CH_2{=}\overset{\cdot}{C}H \; > \; HC{\equiv}C\cdot \; > \; CH_2{=}CH\overset{\cdot}{C}H_2$

(c) $CH_2{=}CH\overset{\cdot}{C}H_2 \; > \; HC{\equiv}C\cdot \; > \; CH_2{=}\overset{\cdot}{C}H$

(d) $CH_2{=}CH\overset{\cdot}{C}H_2 \; > \; CH_2{=}\overset{\cdot}{C}H \; > \; HC{\equiv}C\cdot$

(e) $CH_2{=}\overset{\cdot}{C}H \; > \; CH_2{=}CH\overset{\cdot}{C}H_2 \; > \; HC{\equiv}C\cdot$

10.2 In the radical chlorination of methane, one propagation step is shown as

$$Cl\cdot \; + \; CH_4 \longrightarrow HCl \; + \; \cdot CH_3$$

Why do we eliminate the possibility that this step goes as shown below?

$$Cl\cdot \; + \; CH_4 \longrightarrow CH_3Cl \; + \; H\cdot$$

(a) Because in the next propagation step, H· would have to react with Cl_2 to form Cl· and HCl; this reaction is not feasible.

(b) Because this alternative step has a more endothermic $\Delta H°$ than the first.

(c) Because free hydrogen atoms cannot exist.

(d) Because this alternative step is not consistent with the high photochemical efficiency of this reaction.

Table 10.1 Single-bond homolytic dissociation energies $DH°$ at 25°C

$$A : B \longrightarrow A· + B·$$

Compound	kJ mol^{-1}	Compound	kJ mol^{-1}
H—H	436	$(CH_3)_2CH$—Br	298
D—D	443	$(CH_3)_2CH$—I	222
F—F	159	$(CH_3)_2CH$—OH	402
Cl—Cl	243	$(CH_3)_2CH$—OCH_3	359
Br—Br	193	$(CH_3)_2CHCH_2$—H	422
I—I	151	$(CH_3)_3C$—H	400
H—F	570	$(CH_3)_3C$—Cl	349
H—Cl	432	$(CH_3)_3C$—Br	292
H—Br	366	$(CH_3)_3C$—I	227
H—I	298	$(CH_3)_3C$—OH	400
CH_3—H	440	$(CH_3)_3C$—OCH_3	348
CH_3—F	461	$C_6H_5CH_2$—H	375
CH_3—Cl	352	$CH_2=CHCH_2$—H	369
CH_3—Br	293	$CH_2=CH$—H	465
CH_3—I	240	C_6H_5—H	474
CH_3—OH	387	$HC\equiv C$—H	547
CH_3—OCH_3	348	CH_3—CH_3	378
CH_3CH_2—H	421	CH_3CH_2—CH_3	371
CH_3CH_2—F	444	$CH_3CH_2CH_2$—CH_3	374
CH_3CH_2—Cl	353	CH_3CH_2—CH_2CH_3	343
CH_3CH_2—Br	295	$(CH_3)_2CH$—CH_3	371
CH_3CH_2—I	233	$(CH_3)_3C$—CH_3	363
CH_3CH_2—OH	393	HO—H	499
CH_3CH_2—OCH_3	352	HOO—H	356
$CH_3CH_2CH_2$—H	423	HO—OH	214
$CH_3CH_2CH_2$—F	444	$(CH_3)_3CO$—$OC(CH_3)_3$	157
$CH_3CH_2CH_2$—Cl	354		
$CH_3CH_2CH_2$—Br	294		
$CH_3CH_2CH_2$—I	239	$\underset{\parallel}{\overset{O}{}}\ \underset{\parallel}{\overset{O}{}}$ C_6H_5CO—OCC_6H_5	139
$CH_3CH_2CH_2$—OH	395	CH_3CH_2O—OCH_3	184
$CH_3CH_2CH_2$—OCH_3	355	CH_3CH_2O—H	431
$(CH_3)_2CH$—H	413		
$(CH_3)_2CH$—F	439	$\overset{O}{\underset{\parallel}{}}$ CH_3C–H	364
$(CH_3)_2CH$—Cl	355		

10.3 Pure (*S*)-$CH_3CH_2CHBrCH_3$ is subjected to monobromination to form several isomers of $C_4H_8Br_2$. Which of the following is *not* produced?

(a)   (b)   (c)

(d)   (e)

10.4   Using the data of Table 10.1, calculate the heat of reaction, $DH°$, of the reaction,

$$CH_3CH_3 \ + \ Br_2 \ \longrightarrow \ \diagup\!\!\diagdown_{Br} \ + \ HBr$$

(a) 47 kJ mol^{-1} (b) -47 kJ mol^{-1} (c) 1275 kJ mol^{-1}

(d) -1275 kJ mol^{-1} (e) -157 kJ mol^{-1}

10.5   What is the most stable radical that would be formed in the following reaction?

Cl· +   (isobutane structure) $\longrightarrow$ [ ]   +   HCl

10.6   The reaction of 2-methylbutane with chlorine would yield a total of _____ different monochloro products (including stereoisomers).

11 ALCOHOLS AND ETHERS: SYNTHESIS AND REACTIONS

SOLUTIONS TO PROBLEMS

Note: A mixture of bond-line and condensed structural formulas is used for solutions in this chapter so as to aid your facility in using both types.

11.1 These names mix two systems of nomenclature (functional class and substitutive; see Section 4.3F). The proper names are: isopropyl alcohol (functional class) or 2-propanol (substitutive), and *tert*-butyl alcohol (functional class) or 2-methyl-2-propanol (substitutive). Names with mixed systems of nomenclature should not be used.

11.2 (a)

1-Propanol
or propan-1-ol
(Propyl alcohol)

2-Propanol
or propan-2-ol
(Isopropyl alcohol)

Methoxyethane
(Ethyl methyl ether)

(b)

1-Butanol
or
butan-1-ol
(Butyl alcohol)

2-Methyl-1-propanol
or
2-methylpropan-1-ol
(Isobutyl alcohol)

2-Butanol
or
butan-2-ol
(*sec*-Butyl alcohol)

2-Methyl-2-propanol
or
2-methylpropan-2-ol
(*tert*-Butyl alcohol)

1-Methoxypropane
(Methyl propyl ether)

Ethoxyethane
(Diethyl ether)

2-Methoxypropane
(Isopropyl methyl ether)

11.3 (a)   (b)   (c)

11.4   A rearrangement takes place.

(a)

2,3-Dimethyl-2-butanol
(major product)

(b) (1) $Hg(OAc)_2$/THF-H_2O; (2) $NaBH_4$, HO^-
(oxymercuration-demercuration)

11.5   (a)

Stronger	Stronger	Weaker	Weaker
acid	base	base	acid

(b)

Stronger	Stronger	Weaker	Weaker
acid	base	base	acid

(c)

Weaker	Weaker	Stronger	Stronger
acid	base	base	acid

(d)

Weaker	Weaker	Stronger	Stronger
acid	base	base	acid

11.6

11.7 (a) Tertiary alcohols react faster than secondary alcohols because they form more stable carbocations; that is, 3° rather than 2°:

(b) CH_3OH reacts faster than 1° alcohols because it offers less hindrance to S_N2 attack. (Recall that CH_3OH and 1° alcohols must react through an S_N2 mechanism.)

11.8 (a) CF_3SO_2Cl $\xrightarrow[\text{base (–HCl)}]{}$

(b) CH_3SO_2Cl $\xrightarrow[\text{base (–HCl)}]{}$

(c) H_3C-⟨⟩$-SO_2Cl$ $\xrightarrow[\text{base (–HCl)}]{CH_3OH}$ H_3C-⟨⟩$-SO_2OCH_3$

11.9 (a)

(b)

cis-2-Methyl-
cyclohexanol

11.10 Use an alcohol containing labeled oxygen. If all of the label appears in the sulfonate ester, then one can conclude that the alcohol C—O bond does not break during the reaction:

$$R{-}^{18}O{-}H \;+\; R'{-}SO_2Cl \;\xrightarrow[(-HCl)]{\text{base}}\; R{-}^{18}O{-}SO_2{-}R'$$

11.11

This reaction succeeds because a 3° carbocation is much more stable than a 1° carbocation. Consequently, mixing the 1° alcohol and H_2SO_4 does not lead to formation of appreciable amounts of a 1° carbocation. However, when the 3° alcohol is added, it is rapidly converted to a 3° carbocation, which then reacts with the 1° alcohol that is present in the mixture.

11.12 (a) (1)

$(L = X, OSO_2R, \text{or } OSO_2OR)$

(2)

$(L = X, OSO_2R, \text{or } OSO_2OR)$

(b) Both methods involve S_N2 reactions. Therefore, method (1) is better because substitution takes place at an unhindered methyl carbon atom. In method (2), where substitution must take place at a relatively hindered secondary carbon atom, the reaction would be accompanied by considerable elimination.

11.13 (a)

(b) The —Ö:⁻ group must displace the Cl⁻ from the backside,

trans-2-Chlorocyclohexanol

(c) Backside attack is not possible with the cis isomer (below); therefore, it does not form an epoxide.

cis-2-Chlorocyclohexanol

11.14

11.15 (a)

(b)

11.16 (a)

S_N2 attack by I^- occurs at the methyl carbon atom because it is less hindered; therefore, the bond between the *sec*-butyl group and the oxygen is not broken.

(b)

In this reaction the much more stable *tert*-butyl cation is produced. It then combines with I⁻ to form *tert*-butyl iodide.

11.17

11.18

11.19 (a)

Methyl Cellosolve

(b) An analogous reaction yields ethyl cellosolve,

.

(c)

(d)

(e)

11.20 The reaction is an S_N2 reaction, and thus nucleophilic attack takes place much more rapidly at the primary carbon atom than at the more hindered tertiary carbon atom.

Major product

MeÖ:⁻ + [epoxide structure] slow / MeOH → HO OCH₃ Minor product

11.21 Ethoxide ion attacks the epoxide ring at the primary carbon because it is less hindered, and the following reactions take place.

Cl + :ÖEt ⟶ Cl :Ö:⁻ OEt

⟶ [epoxide structure] OEt

11.22

+ H₃O⁺ ⇌ H₂Ö: [structure with OH⁺ and H₂O⁺]

H₂O / −H₃O⁺ → [structure with OH, OH, H] (plus enantiomer, by attack at the other epoxide carbon)

11.23 A: 2-Butyne
B: H₂, Ni₂B (P-2)
C: *m*CPBA

D: [epoxide structure]

E: MeOH, cat. acid

11.24 (a) [crown ether structure] 15-Crown-5   (b) [crown ether structure] 12-Crown-4

Problems

Nomenclature

11.25 (a) 3,3-Dimethyl-1-butanol or 3,3-dimethylbutan-1-ol

(b) 4-Penten-2-ol or pent-4-en-2-ol

(c) 2-Methyl-1,4-butanediol or 2-methylbutan-1,4-diol

(d) 2-Phenylethanol

(e) 1-Methylcyclopentanol

(f) *cis*-3-Methylcyclohexanol

11.26 (a) [structure: cis-2-buten-1-ol with OH]   (b) [structure: HO, HO, OH, H]

(c) [cyclopentane-1,2-diol, HO, OH]   (d) [cyclobutane with OH and ethyl]

(e) [structure with OH, Cl, alkyne chain]   (f) [tetrahydrofuran, O]

(g) [structure with O, ethyl ether]   (h) [phenyl ethyl ether, O]

(i) [diisopropyl ether, O]   (j) [structure with O, OH]

Reactions and Synthesis

11.27 (a) [pentene structure]   (c) [structure] or [structure] or [structure]

(b) [vinylcyclopentane]   (d) [ethylidenecyclopentane] or [ethylcyclopentene]

11.28 (a) [3,3-dimethyl-1-butene]   (c) [styrene]

(b) [hexene]   (d) [methylcyclopentene]

11.29 (a) 3 [alkene] $\xrightarrow[\text{(hydroboration)}]{\text{BH}_3 : \text{THF}}$ $\left(\text{[chain]}\right)_3 \text{B}$

$\xrightarrow[\text{(oxidation)}]{\text{H}_2\text{O}_2/\text{HO}^-}$ [butanol with OH]

(b) [chain]Cl $\xrightarrow{\text{HO}^-}$ [chain]OH

(c) [structure with Cl] $\xrightarrow{t\text{-BuOK}/t\text{-BuOH}}$ [butene]

$\xrightarrow[\text{ROOR}]{\text{HBr}}$ [chain]Br $\xrightarrow{\text{HO}^-}$ [chain]OH

(or by hydroboration-oxidation of 1-butene formed by the elimination reaction)

(d) [but-1-yne] $\xrightarrow[\text{Ni}_2\text{B (P-2)}]{\text{H}_2}$ [but-1-ene]

$\xrightarrow{\text{[as in (a)]}}$ [butan-1-ol, with OH]

11.30 (a) 3 [2-butanol, with OH] $+$ PBr$_3$ $\longrightarrow$ 3 [2-bromobutane, with Br] $+$ H$_3$PO$_3$

(b) [1-butanol, with OH] $\xrightarrow{\text{PBr}_3}$ [1-bromobutane, with Br] $\xrightarrow[\text{t-BuOH}]{\text{t-BuOK}}$

[1-butene] $\xrightarrow[\text{(no peroxides)}]{\text{HBr}}$ [2-bromobutane, with Br]

(c) See (b) above.

(d) [but-1-yne] $\xrightarrow[\text{Ni}_2\text{B (P-2)}]{\text{H}_2}$ [1-butene] $\xrightarrow[\text{(no peroxides)}]{\text{HBr}}$ [2-bromobutane, with Br]

11.31 (a) [isobutylene] $\xrightarrow[\text{(2) H}_2\text{O}_2,\ \text{HO}^-]{\text{(1) BH}_3 : \text{THF}}$ [isobutyl alcohol, with OH]

(b) [isobutylene] $\xrightarrow[\text{(2)}]{\text{(1) BH}_3 : \text{THF}}$ [isobutane with T]

[acetic acid OT ester structure: CH$_3$C(=O)OT]

(c) [isobutylene] $\xrightarrow{\text{BD}_3 : \text{THF}}$ [organoborane R$_2$B with D] $\xrightarrow{\text{CH}_3\text{C(=O)OT}}$ [product with D and T]

(d) [isobutyl alcohol, with OH] $\xrightarrow{\text{Na}}$ [alkoxide with ONa] $\xrightarrow{\text{Br}}$ [ether product]

[from (a)]

11.32 (a) [cyclohexanol, with OH] $+$ SOCl$_2$ $\longrightarrow$ [cyclohexyl chloride, with Cl] $+$ SO$_2$ $+$ HCl

(b) [cyclohexene] $+$ HCl $\longrightarrow$ [cyclohexyl chloride, with Cl]

(c) $\xrightarrow[\text{(no peroxides)}]{\text{HBr}}$

(d) $\xrightarrow[\text{(2) H}_2\text{O}_2,\text{ HO}^-]{\text{(1) BH}_3 : \text{THF}}$ + enantiomer

(e) $\xrightarrow{}$ $\xrightarrow[\text{(2) H}_2\text{O}_2,\text{ HO}^-]{\text{(1) BH}_3 : \text{THF}}$

11.33 (a) CH_3Br + (c) (e) +

(b) + (d) (2 molar equivalents)

11.34 **A:** **G:**

B: **H:**

C: **I:**

D: + $CH_3SO_3^- Na^+$ **J:**

E: **K:**

F: **L:**

11.35 **A:** **D:** + $CH_3SO_3^- Na^+$

B: **E:**

C: **F:**

G: +

J:

H:

K:

I: +

L:

11.36 (a) $\xrightarrow[\text{heat, } h\nu]{\text{Br}_2}$

(b) $\xrightarrow[\text{EtOH}]{\text{NaOEt}}$

(c) $\xrightarrow[\text{peroxides}]{\text{HBr}}$

(d) $\xrightarrow[\text{acetone}]{\text{KI}}$

(e) $\xrightarrow[\text{H}_2\text{O}]{\text{HO}^-}$

or

$\xrightarrow[\text{(2) H}_2\text{O}_2\text{, HO}^-]{\text{(1) BH}_3\text{:THF}}$

(f) $\xrightarrow[\text{(no peroxides)}]{\text{HBr}}$

(g) $\xrightarrow[\text{CH}_3\text{OH}]{\text{CH}_3\text{ONa}}$

(h) $\xrightarrow{}$

(i) $\xrightarrow{\text{NaCN}}$

(j) $\xrightarrow{\text{CH}_3\text{SNa}}$

(k)

$$\xrightarrow[\text{H}_2\text{O}]{\text{Br}_2}$$

$$\xrightarrow{\text{HO}^-}$$

(l)

$$\xrightarrow{\text{OsO}_4,\ \text{pyr}}$$

$$\xrightarrow[\text{H}_2\text{O}]{\text{NaHSO}_3}$$

(m)

$$\xrightarrow[\text{excess}]{\text{NH}_3}$$

(n)

$$\xrightarrow[\text{EtOH}]{^-\text{OEt}}$$

11.37 (a)

$$\xrightarrow{\begin{array}{l}1.\ \text{PBr}_3\\2.\ \text{NaOC(CH}_3)_3\\3.\ \text{H}_2\text{O, H}_2\text{SO}_4\ \text{cat.}\end{array}}$$

(b)

$$\xrightarrow{\begin{array}{l}1.\ \text{NaOC(CH}_3)_3\\2.\ \text{CH}_3\text{C(O)OOH}\end{array}}$$

(c)

$$\xrightarrow{\begin{array}{l}1.\ \text{NaOC(CH}_3)_3\\2.\ \text{H}_2\text{O, H}_2\text{SO}_4\ \text{cat.}\end{array}}$$

(d)

$$\xrightarrow{\begin{array}{c}\text{NaOH, H}_2\text{O}\\\text{or}\\\text{H}_2\text{O, H}_2\text{SO}_4\ \text{cat.}\end{array}}$$

(e)

$$\xrightarrow{m\text{CPBA}}$$

(f)

$$\xrightarrow{\begin{array}{l}1.\ \text{NaOC(CH}_3)_3\\2.\ \text{Br}_2,\ \text{H}_2\text{O}\end{array}}$$

11.38 (a) $\xrightarrow{\text{SOCl}_2}$ $\equiv$

(b) $\xrightarrow{\text{HBr}}$ $\equiv$

(c) $\xrightarrow{\text{NaNH}_2}$ O^-Na^+ + NH_3

(d) $\xrightarrow{\text{PBr}_3}$

(e) $\xrightarrow[\text{2) NaSCH}_2\text{CH}_3]{\text{1) TsCl, pyridine}}$

(f) $\xrightarrow[\text{H}_2\text{SO}_4]{\text{NaI}}$

11.39 (a) $\xrightarrow{\text{HI (excess)}}$ +

(b) $\xrightarrow{\text{HI (excess)}}$ +

(c) $\xrightarrow{\text{H}_2\text{SO}_4, \text{H}_2\text{O}}$ +

(d) $\xrightarrow{\text{NaOCH}_3}$

(e) $\xrightarrow{\text{HOCH}_3, \text{H}_2\text{SO}_4}$

(f) $\xrightarrow[\text{2. H}_2\text{O}]{\text{1. EtSNa}}$

(g)

HCl (1 equiv)

(h)

MeONa → no rxn

(i)

1. EtONa
2. MeI

(j)

HI

(k)

(1) NaH

(2) Br

(l)

(1) NaOH

(2)

11.40 (a)

1. EtSNa
2. MeI

(b)

1. Na⁺ :≡
2. H₂O
3.

(c)

HBr (excess) → 2

(d)

1. HI
2. NaH
   or
   H$_2$SO$_4$ (cat)

+

11.41

$\xrightarrow[400\ °C]{Cl_2}$ Cl

$\xrightarrow[H_2O]{Cl_2}$

OH
Cl   Cl

$\downarrow$ HO$^-$, 1 eq **(b)**

OH
HO   OH

$\xleftarrow[\text{excess}]{\text{HO}^-}$
(a)

Cl

O

Glycerol   Epichlorohydrin

11.42 (a) **A =** [structure] CH$_3$ / OH + enantiomer   **B =** [structure] CH$_3$ / OTs + enantiomer

C = [structure] CH$_3$ / OH + enantiomer

D = [structure] CH$_3$ / I + enantiomer   **A** and **C** are diastereomers.

(b) **E =** CH$_3$ ... OMs   =

H
OMs

F = CH$_3$ ... H

$\rightleftharpoons$

CH$_3$
H   C≡CH
H

‖‖

CH$_3$

(c) **G** =

H =

I =

J =

H and **J** are enantiomers.

11.43

Epichlorohydrin

Toliprolol

11.44

2,4-D

11.45

Dibenzo-18-crown-6

Mechanisms

11.46 (a)

+ H$_2$O

3° Carbocation
is more stable

+ HA

(b)

(from H₂O and cat. H₂SO₄)

(c) Br—Br

(d)

H₃PO₄ (cat), CH₃CH₂OH

(from cat. H₃PO₄
and EtOH)

Inversion should occur through
interaction of alcohol with carbocation
intermediate blocking back face

11.47 (a)

(b) The trans product because the Cl⁻ attacks anti to the epoxide and an inversion of configuration occurs.

11.48

Two S$_N$2 reactions yield retention of configuration.

11.49 Collapse of the α-haloalcohol by loss of a proton and expulsion of a halide ion leads to the thermodynamically-favored carbonyl double bond. Practically speaking, the position of the following equilibrium is completely to the right.

11.50

The reaction, known as the pinacol rearrangement, involves a 1,2-methanide shift to the positive center produced from the loss of the protonated —OH group.

11.51 The angular methyl group impedes attack by the peroxy acid on the front of the molecule (as drawn in the problem). II results from epoxidation from the back of the molecule—the less hindered side.

11.52 For ethylene glycol, hydrogen bonding provides stabilization of the gauche conformer. This cannot occur in the case of gauche butane.

only van der Waals repulsive forces between methyl groups

Synthesizing the Material

11.53

A (C_9H_{16})

B $(C_9H_{15}Na)$

C $(C_{19}H_{36})$

D $(C_{19}H_{38})$

Disparlure $(C_{19}H_{38}O)$

11.54 (a)

(b)

(c)

11.55

(a)

(3) BH₃:THF
(4) H₂O₂, HO⁻

A

(b)

(1) *m*CPBA → (racemic) → (2) HBr → (racemic)

(3) PBr₃, pyridine

(racemic)

B

(c)

(1) Hg(OAc)₂
(2) NaBH₄, HO⁻ → OH → (3) *p*-TsCl, pyridine → OTs

(4) KCN, 18-crown-6

or   or   OTs

C   **C**   **C**

Note: no reaction in the last step since S$_N$2 cannot happen with a tertiary leaving group. Elimination products could also form and would be an acceptable answer for **C**, with the alkene inside or outside ring.

11.56 (a)

(b)

Note: oxidation of an alkene should proceed faster and under milder conditions than a methyl group in Step 3. You will learn about this other reaction in Chapter 15.

Challenge Problems

11.57 The reactions proceed through the formation of bromonium ions identical to those formed in the bromination of *trans*- and *cis*-2-butene (see Section 8.12A).

meso-2,3-Dibromobutane

(Attack at the other carbon atom of the bromonium ion gives the same product.)

2,3-Dibromobutane (racemic)

11.58

The anion and cation form a close ion pair through which chloride is transferred with retention.

11.59

A and B are enantiomers
A, C, and D are all diastereomers
B, C, and D are all diastereomers
C is meso
D is meso

11.60

QUIZ

11.1 Which set of reagents would effect the conversion,

(a) BH_3:THF, then H_2O_2/HO^- (b) $Hg(OAc)_2$, THF-H_2O, then $NaBH_4/HO^-$

(c) H_3O^+, H_2O (d) More than one of these   (e) None of these

11.2 Which of the reagents in item 11.1 would effect the conversion,

11.3 The following compounds have identical molecular weights. Which would have the lowest boiling point?

(a) 1-Butanol

(b) 2-Butanol

(c) 2-Methyl-1-propanol

(d) 1,1-Dimethylethanol

(e) 1-Methoxypropane

11.4 Complete the following synthesis:

12

ALCOHOLS FROM CARBONYL COMPOUNDS: OXIDATION-REDUCTION AND ORGANOMETALLIC COMPOUNDS

SOLUTIONS TO PROBLEMS

12.1 (a) $LiAlH_4$ (d) $LiAlH_4$

     (b) $LiAlH_4$ (e) $LiAlH_4$

     (c) $NaBH_4$

12.2 (a)

     (b)   No reaction. Ethers can not be oxidized.

     (c)

12.3 (a)

CrO_3Cl^- (PCC)/CH_2Cl_2 or $\dfrac{\text{(1) DMSO, (COCl)}_2\text{, low temp.}}{\text{(2) Et}_3\text{N}}$ ⟶

     (b) $KMnO_4$, HO^-, H_2O, heat; then H_3O^+ [or conditions as in (c) below]

     (c) H_2CrO_4/acetone [or conditions as in (b) above], or Swern oxidation

     (d) (1) O_3 (2) Me_2S

     (e) $LiAlH_4$

12.4 (a)

$MgBr$ + $H\!:\!\ddot{O}H$ ⇌   + $^-\!:\!\ddot{O}H$ + Mg^{2+} + Br^-

                 pK_a 15.7 $pK_a \sim 50$

      Stronger base   Stronger acid   Weaker acid   Weaker base

(b)

pK$_a$ 15.5 pK$_a$ 43-45

Stronger base Stronger acid Weaker acid Weaker base

(c)

pK$_a$ 18 pK$_a$ ~ 50

Stronger base Stronger acid Weaker acid Weaker base

(d)

pK$_a$ 5 pK$_a$ ~ 50

Stronger base Stronger acid Weaker acid Weaker base

12.5 (a)

(b)

12.6

12.7 (a) (1)

$$\text{(structure with OH)} \implies \text{(ketone)} + CH_3MgI$$

$$\text{(2-butanone)} + CH_3MgI \xrightarrow[\text{(2) }^+NH_4/H_2O]{\text{(1) ether}} \text{(tertiary alcohol with OH)}$$

(2)

$$\text{(structure with OH)} \implies \text{(ketone)} + \text{(ethyl)}MgBr$$

$$\text{(acetone)} + \text{(ethyl)}MgBr \xrightarrow[\text{(2) }^+NH_4/H_2O]{\text{(1) ether}} \text{(tertiary alcohol with OH)}$$

(3)

$$\text{(structure with OH)} \implies \text{(ethyl ester)} + 2\ CH_3MgI$$

$$\text{(ethyl propanoate)} + 2\ CH_3MgI \xrightarrow[\text{(2) }^+NH_4/\ H_2O]{\text{(1) ether}} \text{(tertiary alcohol with OH)}$$

(b) (1)

$$\text{(structure with OH)} \implies \text{(ketone)} + CH_3MgI$$

$$\text{(ketone)} + CH_3MgI \xrightarrow[\text{(2) }^+NH_4\ /H_2O]{\text{(1) ether}} \text{(tertiary alcohol with OH)}$$

(2)

$$\text{(structure with OH)} \implies \text{(ketone)} + \text{(propyl)}MgBr$$

$$\text{(2-butanone)} + \text{(propyl)}MgBr \xrightarrow[\text{(2) }^+NH_4\ /H_2O]{\text{(1) ether}} \text{(tertiary alcohol with OH)}$$

(3)

$$\text{(structure with OH)} \implies \text{(methyl ester, }OCH_3) + \text{(propyl)}MgBr$$

$$\text{(methyl acetate, }OCH_3) + 2\ \text{(propyl)}MgBr \xrightarrow[\text{(2) }^+NH_4\ /H_2O]{\text{(1) ether}} \text{(tertiary alcohol with OH)}$$

(c) (1) [structure with OH] $\Longrightarrow$ [aldehyde O=CH] + [structure with MgBr]

[acetaldehyde O=CH] + [structure with MgBr] $\xrightarrow[\text{(2) H}_3\text{O}^+]{\text{(1) ether}}$ [alcohol structure with OH]

(2) [structure with OH] $\Longrightarrow$ CH$_3$MgI + [aldehyde structure with O, H]

CH$_3$MgI + [aldehyde structure with O, H] $\xrightarrow[\text{(2) H}_3\text{O}^+]{\text{(1) ether}}$ [alcohol structure with OH]

(d) (1) [phenyl structure with OH] $\Longrightarrow$ [acetophenone O] + [structure with MgBr]

[acetophenone O] + [structure with MgBr] $\xrightarrow[\text{(2) }^+\text{NH}_4\text{/H}_2\text{O}]{\text{(1) ether}}$ [phenyl alcohol structure with OH]

(2) [phenyl structure with OH] $\Longrightarrow$ [phenyl ketone O] + CH$_3$MgI

[phenyl ketone O] + CH$_3$MgI $\xrightarrow[\text{(2) }^+\text{NH}_4\text{/H}_2\text{O}]{\text{(1) ether}}$ [phenyl alcohol structure with OH]

(3)

(e) (1)

(2)

(f) (1)

$$\text{PhCH}_2\text{CH}_2\text{CH}_2\text{OH} \Longrightarrow \text{PhCH}_2\text{CH}_2\text{MgBr} + \text{HCHO} \Longrightarrow$$

$$\text{PhCH}_2\text{CH}_2\text{Br} \Longrightarrow \text{PhCH}_2\text{CH}_2\text{OH} \Longrightarrow \text{PhMgBr} + \text{(epoxide)}$$

$$\text{PhMgBr} \xrightarrow[\text{(2) H}_3\text{O}^+]{\text{(1) (epoxide), ether}} \text{PhCH}_2\text{CH}_2\text{OH} \xrightarrow{\text{PBr}_3} \text{PhCH}_2\text{CH}_2\text{Br}$$

$$\xrightarrow[\text{(2) HCHO}]{\text{(1) Mg, ether}} \text{PhCH}_2\text{CH}_2\text{CH}_2\text{OMgBr} \xrightarrow{\text{H}_3\text{O}^+} \text{PhCH}_2\text{CH}_2\text{CH}_2\text{OH}$$

(2)

$$\text{PhCH}_2\text{CH}_2\text{CH}_2\text{OH} \Longrightarrow \text{PhCH}_2\text{MgBr} + \text{(epoxide)} \Longrightarrow \text{PhCH}_2\text{Br}$$

$$\Longrightarrow \text{PhCH}_2\text{OH} \Longrightarrow \text{PhMgBr} + \text{HCHO}$$

$$\text{PhMgBr} \xrightarrow[\text{(2) H}_3\text{O}^+]{\text{(1) HCHO, ether}} \text{PhCH}_2\text{OH} \xrightarrow{\text{PBr}_3} \text{PhCH}_2\text{Br}$$

$$\xrightarrow[\text{(2) (epoxide)}]{\text{(1) Mg, ether}} \text{PhCH}_2\text{CH}_2\text{CH}_2\text{OMgBr} \xrightarrow{\text{H}_3\text{O}^+} \text{PhCH}_2\text{CH}_2\text{CH}_2\text{OH}$$

12.8 (a)

Swern Oxidation:
(1) DMSO, (COCl)₂, low temp.
(2) Et₃N
or PCC, CH₂Cl₂

(b)

(1) H—CHO, ether
(2) H₃O⁺

Swern Oxidation:
(1) DMSO, (COCl)₂, low temp.
(2) Et₃N
or PCC, CH₂Cl₂

(c)

ether
[from part (a)]

⁺NH₄
H₂O

(d)

Swern Oxidation:
(1) DMSO, (COCl)₂, low temp.
(2) Et₃N
or PCC, CH₂Cl₂

ether

H₃O⁺

Problems

Reagents and Reactions

12.9 (a) CH_3CH_3

(b) D

(c) OH Ph

(d) OH Ph Ph

(e) OH Ph

(f) OH Ph

(g) CH_3CH_3 + OH

12.10 (a) OH

(b) OH

(c) + OH

(d) + OLi

(e) D + O OLi

12.11 (a) OH +

(b) N

(c)

(d) OCH_3 +

(e) OH

(f) OH

(g) OH

(h) OH

(i) OH

(j) + CH_3OLi

(k) + H≡Li

12.12 (a) LiAlH$_4$ (d) (1) KMnO$_4$, HO$^-$, heat (2) H$_3$O$^+$

(b) NaBH$_4$ (or H$_2$CrO$_4$/acetone)

(c) LiAlH$_4$ (e) PCC/CH$_2$Cl$_2$ or Swern oxidation

12.13 (a) Myrtenol $\xrightarrow[\text{(2) Et}_3\text{N}]{\text{(1) DMSO, (COCl)}_2\text{, low temp.}}$

(b) Chrysanthemyl alcohol $\xrightarrow[\text{(2) Et}_3\text{N}]{\text{(1) DMSO, (COCl)}_2\text{, low temp.}}$

12.14 (a) EtMgBr +

(b) EtMgBr +

12.15 (a)

$$\xrightarrow[\text{2. H}_2\text{O}]{\text{1. NaBH}_4\text{, EtOH}}$$

(b)

$$\xrightarrow[\text{2. H}_2\text{O/H}_2\text{SO}_4]{\text{1. LiAlH}_4\text{, THF}}$$

(c)

$$\xrightarrow[\text{2. H}_2\text{O}]{\text{1. NaBH}_4\text{, EtOH}}$$

12.16 (a)

$$\xrightarrow[\text{2. H}_3\text{O}^+]{\text{1. KMnO}_4\text{, NaOH}}$$

(b)

$$\xrightarrow[\text{CH}_2\text{Cl}_2]{\text{PCC}}$$

(c)

$$\xrightarrow[\text{(2) Et}_3\text{N}]{\text{(1) DMSO, (COCl)}_2\text{, low temp.}}$$

(d)

$$\xrightarrow{\text{H}_2\text{CrO}_4}$$ No Rxn
3° Alcohols do not oxidize

(e)

$$\xrightarrow{\text{H}_2\text{CrO}_4}$$

12.17 (a)

$$\xrightarrow[\text{CH}_2\text{Cl}_2]{\substack{\text{Swern Oxidation (as in 12.18(c))} \\ \text{or} \\ \text{PCC}}}$$

(b)

HO—⬡—OH   →[H$_2$CrO$_4$]   HO$_2$C—⬡—CO$_2$H

(c)

Swern Oxidation
or
PCC
CH$_2$Cl$_2$

(d)

(1) LiAlH$_4$
(2) aq. H$_2$SO$_4$

(e)

NaBH$_4$

12.18 (a)

(1) CH$_3$MgBr
(2) H$_3$O$^+$

(b)

(1) ⬭MgBr
(2) NH$_4$Cl, H$_2$O

(c)

(1) ⬡MgBr
(2) H$_3$O$^+$

(d)

(1) CH$_3$CH$_2$Li (excess)
(2) NH$_4$Cl, H$_2$O

Mechanisms

12.19

12.20

Workup in aqueous acid

12.21

Workup in aqueous acid

12.22 The three- and four-membered rings are strained, and so they open on reaction with RMgX or RLi. THF possesses an essentially unstrained ring and hence is far more resistant to attack by an organometallic compound.

Synthesis

12.23 (a) CH_4 + Li≡

(b)

(c)

(d) + $Na^{+-}OSO_2Me$

(e)

(f)

(g) + $Na^{+-}OSO_2Me$

(h)

12.24 (a) 3 + PBr_3 ⟶ 3 + H_3PO_3

+ Mg $\xrightarrow{ether}$

+ $\xrightarrow[\text{(2) } H_3O^+]{\text{(1) ether}}$

(b) [from part (a)] + $\xrightarrow[\text{(2) } H_3O^+]{\text{(1) ether}}$

(c) [from part (a)] + $\xrightarrow[\text{(2) } H_3O^+]{\text{(1) ether}}$

$\xleftarrow{SOCl_2}$

(d) [structure with MgBr] + [isobutyraldehyde] $\xrightarrow[\text{(2) H}_3\text{O}^+]{\text{(1) ether}}$ [2,4-dimethyl-3-pentanol]

[from part (a)]

(e) [structure with MgBr] + D_2O $\longrightarrow$ [isopropyl-D]

[from part (a)]

12.25 (a) [pentanol] $\xrightarrow{\text{PBr}_3}$ [1-bromopentane with Br]

(b) [bromopentane] $\xrightarrow[t\text{-BuOH}]{t\text{-BuOK}}$ [1-pentene]

[from (a)]

(c) [1-pentene] $\xrightarrow[\text{(2) NaBH}_4/\text{HO}^-]{\text{(1) Hg(OAc)}_2/\text{THF-H}_2\text{O}}$ [2-pentanol with OH]

[from (b)]

(d) [1-pentene] $\xrightarrow[\text{pressure}]{\text{H}_2/\text{Pt, Pd, or Ni}}$ [pentane]

[from (b)]

(e) [2-pentanol with OH] $\xrightarrow{\text{PBr}_3}$ [2-bromopentane with Br]

[from (c)]

(f) [bromopentane with Br] $\xrightarrow{\text{Mg}}{\text{Et}_2\text{O}}$ [pentyl MgBr] $\xrightarrow[\text{(2) H}_3\text{O}^+]{\text{(1) H}_2\text{C=O}}$ [hexanol with OH]

[from (a)]

(g) [pentyl MgBr] $\xrightarrow[\text{(2) H}_3\text{O}^+]{\text{(1) ethylene oxide}}$ [heptanol with OH]

[from (f)]

(h) [pentanol with OH] $\xrightarrow[\text{or PCC, CH}_2\text{Cl}_2]{\begin{array}{l}\text{(1) DMSO, (COCl)}_2\text{, low temp.}\\\text{(2) Et}_3\text{N}\end{array}}$ [pentanal]

(i)   [structure: pentan-2-ol with OH] $\xrightarrow[\text{acetone, H}_2\text{O}]{\text{H}_2\text{CrO}_4}$ [structure: pentan-2-one with O]

[from (c)]

(j)   [structure: 1-pentanol with OH] $\xrightarrow[\text{(2) H}_3\text{O}^+]{\text{(1) KMnO}_4\text{, HO}^-\text{, heat}}$ [structure: carboxylic acid with O, OH]

(k) (1)   [structure: 1-pentanol OH] $\xrightarrow[\text{140 °C}]{\text{H}_2\text{SO}_4}$ $\left(\text{[chain]}\right)_2\text{O}$

(2)   [structure: 1-pentanol OH] $\xrightarrow[(-\text{H}_2)]{\text{NaH}}$ [structure] ONa

$\xrightarrow[\text{[from (a)]}]{\text{[chain] Br}}$ $\left(\text{[chain]}\right)_2\text{O}$

(l)   [structure: pentene] $\xrightarrow{\text{Br}_2}$ [structure with Br, Br] $\xrightarrow[\text{heat}]{3 \text{ NaNH}_2}$

[from (b)]

[structure: alkyne] :⁻⁺Na $\xrightarrow{\text{H}_3\text{O}^+}$ [structure: alkyne] ≡—H

(m)   [structure: alkyne]≡—H $\xrightarrow[\text{(no peroxides)}]{\text{HBr (1 equiv.)}}$ [structure with Br]

[from (l)]

(n)   [structure] Br $\xrightarrow[\text{Et}_2\text{O}]{\text{Li}}$ [structure] Li

[from (a)]

(o)   [structure] MgBr $\xrightarrow[\text{(2) }^+\text{NH}_4/\text{H}_2\text{O}]{\text{(1)}}$ [from (i)]   [structure with OH]

[from (f)]

12.26 (a) NaBH₄, HO⁻

(b) 85% H₃PO₄, heat

(c) H₂/Ni₂B (P-2)

(d) (1) BH₃:THF
(2) H₂O₂, HO⁻

(e) (1) NaH   (2) MeI

(f) (1) PBr₃   (2) Mg   (3) [epoxide O]   (4) HA

(g) (1) CH₃MgBr/Et₂O

(2) [acetone structure O]

(3) NH₄Cl/H₂O

12.27   (a) *Retrosynthetic Analysis*

Synthesis

(b) *Retrosynthetic Analysis*

Synthesis

(c) *Retrosynthetic Analysis*

Synthesis

(d) *Retrosynthetic Analysis*

Synthesis

(e) *Retrosynthetic Analysis*

Synthesis

(f) *Retrosynthetic Analysis*

Synthesis

(g) *Retrosynthetic Analysis*

Synthesis

(h) *Retrosynthetic Analysis*

Synthesis

12.28 (a)

(b)

(c)

12.29 There are multiple strategies to approach this these problems.

(a)

(b)

(c)

12.30 *Retrosynthetic Analysis*

Synthesis

12.31 *Retrosynthetic Analysis*

Synthesis

12.32 *Retrosynthetic Analysis*

Synthesis

Before converting the reactant to a Grignard reagent it is first necessary to mask the alcohol, such as by converting it to a *tert*-butyldimethylsilyl ether. After the Grignard reaction is over, the protecting group is removed.

Synthesizing the Material

12.33 (a)

(1) MeMgBr (excess)
(2) NH₄Cl, H₂O

(b)

(1) Mg
(2)H—CHO
(3) H₃O⁺

(c)

(1) PBr₃
(2) Mg
(3) H₃O⁺

(d)

(1) PCC or Swern oxidation
(2) MgBr
(3) H₃O⁺

(e)

(1) EtMgBr
(2) H₃O⁺
(3) NaH
(4) CH₃Br

(f)

(1)

MgBr
(excess)
(2) NH₄Cl, H₂O
(3) PCC

(g)

(1) BrMg   (1 equiv)   MgBr
(2) H₃O⁺

+   MeOH

12.34

Br₂, hv

H —— H
NaNH₂

(1) BH₃
(2) NaOH, H₂O₂

Mg
Et₂O

(1)

(2) NH₄Cl, H₂O

NaH, EtI

12.35 (a)

(b)

(c)

12.36 (a)

(5) Lindlar's cat., H$_2$

(4) KOt-Bu, t-BuOH
(3) MsCl, pyr.

(2) H$_3$O$^+$
(1) BrMg—≡—

A

(b)

(6) PBr$_3$, pyr

(5) H$_3$O$^+$
(4) PhMgBr

Swern [O]

(2) H$_3$O$^+$
(1) MeMgBr

B

(c)

(4) H$_2$, Pd/C

(3) D$_3$O$^+$
(2) Mg, Et$_2$O

(1) NBS, ROOR, Δ

Either isomer of **C** could lead to the subsequent products. The products shown are assumed to have formed from allylic intermediates under kinetic control, since the double bond does not become conjugated to the benzene ring. Worth noting as well is that the resonance contributor that places a radical next to both the benzene ring and the double bond should be preferred in step 1.

C   or   **C**

Challenge Problems

12.37 2-Phenylethanol, 1,2-diphenylethanol, and 1,1-diphenylethanol are distinct from 2,2-diphenylethanoic acid and benzyl 2-phenylethanoate in that they do not have carbonyl groups. IR spectroscopy can be used to segregate these five compounds into two groups according to those that do or do not exhibit carbonyl absorptions.

^{1}H NMR can differentiate among all of the compounds. In the case of the alcohols, in the ^{1}H NMR spectrum of 2-phenylethanol there will be two triplets of equal integral value, whereas for 1,2-diphenylethanol there will be a doublet and a triplet in a 2:1 area ratio. The triplet will be at higher frequency than the doublet. 1,1-Diphenylethanol will exhibit a singlet for the unsplit methyl hydrogens.

The broadband proton-decoupled ^{13}C NMR spectrum of 2-phenylethanol should show 6 signals (assuming no overlap), four of which are in the chemical shift region for aromatic carbons. 1,2-Diphenylethanol should exhibit 10 signals (assuming no overlap), 8 of which are in the aromatic region. 1,1-Diphenylethanol should show 6 signals (assuming no overlap), four of which would be in the aromatic region. The DEPT ^{13}C NMR spectra would give direct evidence as to the number of attached hydrogens on each carbon.

Regarding the carbonyl compounds, both 2,2-diphenylethanoic acid and benzyl 2-phenylethanoate will show carbonyl absorptions in the IR, but only the former will also have a hydroxyl absorption. The ^{1}H NMR spectrum of 2,2-diphenylethanoic acid should show a broad absorption for the carboxylic acid hydrogen and a sharp singlet for the unsplit hydrogen at C2. Their integral values should be the same, and approximately one-tenth the integral value of the signals in the aromatic region. Benzyl 2-phenylethanoate will exhibit two singlets, one for each of the unsplit CH$_2$ groups. These signals will have an area ratio of 2:5 with respect to the signal for the 10 aromatic hydrogens. The broadband ^{1}H decoupled ^{13}C NMR spectrum for 2,2-diphenylethanoic acid should show four aromatic carbon signals, whereas that for benzyl 2-phenylethanoate (assuming no overlapping signals) would show 8 signals in the aromatic carbon region. Aside from the carbonyl and aromatic carbon signals, benzyl 2-phenylethanoate would show two additional signals, whereas 2,2-diphenylethanoic acid would show only one. DEPT ^{13}C NMR spectra for these two compounds would also distinguish them directly.

12.38 The IR indicates the presence of OH and absence of C=C and C=O. The MS indicates a molecular weight of 116 amu and confirms the presence of hydroxyl. The reaction data indicate **X** contains 2 protons per molecule that are acidic enough to react with a Grignard reagent, meaning two hydroxyl groups per molecule. (This analytical procedure, the Zerewitinoff determination, was routinely done before the advent of NMR.)

Thus **X** has a partial structure such as:

C$_6$H$_{10}$(OH)$_2$ with one ring, or

C$_5$H$_6$O(OH)$_2$ with two rings, or (less likely)

C$_4$H$_2$O$_2$(OH)$_2$ with three rings.

12.39

The alkylmagnesium group of the alkoxide can go on to react as a nucleophile with another carbonyl group.

QUIZ

12.1 Which of the following could be employed to transform ethanol into $\diagup\!\!\!\!\diagdown\!\!\!\!\diagup$OH?

(a) Ethanol + HBr, then Mg/diethyl ether, then H_3O^+

(b) Ethanol + HBr, then Mg/diethyl ether, then $\overset{\displaystyle O}{\underset{H \quad H}{\parallel}}$, then H_3O^+

(c) Ethanol + H_2SO_4/140 °C

(d) Ethanol + Na, then $\overset{\displaystyle O}{\underset{H \quad H}{\parallel}}$, then H_3O^+

(e) Ethanol + H_2SO_4/180 °C, then $\triangle$

12.2 The principal product(s) formed when 1 mol of methylmagnesium iodide reacts with 1 mol of

$\overset{\displaystyle O}{\parallel}\diagup\!\!\!\!\diagdown\!\!\!\!\diagup\!\!\!\!\diagdown$OH

(a) CH_4 + $\overset{\displaystyle O}{\parallel}\diagup\!\!\!\!\diagdown\!\!\!\!\diagup\!\!\!\!\diagdown$OMgI

(b) $\overset{\displaystyle OMgI}{\diagup\!\!\!\!\diagdown\!\!\!\!\diagup\!\!\!\!\diagdown}$OH

(c) $\overset{\displaystyle O}{\parallel}\diagup\!\!\!\!\diagdown\!\!\!\!\diagup\diagdown_O\diagup CH_3$

(d) $\overset{\displaystyle OH}{\diagup\!\!\!\!\diagdown\!\!\!\!\diagup\diagdown}_O\diagup CH_3$

(e)   None of the above

12.3 Supply the missing reagents.

$CH_3{-}O\diagdown\!\!\!\diagup\!\!\!\equiv$

12.4 Supply the missing reagents and intermediates.

12.5 Supply the missing starting compound.

Ion-Electron Half-Reaction Method for Balancing Organic Oxidation-Reduction Equations

Only two simple rules are needed:

Rule 1 Electrons (e^-) together with protons (H^+) are arbitrarily considered the reducing agents in the half-reaction for the reduction of the oxidizing agent. Ion charges are balanced by *adding electrons to the left-hand side.* (If the reaction is run in neutral or basic solution, add an equal number of HO^- ions to both sides of the balanced half-reaction to neutralize the H^+, and show the resulting $H^+ + HO^-$ as H_2O.)

Rule 2 Water (H_2O) is arbitrarily taken as the formal source of oxygen for the oxidation of the organic compound, producing *product, protons,* and *electrons* on the right-hand side. (Again, use HO^- to neutralize H^+ in the *balanced* half-reaction in neutral or basic media.)

EXAMPLE 1

Write a balanced equation for the oxidation of RCH_2OH to RCO_2H by $Cr_2O_7^{2-}$ in acid solution.

Reduction half-reaction:

$$Cr_2O_7^{2-} + H^+ + e^- = 2Cr^{3+} + 7H_2O$$

Balancing atoms and charges:

$$Cr_2O_7{}^{2-} + 14H^+ + 6e^- = 2Cr^{3+} + 7H_2O$$

Oxidation half-reaction:

$$RCH_2OH + H_2O = RCO_2H + 4H^+ + 4e^-$$

The least common multiple of a 6-electron uptake in the reduction step and a 4-electron loss in the oxidation step is 12, so we multiply the first half-reaction by 2 and the second by 3, and add:

$$3RCH_2OH + 3H_2O + 2Cr_2O_7{}^{2-} + 28H^+ = 3RCO_2H + 12H^+ + 4Cr^{3+} + 14H_2O$$

Canceling common terms, we get:

$$3RCH_2OH + 2Cr_2O_7{}^{2-} + 16H^+ \longrightarrow 3RCO_2H + 4Cr^{3+} + 11H_2O$$

This shows that the oxidation of 3 mol of a primary alcohol to a carboxylic acid requires 2 mol of dichromate.

EXAMPLE 2

Write a balanced equation for the oxidation of styrene to benzoate ion and carbonate ion by $MnO_4{}^-$ in alkaline solution.

Reduction half-reaction:

$$MnO_4{}^- + 4H^+ + 3e^- = MnO_2 + 2H_2O \text{ (in acid)}$$

Since this reaction is carried out in basic solution, we must add 4 HO^- to neutralize the $4H^+$ on the left side, and, of course, 4 HO^- to the right side to maintain a balanced equation.

$$MnO_4{}^- + 4H^+ + 4HO^- + 3e^- = MnO_2 + 2H_2O + 4HO^-$$

or, $$MnO_4{}^- + 2H_2O + 3e^- = MnO_2 + 4HO^-$$

Oxidation half-reaction:

$$ArCH{=}CH_2 + 5H_2O = ArCO_2{}^- + CO_3{}^{2-} + 13H^+ + 10e^-$$

We add 13 HO^- to each side to neutralize the H^+ on the right side,

$$ArCH{=}CH_2 + 5H_2O + 13HO^- = ArCO_2{}^- + CO_3{}^{2-} + 13H_2O + 10e^-$$

The least common multiple is 30, so we multiply the reduction half-reaction by 10 and the oxidation half-reaction by 3 and add:

$$3ArCH{=}CH_2 + 39HO^- + 10MnO_4{}^- + 20H_2O = 3ArCO_2{}^- + 3CO_3{}^{2-} +$$
$$24H_2O + 10MnO_2 + 40HO^-$$

Canceling:

$$3ArCH{=}CH_2 + 10MnO_4{}^- \longrightarrow 3ArCO_2{}^- + 3CO_3{}^{2-} + 4H_2O + 10MnO_2 + HO^-$$

SAMPLE PROBLEMS

Using the ion-electron half-reaction method, write balanced equations for the following oxidation reactions.

(a) Cyclohexene + MnO_4^- + H^+ $\xrightarrow{\text{(hot)}}$ $HO_2C(CH_2)_4CO_2H$ + Mn^{2+} + H_2O

(b) Cyclopentene + MnO_4^- + H_2O $\xrightarrow{\text{(cold)}}$ cis-1,2-cyclopentanediol + MnO_2 + HO^-

(c) Cyclopentanol + HNO_3 $\xrightarrow{\text{(hot)}}$ $HO_2C(CH_2)_3CO_2H$ + NO_2 + H_2O

(d) 1,2,3-Cyclohexanetriol + HIO_4 $\xrightarrow{\text{(cold)}}$ $OCH(CH_2)_3CHO$ + HCO_2H + HIO_3

SOLUTIONS TO SAMPLE PROBLEMS

(a) Reduction:

$$MnO_4^- + 8\,H^+ + 5e^- = Mn^{2+} + 4\,H_2O$$

Oxidation:

$$+ \ 4\,H_2O = \quad + \ 8\,H^+ \ + \ 8\,e^-$$

The least common multiple is 40:

$$8\,MnO_4^- + 64\,H^+ + 40\,e^- = 8Mn^{2+} + 32\,H_2O$$

$$5 \quad + \ 20\,H_2O = \ 5 \quad + \ 40\,H^+ \ + \ 40\,e^-$$

Adding and canceling:

$$5 \quad + 8\,MnO_4^- \ + \ 24\,H^+ \longrightarrow 5 \quad + 8\,Mn^{2+} + 12\,H_2O$$

(b) Reduction:

$$MnO_4^- + 2\,H_2O + 3e^- = MnO_2 + 4\,HO^-$$

Oxidation:

$$+ \ 2\,HO^- = \quad + \ 2\,e^-$$

The least common multiple is 6:

$$2\,MnO_4^- + 4\,H_2O + 6\,e^- = 2\,MnO_2 + 8\,HO^-$$

$$3 \bigcirc\!\!\!\| \;+\; 6\,HO^- \;=\; 3 \bigcirc\!\!\!\!\begin{smallmatrix}OH\\[2pt]OH\end{smallmatrix} \;+\; 6\,e^-$$

Adding and canceling:

$$3 \bigcirc\!\!\!\| \;+\; 2\,MnO_4^- \;+\; 4\,H_2O \longrightarrow 3 \bigcirc\!\!\!\!\begin{smallmatrix}OH\\[2pt]OH\end{smallmatrix} \;+\; 2\,MnO_2 \;+\; 2\,HO^-$$

(c) Reduction:

$$HNO_3 + H^+ + e^- = NO_2 + H_2O$$

Oxidation:

$$\bigcirc\!\!-OH \;+\; 3\,H_2O \;=\; \bigcirc\!\!\!\begin{smallmatrix}CO_2H\\[2pt]CO_2H\end{smallmatrix} \;+\; 8\,H^+ \;+\; 8\,e^-$$

The least common multiple is 8:

$$8\,HNO_3 + 8\,H^+ + 8\,e^- = 8\,NO_2 + 8\,H_2O$$

$$\bigcirc\!\!-OH \;+\; 3\,H_2O \;=\; \bigcirc\!\!\!\begin{smallmatrix}CO_2H\\[2pt]CO_2H\end{smallmatrix} \;+\; 8\,H^+ \;+\; 8\,e^-$$

Adding and canceling:

$$\bigcirc\!\!-OH \;+\; 8\,HNO_3 \longrightarrow \bigcirc\!\!\!\begin{smallmatrix}CO_2H\\[2pt]CO_2H\end{smallmatrix} \;+\; 8\,NO_2 \;+\; 5\,H_2O$$

(d) Reduction:

$$HIO_4 + 2\,H^+ + 2\,e^- = HIO_3 + H_2O$$

Oxidation:

$$\text{(cyclohexane diol)} \;+\; H_2O \;=\; \begin{smallmatrix}CHO\\[6pt]CHO\end{smallmatrix} \;+\; HC\!-\!OH \;+\; 4\,H^+ \;+\; 4\,e^-$$

The least common multiple is 4:

$$2\,HIO_4 + 4\,H^+ + 4\,e^- = 2\,HIO_3 + 2H_2O$$

Adding and canceling:

WileyPLUS

ANSWERS TO
FIRST REVIEW PROBLEM SET

1. (a)

2° Carbocation

3° Carbocation   +   HA

(b)

then,

(c) The enantiomer of the product given would be formed in an equimolar amount via the following reaction:

The *trans*-1,2-dibromocyclopentane would be formed as a racemic form via the reaction of the bromonium ion with a bromide ion:

Racemic *trans*-1,2-dibromocyclopentane

And, *trans*-2-bromocyclopentanol (the bromohydrin) would be formed (as a racemic form) via the reaction of the bromonium ion with water.

Racemic *trans*-2-bromocyclopentanol

2. (a) CHCl$_3$ (b) The cis isomer   (c) CH$_3$Cl

3. (a)   (b)

(c)

4. (a) 4-Chlorophenol will dissolve in aqueous NaOH; 4-chloro-1-methylbenzene will not.

(b) 4-Methylbenzoic acid will dissolve in aqueous NaHCO$_3$; 4-methylphenol will not.

(c) 2,4,6-Trinitrophenol, because it is so exceptionally acidic (pK_a = 0.38), will dissolve in aqueous NaHCO$_3$; 4-methylphenol (pK_a = 10.17) will not.

(d) 4-Ethylphenol will dissolve in aqueous NaOH; ethyl phenyl ether will not.

5. This indicates that the bonds in BF$_3$ are geometrically arranged so as to cancel each others' polarities in contrast to the case of NF$_3$. This, together with other evidence, indicates that BF$_3$ has trigonal planar structure and NF$_3$ has trigonal pyramidal structure.

6. (a)

(1) Br$_2$, H$_2$O
(2) NaOH
(3) NaOMe, MeOH

(b)

(1) BH$_3$:THF
(2) NaOH, H$_2$O$_2$
(3) NaH
(4) $\bigwedge$—Cl

(c)

(1) PBr$_3$
(2) NaSCH$_3$, HSCH$_3$

(d)

(1) mCPBA
(2) EtOH, H$_2$SO$_4$

(e)

(1) TsCl
(2) NaN$_3$

(f)

concd. H$_2$SO$_4$, EtOH

(g)

NaNH$_2$

$\bigwedge\!\!\!\bigwedge$—OH $\longrightarrow$ $\bigwedge\!\!\!\bigwedge$—ONa + NH$_3$

7. (a)

TS = transition state; E$_a$ = activation energy

(b)

8. **A** =

B =

C =

Muscalure =

9.

(*E*)-2,3-Diphenyl-2-butene   (*Z*)-2,3-Diphenyl-2-butene

Because catalytic hydrogenation is a syn addition, catalytic hydrogenation of the (*Z*) isomer would yield a meso compound.

(*Z*)   →   (by addition at one face)   ≡   (by addition at the other face)

A meso compound

Syn addition of hydrogen to the (*E*) isomer would yield a racemic form:

(*E*)   →   (by addition at one face)   +   (by addition at the other face)

Enantiomers - a racemic form

10. From the molecular formula of **A** and of its hydrogenation product **B**, we can conclude that **A** has two rings and a double bond. (**B** has two rings.)

   From the product of strong oxidation with $KMnO_4$ and its stereochemistry (i.e., compound **C**), we can deduce the structure of **A**.

meso-1,3-Cyclopentane-
dicarboxylic acid

Compound **B** is bicyclo[2.2.1]heptane and **C** is a glycol.

Notice that **C** is also a meso compound.

11. (a)

(b)

(c)

(d)

(e)

(f)

(1) concd. H_2SO_4
(2) BH_3:THF
(3) NaOH, H_2O_2

12. (a)

$\xrightarrow{\text{PBr}_3}$

(b)

$\xrightarrow{\text{SOCl}_2}$

(c)

$\xrightarrow[\text{or SOCl}_2]{\text{HCl}}$

(d)

$\xrightarrow[\text{or PBr}_3 \text{ (2 eq)}]{\text{HBr (2 eq)}}$

(e)

$\xrightarrow[\text{(2) } t\text{-BuOK}]{\text{(1) TsCl, Pyr}}$

(f)

$\xrightarrow{\text{H}_2\text{SO}_4}$

13. (a)

$\xrightarrow[\text{liq. NH}_3]{\text{NaNH}_2}$ $\xrightarrow{\text{CH}_3\text{I}}$

(b)

$\xrightarrow[\text{Ni}_2\text{B (P-2)}]{\text{H}_2}$

[from (a)]

(c)

$\xrightarrow[\text{liq. NH}_3]{\text{Li}}$

[from (a)]

(d)

$$\text{---}\!\equiv\quad \xrightarrow[\text{(2) CH}_3\text{Br}]{\text{(1) NaNH}_2}\quad \text{---}\!\equiv\!\text{---}\quad \xrightarrow[\substack{\text{Ni}_2\text{B(P-2) or}\\ \text{(1) Na, liq. NH}_3\\ \text{(2) NH}_4\text{Cl}}]{\text{H}_2}\quad \text{(cis)} \quad \text{or} \quad \text{(trans)}$$

| HBr

$$\xleftarrow[\substack{t\text{-BuOH}\\ \text{heat}}]{t\text{-BuOK}}$$

Br

(e)

[from (d)]

$$\xrightarrow{\text{NBS}}$$

Br

+

Br

$$\xrightarrow[t\text{-BuOH}]{t\text{-BuOK}}$$

(f)

[from (d)]

$$\xrightarrow[\text{ROOR}]{\text{HBr}}$$

Br

(g)

or

[from (b) or (c)]

or

[from (d)]

$$\xrightarrow[\text{no peroxides}]{\text{HBr}}$$

Br

racemic

$$\xrightarrow[\text{no peroxides}]{\text{HBr}}$$

Br

racemic

(h)

[from (c)]   (anti addition)

Br   H

H   Br

(2R,3S)

A meso compound

(cf. Section 8.11)

(i)

[from (b)]   (anti addition)

Br   H

H   Br

(2R,3R)

+

H   Br

Br   H

(2S,3S)

A racemic form

(j) [structure] $\xrightarrow[\substack{(2)\ NaHSO_3 \\ (syn\ addition)}]{(1)\ OsO_4}$ [structure: H, H, HO, OH]   (cf. Section 8.15)

[from (b)]

or

[structure] $\xrightarrow[\substack{(2)\ H_3O^+,\ heat \\ (anti\ addition)}]{(1)\ R\overset{O}{\underset{}{C}}OOH}$ [structure: H, H, HO, OH]   (cf. Section 11.15)

[from (c)]

(k) —≡— $\xrightarrow[\substack{O \\ \diagup\!\!\!\diagdown\!OH}]{HBr,\ Br^-}$ [structure with Br]   (cf. Section 8.18)

14. [structure] $\xrightarrow[hv,\ heat]{Br_2}$ [structure with Br]   (cf. Section 10.5)

(a) [structure with Br] $\xrightarrow[\substack{EtOH \\ heat}]{NaOEt}$ [structure]

(b) [structure] $\xrightarrow[H_2O]{H_3O^+}$ [structure with OH]

[from (a)]

(c) [structure] $\xrightarrow[(2)\ H_2O_2,\ HO^-]{(1)\ BH_3:THF}$ [structure with OH]

[from (a)]

(d) [structure] $\xrightarrow[\substack{ROOR \\ heat}]{HBr}$ [structure with Br] $\xrightarrow[t\text{-}BuOH]{t\text{-}BuOK}$

[from (a)]

[structure] $\xrightarrow{Br_2}$ [structure with Br, Br] $\xrightarrow[heat]{3\ NaNH_2}$

[structure] ≡:⁻Na⁺ $\xrightarrow{H_3O^+}$ [structure] ≡—H

(e) [structure] $\xrightarrow[\substack{ROOR \\ heat}]{HBr}$ [structure with Br]

[from (d)]

(f) [from (d)] $\xrightarrow{\text{HCl}}$

(g) [from (a)] $\xrightarrow{\text{HCl}}$

(h) [from (e)] Br $\xrightarrow[\substack{\text{acetone} \\ S_N 2}]{\text{NaI}}$ I

(i) [from (a)] $\xrightarrow[\text{(2) Me}_2\text{S}]{\text{(1) O}_3}$ + H

(j) [from (d)] $\xrightarrow[\text{(2) Me}_2\text{S}]{\text{(1) O}_3}$ + H H

15. The solvolysis reaction is $S_N 1$. The intermediate allylic cation is achiral and therefore reacts with water to give the enantiomeric 3-buten-2-ols in equal amounts and to give some of the achiral 2-buten-1-ol:

Cl $\xrightarrow{\text{H}_2\text{O}}$ + $\longrightarrow$ OH

(R) or (S)   Achiral   Racemic

+

HO

Achiral
(but two diastereomers
are possible)

16. (a) In concentrated base and ethanol (a relatively nonpolar solvent), the $S_N 2$ reaction is favored. Thus, the rate depends on the concentration of both the alkyl halide and EtONa. Since no carbocation is formed, the only product is

EtO

(b) When the concentration of EtONa is small or zero, the reaction occurs through the S_N1 mechanism. The carbocation that is produced in the first step of the S_N1 mechanism is a resonance hybrid.

This ion reacts with the nucleophile (EtONa or EtOH) to produce two isomeric ethers

17. Compounds that undergo reactions by an S_N1 path must be capable of forming relatively stable carbocations. Primary halides of the type $ROCH_2X$ form carbocations that are stabilized by resonance:

18. The relative rates are in the order of the relative stabilities of the carbocations:

The solvolysis reaction involves a carbocation intermediate.

19.

20. (a) would be the most reactive in an S_N2 reaction because it is a 1° allylic halide. There would, therefore, be less steric hindrance to the attacking nucleophile.

(b) would be the most reactive in an S_N1 reaction because it is a 3° allylic halide. The carbocation formed in the rate-determining step, being both 3° and allylic, would be the most stable.

21.

B cannot undergo dehydrohalogenation because it has no β hydrogen; however, **C** and **D** can, as shown next.

22.

23. The eliminations are anti eliminations, requiring an anti coplanar arrangement of the bromine atoms.

meso-2,3-Dibromobutane   *trans*-2-Butene

(2*S*,3*S*)-2,3-Dibromobutane   *cis*-2-Butene

(2*R*,3*R*)-2,3-Dibromobutane   *cis*-2-Butene

24. The eliminations are anti eliminations, requiring an anti coplanar arrangement of the —H and —Br.

meso-1,2-Dibromo-   (*E*)-1-Bromo-1,2-
1,2-diphenylethane   diphenylethene

(2*R*,3*R*)-1,2-Dibromo-   (*Z*)-1-Bromo-1,2-
1,2-diphenylethane   diphenylethene

(2*S*,3*S*)-1,2-Dibromo-1,2-diphenylethane will also give (*Z*)-1-bromo-1,2-diphenylethene in an anti elmination.

25. In all the following structures, notice that the large *tert*-butyl group is equatorial.

(a)

+ enantiomer
as a racemic form

(bromine addition is anti; cf. Section 8.11)

(b)

+ enantiomer
as a racemic form

(syn hydroxylation; cf. Section 8.15)

(c)

+ enantiomer
as a racemic form

(anti hydroxylation; cf. Section 11.15)

(d)

+ enantiomer
as a racemic form

(anti Markovnikov syn addition of —H and —OH; cf. Section 8.6)

(e)

(Markovnikov addition of —H and —OH; cf. Section 8.4, 8.5)

(f)

+ enantiomer
as a racemic form

(anti addition of —Br and —OH, with —Br and —OH placement resulting from the more stable partial carbocation in the intermediate bromonium ion; cf. Section 8.13)

(g)

+ enantiomer
as a racemic form

(anti addition of —I and —Cl, following
Markovnikov's rule; cf. Section 8.11)

(h)

(i)

+ enantiomer
as a racemic form

(syn addition of deuterium;
cf. Section 7.15)

(j)

+ enantiomer
as a racemic form

(anti Markovnikov syn addition of —D and

, with being replaced by —T

where it stands; cf. Section 8.10)

26. **A** =   **B** =   **C** =

(racemic)

27. (a) The following products are diastereomers. They would have different boiling points
and would be in separate fractions. Each fraction would be optically active.

(*R*)-3-Methyl-1-pentene   (optically active)   (optically active)

Diastereomers

(b) Only one product is formed. It is achiral, and, therefore, it would not be optically
active.

(optically inactive)

(c) Two diastereomeric products are formed. Two fractions would be obtained. Each fraction would be optically active.

$$\xrightarrow[\text{(2) NaHSO}_3]{\text{(1) OsO}_4}$$

(optically active)   +   (optically active)

Diastereomers

(d) One optically active compound is produced.

$$\xrightarrow[\text{(2) H}_2\text{O}_2, \text{HO}^-]{\text{(1) BH}_3\text{:THF}}$$

(optically active)

(e) Two diastereomeric products are formed. Two fractions would be obtained. Each fraction would be optically active.

$$\xrightarrow[\text{(2) NaBH}_4, \text{HO}^-]{\text{(1) Hg(OAc)}_2, \text{THF–H}_2\text{O}}$$

(optically active)   +   (optically active)

Diastereomers

(f) Two diastereomeric products are formed. Two fractions would be obtained. Each fraction would be optically active.

$$\xrightarrow[\text{(2) H}_3\text{O}^+, \text{H}_2\text{O}]{\text{(1) }m\text{CPBA}}$$

28.

A → (NaOEt / EtOH) → D → (1) O$_3$ (2) Me$_2$S → diketone

B + C (a meso compound)

C ← (H$_2$ / Pt) ← D

29. (a)

(1) CH$_3$MgBr
(2) H$_3$O$^+$

(b)

(1) EtLi
(2) H$_3$O$^+$

(c)

(1) CH$_3$MgBr (excess), ether
(2) H$_3$O$^+$

+ CH$_3$OH

30.

(1) Br$_2$, hv
(2) Mg, ether
(3) O
(4) H$_3$O$^+$

31. (a)

$$\xrightarrow[\text{pyridine}]{\text{SOCl}_2}$$

(b)

$$\xrightarrow[\text{(2) H}_3\text{O}^+]{\text{(1) LiAlH}_4 \text{ (excess)}}$$

(c)

$$\xrightarrow[\text{(2) H}_3\text{O}^+]{\text{(1) LiAlH}_4}$$

(d)

$$\xrightarrow[\text{(2) NaHSO}_3]{\text{(1) OsO}_4}$$

(e)

$$\xrightarrow{\text{NaH}}$$

(f)

$$\xrightarrow{\text{H}_2\text{CrO}_4}$$

(g)

$$\xrightarrow[\text{(2) H}_3\text{O}^+]{\text{(1)} \quad \text{MgBr (1 equiv.)}}$$

(h)

$$\xrightarrow[\text{(2) H}_3\text{O}^+]{\text{(1) NaBH}_4}$$

32.

33.

34.

Y

^{1}H NMR

Singlet at δ 1.4   Presence of only one proton signal suggests
that all 18 protons are equivalent. The chemical
shift suggests that an electronegative
atom is not bonded at the carbon bearing the
protons.

^{13}C NMR

δ 87

$$-\overset{|}{\underset{|}{C}}-O$$

δ 151

$$\overset{\backslash}{\underset{/}{C}}=O$$

IR

1750–1800 cm^{-1} C=O
2 carbonyls indicated
by split peak

35.

m/z 120 = M$^{\overset{+}{\cdot}}$

105 = M$^{\overset{+}{\cdot}}$ − 15(CH$_3$·) =

77 = M$^{\overset{+}{\cdot}}$ − 43(i-Pr·) = C$_6$H$_5^+$

δ7.2–7.6	5 ring protons
2.95	CH of isopropyl group
1.29	equivalent CH$_3$s of isopropyl group

36. C$_5$H$_{10}$O has IHD = 2

IR absorption indicates C=O

^{13}C NMR spectrum for **X** is consistent with structure

37. (a)

1 meso   **2** meso   **3** meso   **4** meso

5 meso   **6** meso   **7**   **8**

Enantiomers

9 meso

(b) Isomer **9** is slow to react in an E2 reaction because in its more stable conformation (see following structure) all the chlorine atoms are equatorial and an anti coplanar transition state cannot be achieved. All other isomers **1–8** can have a —Cl axial and thus achieve an anti coplanar transition state.

9

38. (a)

F_2

1   **2**

Enantiomers
(obtained in one fraction as an
optically inactive racemic form)

+   **3**   +   **4**   +   **5**   +   **6**

3
(achiral and, therefore,
optically inactive)

4   **5**
Enantiomers
(obtained in one fraction as an
optically inactive racemic form)

6
(achiral and, therefore,
optically inactive)

(b) Four fractions. The enantiomeric pairs would not be separated by fractional distillation because enantiomers have the same boiling points.

(c) All of the fractions would be optically inactive.

(d) The fraction containing **1** and **2** and the fraction containing **4** and **5**.

39. (a)

F_2

(*R*)-2-Fluorobutane

1
(optically active)

2
(achiral and, therefore,
optically inactive)

3
(optically active)

4
meso compound
(optically inactive)

5
(optically active)

(b) Five. Compounds **3** and **4** are diastereomers. All others are constitutional isomers of each other.

(c) See above.

40.

(R)   (S)   (R)   (S)

meso   meso

Each of the two structures just given has a plane of symmetry (indicated by the dashed line), and, therefore, each is a meso compound. The two structures are not superposable one on the other; therefore, they represent molecules of different compounds and are diastereomers.

41. Only a proton or deuteron anti to the bromine can be eliminated; that is, the two groups undergoing elimination (H and Br or D and Br) must lie in an anti coplanar arrangement. The two conformations of *erythro*-2-bromobutane-3-*d* in which a proton or deuteron is anti coplanar to the bromine are **I** and **II**.

Conformation **I** can undergo loss of HBr to yield *cis*-2-*d*-2-butene. Conformation **II** can undergo loss of DBr to yield *trans*-2-butene.

To a minor extent, a proton of the methyl group can be eliminated with the bromine.

13

CONJUGATED UNSATURATED SYSTEMS

SOLUTIONS TO PROBLEMS

13.1 The following two allylic radicals are possible, differing only because of the isotopic label. Together they allow for four constitutional isomers with respect to the ^{13}C label. (In the absence of the isotopic label, only one constitutional isomer (as a racemic mixture) would be possible.)

($\star = {}^{13}C$ labeled position)

13.2 (a)

(b) because it represents a 2° carbocation.

(c) and (racemic)

13.3 (a)

(b)

(c)

(d)

(e)

(f)

(g)

(h)

(i)

(j)

(minor)

13.4 (a) because the positive charge is on a tertiary carbon atom rather than a primary one (rule 8).

(b) because the positive charge is on a secondary carbon atom rather than a primary one (rule 8).

(c) $=\text{N}+$ because all atoms have a complete octet (rule 8b), and there are more covalent bonds (rule 8a).

(d) because it has no charge separation (rule 8c).

(e) because the radical is on a secondary carbon atom rather than a primary one (rule 8).

(f) $H_2\ddot{N}$⸺≡N: because it has no charge separation (rule 8c).

13.5 In resonance structures, the positions of the nuclei must remain the same for all structures (rule 2). The keto and enol forms shown differ not only in the positions of their electrons, but also in the position of one of the hydrogen atoms. In the enol form, it is attached to an oxygen atom; in the keto form, it has been moved so that it is attached to a carbon atom.

13.6 (a) (3Z)-Penta-1,3-diene, (2E,4E)-2,4-hexadiene, (2Z,4E)-hexa-2,4-diene, and 1,3-cyclohexadiene are conjugated dienes.

(b) 1,4-Cyclohexadiene and 1,4-pentadiene are isolated dienes.

(c) Pent-1-en-4-yne (1-penten-4-yne) is an isolated enyne.

13.7 The formula, C_6H_8, tells us that **A** and **B** have six hydrogen atoms less than an alkane. This unsaturation may be due to three double bonds, one triple bond and one double bond, or combinations of two double bonds and a ring, or one triple bond and a ring. Since both **A** and **B** react with 2 mol of H_2 to yield cyclohexane, they are either cyclohexyne or cyclohexadienes. The absorption maximum of 256 nm for **A** tells us that it is conjugated. Compound **B**, with no absorption maximum beyond 200 nm, possesses isolated double bonds. We can rule out cyclohexyne because of ring strain caused by the requirement of linearity of the —C≡C— system. Therefore, **A** is 1,3-cyclohexadiene; **B** is 1,4-cyclohexadiene. **A** has three signals in its ^{13}C NMR spectrum. With its higher symmetry, **B** shows only two ^{13}C NMR signals.

13.8 All three compounds have an unbranched five-carbon chain, because the product of hydrogenation is unbranched pentane. The formula, C_5H_6, suggests that they have one double bond and one triple bond. Compounds **D**, **E**, and **F** must differ, therefore, in the way the multiple bonds are distributed in the chain. Compounds **E** and **F** have a terminal —C≡CH [IR absorption at ~3300 cm^{-1}]. The UV absorption maximum near 230 nm for **D** and **E** suggests that in these compounds, the multiple bonds are conjugated. Absence of UV absorption beyond 200 nm indicates that the unsaturation sites are isolated in **F**. The structures are

13.9 (a)

(b)

13.10 Addition of the proton gives the resonance hybrid.

(a)

The inductive effect of the methyl group in **I** stabilizes the positive charge on the adjacent carbon. Such stabilization of the positive charge does not occur in **II**. Because **I** contributes more heavily to the resonance hybrid than does **II**, C2 bears a greater positive charge and reacts faster with the bromide ion.

(b) In the 1,4-addition product, the double bond is more highly substituted than in the 1,2-addition product; hence it is the more stable alkene.

13.11 Endo and exo product formation in the Diels–Alder reaction of cyclopentadiene and maleic anhydride.

13.12 (a) The dienophile can approach the diene in an endo fashion from above or below the diene. Approach from above leads to one enantiomer. Approach from the bottom leads to the other.

(b) They are diastereomers.

13.13

Major-by endo addition
(plus enantiomer)

Minor-by exo addition
(plus enantiomer)

13.14 (a)

(b)

(c)

(major product)

(minor product)

(d)

+ enantiomer

(e)

13.15 (a) Use the trans diester because the stereochemistry is retained in the adduct.

(b) Here, the cis relationship of the acetyl groups requires the use of the cis dienophile.

13.16

13.17

(diene underneath the plane of the dienophile)

(diene above the plane of the dienophile)

(Or, in each case, the other face of the dienophile could present itself to the diene, resulting in the respective enantiomer.)

Problems

Conjugated Systems

13.18 (a) Br⌒⌒⌒Br $\xrightarrow[t\text{-BuOH, heat}]{t\text{-BuOK (2 equiv.)}}$ ⌒⌒⌒

(b) HO⌒⌒⌒OH $\xrightarrow[\text{heat}]{\text{concd } H_2SO_4}$ ⌒⌒⌒

(c) ⌒⌒OH $\xrightarrow[\text{heat}]{\text{concd } H_2SO_4}$ ⌒⌒⌒

(d) ⌒⌒Cl $\xrightarrow[t\text{-BuOH, heat}]{t\text{-BuOK}}$ ⌒⌒⌒

(e) ⌒⌒(Cl) $\xrightarrow[t\text{-BuOH, heat}]{t\text{-BuOK}}$ ⌒⌒⌒

(f) ⌒⌒(OH) $\xrightarrow[\text{heat}]{\text{concd } H_2SO_4}$ ⌒⌒⌒

(g) ⌇⌒ + H_2 $\xrightarrow{Ni_2B \text{ (P-2)}}$ ⌒⌒⌒

13.19

13.20 (a) Cl⌒⌒(Cl)⌒ (racemic) + Cl⌒⌒⌒Cl (E) + (Z)

(b) Cl⌒(Cl)⌒(Cl)⌒Cl (3 stereoisomers)

(c) Br⌒(Br)⌒(Br)⌒Br (3 stereoisomers)

(d) ⌒⌒⌒

(e) + (E) + (Z)

(racemic)

[+ (racemic) + (E) + (Z)]

(f) 4 CO$_2$ (*Note:* KMnO$_4$ oxidizes HO$_2$C—CO$_2$H to 2 CO$_2$.)

(g) + (E) + (Z)

(racemic)

13.21 (a) $\xrightarrow{\text{H}_2,\ \text{Pd/C}}$

(b) $\xrightarrow{\substack{(1)\ \text{O}_3 \\ (2)\ \text{Me}_2\text{S}}}$

(c) $\xrightarrow{\text{Br}_2,\ h\nu}$

(d) $\xrightarrow{\text{HBr, heat}}$

13.22 (a) + $\xrightarrow[\Delta]{\text{ROOR}}$ (racemic) (+ (E) + (Z))

(NBS)

$\xrightarrow[\text{$t$-BuOH, heat}]{\text{$t$-BuOK}}$

Note: In the second step, both allylic halides undergo elimination of HBr to yield 1,3-butadiene; therefore, separating the mixture produced in the first step in unnecessary. The BrCH$_2$CH=CHCH$_3$ undergoes a 1,4-elimination (the opposite of a 1,4-addition).

(b) + NBS $\xrightarrow[\Delta]{ROOR}$

(racemic)

$\left(\begin{array}{c} \text{Br} \\ + \\ (E)+(Z) \end{array}\right)$ $\xrightarrow[\text{t-BuOH, heat}]{\text{t-BuOK}}$

Here again both products undergo elimination of HBr to yield 1,3-pentadiene.

(c) OH $\xrightarrow[\text{heat}]{\text{concd H}_2\text{SO}_4}$ + [as in (a)]

$\left(+ \text{}\right)$

BrBr $\xleftarrow[\text{heat}]{\text{Br}_2}$

$(E)+(Z)$

(d) + NBS $\xrightarrow[\Delta]{ROOR}$ Br +

$(E)+(Z)$ (racemic)

(e) + Br$_2$ $\xrightarrow[\text{heat}]{\text{light}}$ $\xrightarrow[\text{t-BuOH, heat}]{\text{t-BuOK}}$ $\xrightarrow[\Delta]{\text{NBS, ROOR}}$

(excess)   Br   (racemic)

(f) $\xrightarrow[\text{t-BuOH, heat}]{\text{t-BuOK}}$ $\left(\text{same as} \text{}\right)$

13.23 $R-\ddot{\text{O}}-\ddot{\text{O}}-R \xrightarrow[\text{or light}]{\text{heat}} 2\,R-\ddot{\text{O}}\cdot$

$R-\ddot{\text{O}}\cdot + H-\ddot{\text{Br}}: \longrightarrow R-\ddot{\text{O}}-H + \cdot\ddot{\text{Br}}:$

+ $\cdot\ddot{\text{Br}}: \longrightarrow \left[\text{} \longleftrightarrow \text{}\right]$

$\xrightarrow{\text{HBr}}$ + [+ (Z) isomer] + $\cdot\ddot{\text{Br}}:$

13.24 (a)

^{13}C NMR	2 signals	4 signals
UV-Vis	217 nm (s) [conj. system]	185 nm (w)
IR	~1600 cm^{-1} (s) [conj. system]	~3300 cm^{-1} (s) [$\equiv$ C—H] 2100-2260 cm^{-1} (w) [C$\equiv$C]

(b)

UV-Vis	217 nm (s) [conj. system]	transparent in UV-Vis
^{1}H NMR	$\delta \sim 5.0$ [$=CH_2$] $\delta \sim 6.5$ [—CH$=$]	$\delta \sim 1.0$ (t) [CH$_3$] $\delta \sim 1.4$ (q) [CH$_2$]
^{13}C NMR	$\delta \sim 117$ [$=CH_2$] $\delta \sim 137$ [$=CH$—]	$\delta \sim 13$ [CH$_3$] $\delta \sim 25$ [CH$_2$]

(c)

IR	2800–3300 cm^{-1} (s, sharp) [C—H]	3200–3550 cm^{-1} (s, broad) [O—H] ~900 cm^{-1} (s) [$=CH$—H] ~1000 cm^{-1} (s) [$=C$—H]
^{13}C NMR	2 signals	4 signals
MS	m/z 58 (M$^+$)	m/z 72 (M$^+$), 54 (M$^+$ − 18)

(d)

MS	m/z 54 (M$^+$)	m/z 134 (M$^+$), 136 (M$^+$ +2)
^{13}C NMR	2 signals	4 signals
UV-Vis	217 nm (s) [conj. system]	<200 nm

(e)

^{1}H NMR	2 signals	1 signal
IR	960–980 cm^{-1} [trans C—H]	no alkene C—H

s = strong; w = weak; q = quartet; t = triplet

13.25

The resonance hybrid, **I**, has the positive charge, in part, on the tertiary carbon atom; in **II**, the positive charge is on primary and secondary carbon atoms only. Therefore, hybrid **I** is more stable and will be the intermediate carbocation. A 1,4-addition to **I** gives

13.26 The products are (racemic) and Br [also (Z)]. They are formed from an allylic radical in the following way:

13.27 (a) Because a highly resonance-stabilized radical is formed:

(b) Because the carbanion is more stable:

That is, for the carbanion derived from the diene we can write more resonance structures of nearly equal energies.

13.28

Where did the Br_2 come from? It was formed from NBS, as follows.

13.29

A

Kinetic product
(1,2-Addition)

C

Thermodynamic product
(1,4-Addition)

D

C is the kinetic product because it has a smaller E_{act} from B than D.

D is the thermodynamic product because it is more stable.

Reaction Coordinate

13.30 (a) [structure] $\xrightarrow[\text{−15 °C}]{\text{HBr}}$ [structure with Br]

(b) [structure] $\xrightarrow[\text{40 °C}]{\text{HBr}}$ [structure with Br] + [structure with Br]

　　　　　　　　　　　　　　　　Major

(c) [structure] $\xrightarrow[\text{heat}]{\text{NBS, } hv}$ [structure with Br] + [structure with Br]

　　　　　　　　　　　　　　　　Major

13.31 Protonation of the alcohol and loss of water leads to an allylic cation that can react with a chloride ion at either C1 or C3.

[structure] OH $\xrightarrow{\text{H}_3\text{O}^+}$ [structure] $\xrightarrow{-\text{H}_2\text{O}}$

[allylic cation] $\longleftrightarrow$ [allylic cation] $\xrightarrow{\text{Cl}^-}$ [structure] Cl + [structure] Cl

　　　　　　　　　　　　　　　　　　[also (Z)]　　　(racemic)

13.32 (1)

(2)

13.33 (a) The same carbocation (a resonance hybrid) is produced in the dissociation step:

(b) Structure **I** contributes more than **II** to the resonance hybrid of the carbocation (rule 8). Therefore, the hybrid carbocation has a larger partial positive charge on the tertiary carbon atom than on the primary carbon atom. Reaction of the carbocation with water will therefore occur more frequently at the tertiary carbon atom.

13.34 A six-membered ring cannot accommodate a triple bond because of the strain that would be introduced.

Too highly strained

13.35 (a) Propyne. (b) Base (:B⁻) removes a proton, leaving the anion whose resonance structures are shown:

$$CH_2 = C = CH_2 + :B^- \rightleftharpoons H:B + \quad I \longleftrightarrow II$$

Reaction with H:B may then occur at the CH₂ carbanion (**II**). The overall reaction is

$$CH_2 = C = CH_2 + :B^- \rightleftharpoons [CH_2 = C = \ddot{C}H \longleftrightarrow \ddot{C}H_2 - C \equiv CH] + H:B$$

$$CH_3 - C \equiv CH + :B^-$$

13.36 The product formed when 1-bromobutane undergoes elimination is 1-butene, a simple monosubstituted alkene. When 4-bromo-1-butene undergoes elimination, the product is 1,3-butadiene, a conjugated diene, and therefore, a more stable product. The transition states leading to the products reflect the relative stabilities of the products. Since the transition state leading to 1,3-butadiene has the lower free energy of activation of the two, the elimination reaction of 4-bromo-1-butene will occur more rapidly.

Diels-Alder Reactions

13.37

LUMO

HOMO

Energy

13.38 The diene portion of the molecule is locked into an s-trans conformation. It cannot, therefore, achieve the s-cis conformation necessary for a Diels-Alder reaction.

13.39 s-cis conformation is disfavored (steric strain) → No steric strain in s-cis conformation ←

<u>4</u>
"Locked"
s-cis conformation

<u>2</u>

<u>3</u>

<u>1</u>
Cannot achieve s-cis conformation for Diels-Alder reaction.

13.40 (a)
$$\begin{array}{c}O\\ \parallel\end{array}$$ OMe + enantiomer

(b)
$$\begin{array}{c}O\\ \parallel\end{array}$$ OMe + enantiomer

(c) CN + enantiomer

(d) CN + enantiomer

13.41 (a)
$$\begin{array}{c}O\\ \parallel\end{array}$$ OMe OMe
$$\begin{array}{c}\parallel\\ O\end{array}$$

(b) CF$_3$ CF$_3$

13.42 (a)
$$\vartriangle$$ +
$$\begin{array}{c}O\\ \parallel\end{array}$$ OMe
$$\xrightarrow{\Delta}$$
H MeO $$\begin{array}{c}\\ O\end{array}$$ + OMe H

Major

(b)

(c)

(d)

(e)

(f)

(g)

(h)

13.43 (a)

(b)

(c)

(d)

(e)

(f)

(g)

13.44 The endo adduct is less stable than the exo, but is produced at a faster rate at 25°C. At 90°C the Diels–Alder reaction becomes reversible; an equilibrium is established, and the more stable exo adduct predominates.

13.45

Aldrin

Dieldrin

13.46 (a)

Norbornadiene

(b)

EtONa
EtOH
heat

13.47

Cl₂
(dark)

Chlordan

Note: The other double bond is less reactive because of the presence of the two chlorine substituents.

allylic
chlorination

Heptachlor

Synthesizing the Material

13.48 (a)

Δ

(b)

(c)

(1) pyridine (a weak base)

(2) ⟍⟍ , Δ

(d)

(1) *t*-BuOK

(2) HOOC⟍⟍COOH , Δ

(e)

(1) ⟍⟍ , Δ

(2) O₃

(3) Me₂S

(f)

(1) NBS
(1 equiv),
ROOR, Δ

(2) *t*-BuOK,
t-BuOH

(3) ⟍⟍ , Δ

(4) NaBH₄
(excess)

(5) H₃O⁺

Note: if NaBH₄ was
used in MeOH
you could avoid
quench step

13.49 (a)

(5) H₃O⁺
(4) NaBH₄ (excess)

(3) Me₂S
(2) O₃

(1) ethylene,
heat

A

(b)

Key to this question is to think about how to use the Diels–Alder reaction, noting that an electron-deficient dienophile works best, which is why an aldehyde would be the best precursor for the Jones oxidation step

(c)

13.50 (a)

(b)

+ enantiomer

Challenge Problems

13.51

Good polarity interactions lead to major product.

Major (endo) + Minor (endo)

13.52 The first crystalline solid is the Diels–Alder adduct below, mp 125°C,

On melting, this adduct undergoes a reverse Diels–Alder reaction, yielding furan (which vaporizes) and maleic anhydride, mp 56°C,

Furan + Maleic anhydride (mp 56°C)

13.53 MeO$_2$C

OSi(*t*-Bu)Ph$_2$

13.54

The map of electrostatic potential for the pentadienyl carbocation shows greater electron density or less positive charge (less blue color) in the vicinity of the C1—C2 and C4—C5 bonds, suggesting that the most important contributing resonance structure is the one with the most positive charge near C3, a secondary carbon. The other contributing resonance structures have a positive charge at primary carbon atoms.

QUIZ

13.1   Which is the 1,4-addition product from the following reaction:

(a)
(b)

(c)
(d)

(e)

13.2   Which diene and dienophile could be used to synthesize the following compound?

(a)   (b)   (c)

(d)   (e)

13.3 Which reagent(s) could be used to carry out the following reaction?

(a) NBS, ROOR, Δ (NBS = [structure])   (b) NBS, ROOR, Δ, then Br$_2$/hv

(c) Br$_2$, hv, then t-BuOK /t-BuOH, Δ, then NBS, ROOR, Δ

(d) t-BuOK /t-BuOH, then NBS, ROOR, Δ

13.4 Which of the following structures does not contribute to the hybrid for the carbocation formed when 4-chloro-2-pentene ionizes in an S$_N$1 reaction?

(a)   (b)   (c)

(d) All of these contribute to the resonance hybrid.

13.5 Which of the following resonance structures accounts, at least in part, for the lack of S$_N$2 reactivity of vinyl chloride?

(a)   (b)   (c) Neither   (d) Both

SUMMARY OF REACTIONS BY TYPE
CHAPTERS 1–13

I. SUBSTITUTION REACTIONS

Type	Stereochemical Result	Favoring Conditions
S_N2 Chapter 6 See Section 6.14 for a summary of nucleophiles mentioned in early chapters	inversion	1°, 2°, benzylic (1° or 2°), or allylic (1° or 2°) leaving group (e.g., halide, tosylate, mesylate); strong nucleophile; polar aprotic solvent
S_N1 Chapter 6	racemization (via carbocation)	3°, benzylic, or allylic leaving group
Radical substitution Chapter 10	racemization (via radical)	3°, benzylic, or allylic hydrogen; peroxide, heat/light

II. ELIMINATION REACTIONS

Type	Stereochemical/Regiochemical Result	Favoring Conditions
E2 (dehydrohalogenation) Section 7.7	elimination to form the most substituted alkene (Zaitsev elimination) with small bases formation of the less substituted alkene with use of a bulky base (e.g., *tert*-BuOK/*tert*-BuOH)	strong base (e.g., NaOEt/EtOH, KOH/EtOH, *tert*-BuOK/*tert*-BuOH); 2° or 3° leaving group (e.g., halide, tosylate, mesylate, etc.); heat
E1 (dehydrohalogenation) Section 7.8	formation of most substituted alkene; may occur with carbocation rearrangement	3° leaving group; weak base; heat
Dehydration Section 7.10	formation of most substituted alkene; may occur with carbocation rearrangement	catalytic acid (HA, e.g., concd H_2SO_4, H_3PO_4); heat

III. MECHANISTIC SUMMARY OF ALKENE AND ALKYNE ADDITION REACTIONS

Reactant	Stereochemical Result		Regiochemical Result	
	syn	**anti**	**Markovnikov**	**anti-Markovnikov**
Alkenes	H$_2$/Pt, Pd or Ni (Sections 4.16, 7.15, 7.16)	*X$_2$ (Section 8.11)	*H$_2$O, HA (hydration) (Section 8.4)	(i) BH$_3$:THF (ii) HO$^-$, H$_2$O$_2$ (Section 8.6)
	(i) OsO$_4$, (ii) NaHSO$_3$ (Section 8.15)	* X$_2$,H$_2$O (Section 8.13)	* (i) Hg(OAc)$_2$, H$_2$O, THF (ii) NaBH$_4$, HO$^-$ (Section 8.5)	
	RCO$_3$H (e.g., *m*CPBA) (Section 11.13)	(i) RCO$_3$H, (ii) H$_3$O$^+$ or HO$^-$ (Section 11.14)	* HX (no peroxides) (Section 8.2)	HBr (w/peroxides) (Sections 8.2D, 10.10)
	* (i) BH$_3$:THF (ii) H$_2$O$_2$, HO$^-$ (Sections 8.6 and 8.7)	* X$_2$/ROH (Section 8.13)	* additon of other alkenes (Section 10.11, Special Topic C)	
		* X$_2$, Nucleophile (e.g., RO$^-$, RMgX) (Section 8.13)		
	∕ :C ∖ Addition of carbenes (Section 8.14)			

* Shares mechanistic themes with other reactions denoted by the same symbol.

Reactant			
Alkynes	H$_2$, Pd on CaCO$_3$ w/quinoline (Lindlar's cat.) (Section 7.17A)	(i) Li, EtNH$_2$, -78 °C (ii) NH$_4^+$ Cl$^-$ (Section 7.17B)	HX (one or two molar equivalents) (Section 8.18)
	H$_2$, Ni$_2$B (P-2) (Section 7.17A)	X$_2$ (2 equiv.) (Section 8.17)	
	H$_2$ (2 equiv.), Pt or Ni (complete hydrogenation) (Sections 4.16A, 7.17)		

IV. ALKENE AND ALKYNE CLEAVAGE WITH OXIDATION

Conditions	Reactant		Product(s)
Alkenes			
(i) KMnO$_4$/HO$^-$, heat; (ii) H$_3$O$^+$	tetrasubstituted alkene	$\longrightarrow$	two ketones
	trisubstituted alkene	$\longrightarrow$	one ketone, one carboxylic acid
(Section 8.16A)	monosubstituted alkene	$\longrightarrow$	one carboxylic acid and CO$_2$
(i) O$_3$, (ii) Me$_2$S	tetrasubstituted alkene	$\longrightarrow$	two ketones
	trisubstituted alkene	$\longrightarrow$	one ketone, one aldehyde
(Section 8.16B)	monosubstituted alkene	$\longrightarrow$	one aldehyde and formaldehyde
Alkynes			
(i) KMnO$_4$/HO$^-$, heat; (ii) H$_3$O$^+$	terminal alkyne	$\longrightarrow$	a carboxylic acid and formic acid
	internal alkyne	$\longrightarrow$	two carboxylic acids
(Section 8.19)			
(i) O$_3$, (ii) HOAc	terminal alkyne	$\longrightarrow$	a carboxylic acid and formic acid
	internal alkyne	$\longrightarrow$	two carboxylic acids
(Section 8.19)			

V. CARBON–CARBON BOND-FORMING REACTIONS

(a) Alkylation of alkynide anions (with 1° alkyl halides, epoxides, and aldehydes or ketones) (Sections 7.18, 8.20 and 12.8D)

(b) Grignard reaction (with aldehydes and ketones, or epoxides) (Sections 12.7B, C, and 12.8)

(c) Carbocation addition to alkenes (e.g., polymerization) (Special Topic C)

(d) Diels-Alder reaction (Section 13.10)

(e) Addition of a carbene to an alkene (Section 8.14)

(f) Heck, Suzuki, Stille, Sonogashira, Grubbs, and other transition metal organometallic reactions (Chapter 21)

VI. REDUCTIONS/OXIDATIONS (NOT INCLUDING ALKENES/ALKYNES)

(a) 2 R—X (w/Zn/H$_3$O$^+$) $\longrightarrow$ 2 R—H + ZnX$_2$ (Section 5.15)

(b) R—X $\xrightarrow[\text{ether}]{\text{Mg}}$ R—MgX $\xrightarrow{\text{H}_3\text{O}^+}$ R—H (Section 12.7A)

(c) Lithium aluminum hydride (LiAlH$_4$) reduction of alkyl halides to hydrocarbons (Section 12.3D)

(d) Lithium aluminum hydride (LiAlH$_4$) and sodium borohydride (NaBH$_4$) reduction of carbonyl compounds (Section 12.3A, B, C)

(e) Oxidation of alcohols by the Swern oxidation (Section 12.4B), with chromic acid (Section 12.4C), pyridinium chlorochromate (PCC, Section 12.4D), or hot potassium permanganate (Section 12.4E)

VII. MISCELLANEOUS

(a) R—CO—R' + PCl_5 $\longrightarrow$ R—CCl_2—R' $\longrightarrow$ alkynes (Section 7.13)

(b) R—$COOH$ + $SOCl_2$ or PCl_5 $\longrightarrow$ R—$COCl$ $\longrightarrow$ acyl chlorides for Friedel-Crafts reactions, esters, amides, etc. (Sections 15.6 and 17.5)

(c) Terminal alkynes + $NaNH_2$ in liq. NH_3 $\longrightarrow$ alkynide anions (Sections 8.20, 12.8D)

(d) ROH + TsCl, MsCl, or TfCl (with pyridine) $\longrightarrow$ R—OTs, R—OMs, or R—OTf (Section 11.10)

(e) R—OH + Na (or NaH) $\longrightarrow$ R—O^-Na^+ + H_2 (Section 7.6A, 11.11B)

VIII. CHEMICAL TESTS

(a) Alkenes/Alkynes: Br_2 (Section 8.11)

(b) Rings/Unsaturation/etc.: Index of Hydrogen Deficiency (Section 4.17)

(c) Position of Unsaturation: $KMnO_4$ (Section 8.16A); O_3 (Section 8.16B)

METHODS FOR FUNCTIONAL GROUP PREPARATION

CHAPTERS 1–13

I. ALKYL HALIDES

(a) HX addition to alkenes [Markovnikov (Section 8.2) and HBr anti-Markovnikov (Sections 8.2D, 10.10)]

(b) X_2 addition to alkenes (Section 8.11)

(c) Radical halogenation

   (i) X_2/light/heat for alkanes (Sections 10.3–10.5, 10.8, 10.9)

   (ii) *N*-Bromosuccinimide (NBS)/heat/light for allylic and benzylic substitution (Sections 10.8, 10.9)

(d) R—OH + $SOCl_2$ (with pyridine) $\longrightarrow$ R—Cl + SO_2 + pyridinium hydrochloride (Section 11.9)

(e) 3 R—OH + PBr_3 $\longrightarrow$ 3 R—Br + $P(OH)_3$ (Section 11.9)

(f) R—OH + HX $\longrightarrow$ R—X + H_2O (Sections 11.7, 11.8)

II. ALCOHOLS

(a) Markovnikov addition of H_2O with catalytic HA (Section 8.4)

(b) Markovnikov addition of H_2O via (i) $Hg(OAc)_2$, H_2O, THF; (ii) $NaBH_4$, HO^- (Section 8.5)

(c) Anti-Markovnikov addition of H_2O via (i) BH_3:THF; (ii) HO^-, H_2O_2 (Section 8.6)

(d) OsO_4 addition to alkenes (syn) (Section 8.15)

(e) (i) RCO_3H (peroxycarboxylic acids, e.g., *m*CPBA); (ii) H_3O^+ (Sections 11.14, 11.15)

(f) Cleavage of ethers with HX (Section 11.12)

(g) Opening of epoxides by a nucleophile (Sections 11.14, 11.15, 12.7B)

(h) Lithium aluminum hydride or sodium borohydride reduction of carbonyl compounds (Section 12.3)

(i) Grignard reaction with aldehydes, ketones, and epoxides (Sections 12.7B, C, 12.8)

(j) Cleavage of silyl ethers (Section 11.11F)

III. ALKENES

(a) E2 Dehydrohalogenation (preferred over E1 for alkene synthesis) (Section 7.7)

(b) Dehydration of alcohols (Section 7.10)

(c) Hydrogenation of alkynes: (Z) by catalytic hydrogenation, (E) by dissolving metal reduction (Sections 7.17A, B)

(d) Diels-Alder reaction (forms one new double bond with ring formation) (Section 13.10)

(e) Heck, Suzuki, Stille, Sonogashira cross coupling reactions (Chapter 21)

(f) Grubbs olefin metathesis (Chapter 21)

IV. ALKYNES

(a) Alkylation of alkynide anions (Sections 7.18, 8.20, 12.8D)

(b) Double dehydrohalogenation of vicinal or geminal dihalides (Section 7.13)

(c) Sonogashira cross coupling (Chapter 21)

V. CARBON—CARBON BONDS

(a) Alkylation of alkynide anions (Sections 7.18, 8.20, 12.8D)

(b) Organometallic addition to carbonyl compounds and epoxides (e.g., Grignard or RLi reactions) (Sections 12.7B, C, 12.8)

(c) Diels-Alder reaction (Section 13.10)

(d) Addition of alkenes to other alkenes (e.g., polymerization) (Section 10.11 and Special Topic C)

(e) Addition of a carbene to an alkene (Section 8.14)

(f) Heck, Suzuki, Stille, Sonogashira, Grubbs, and other transition metal organometallic reactions (Chapter 21)

VI. ALDEHYDES

(a) (i) O_3; (ii) Me_2S with appropriate alkenes (Section 8.16B)

(b) Swern oxidation of 1° alcohols (Section 12.4B)

(c) Pyridinium chlorochromate (PCC) oxidation of 1° alcohols (Section 12.4D)

VII. KETONES

(a) (i) O_3; (ii) Me_2S with appropriate alkenes (Section 8.16B)

(b) $KMnO_4/HO^-$ cleavage of appropriate alkenes (Section 8.16A)

(c) Swern oxidation of 2° alcohols (Section 12.4B)

(d) H_2CrO_4 oxidation of 2° alcohols (Section 12.4C)

(e) Stille carbonylation by cross coupling with carbon monoxide (Chapter 21)

VIII. CARBOXYLIC ACIDS

(a) (i) $KMnO_4/HO^-$; (ii) H_3O^+ with 1° alcohols (Section 12.4E)

(b) (i) O_3; (ii) HOAc with alkynes (Section 8.19)

(c) (i) $KMnO_4/HO^-$, heat; (ii) H_3O^+ with alkynes and alkenes (Sections 8.19 and 8.16A)

(d) H_2CrO_4 oxidation of 1° alcohols (Section 12.4C, E)

IX. ETHERS (INCLUDING EPOXIDES)

(a) $RO^- + R'X \longrightarrow ROR' + X^-$ (Williamson synthesis) (Section 11.11B, C)

(b) 2 ROH (cat. H_2SO_4, heat) $\longrightarrow$ ROR + HOH (for symmetrical ethers only) (Section 11.11A)

(c) Alkene + RCO_3H (a peroxycarboxylic acid, e.g., *m*CPBA) $\longrightarrow$ an epoxide (Section 11.13)

(d) An epoxide + $RO^- \longrightarrow$ an α-hydroxy ether (also acid catalyzed) (Section 11.14)

(e) $ROH + ClSiR'_3 \longrightarrow ROSiR'_3$ (silyl ether protecting groups, e.g., TBDMS ethers) (Section 11.11F)

14 AROMATIC COMPOUNDS

SOLUTIONS TO PROBLEMS

14.1 (a) 4-Bromobenzoic acid (or *p*-bromobenzoic acid)

(b) 2-Benzyl-1.3-cyclohexadiene

(c) 2-Chloro-2-phenylpentane

(d) Phenyl propyl ether

14.2 Compounds (a) and (b) would yield only one monosubstitution product.

14.3 Resonance structures are defined as being structures that differ *only* in the positions of the electrons. In the two 1,3,5-cyclohexatrienes shown, the carbon atoms are in different positions; therefore, they cannot be resonance structures.

14.4 The cyclopentadienyl cation would be a diradical. We would not expect it to be aromatic.

14.5 (a) No, the cycloheptatrienyl anion (below) would be a diradical.

(b) The cycloheptatrienyl cation (below) would be aromatic because it would have a closed bonding shell of delocalized π electrons.

14.6 If the 1,3,5-cycloheptatrienyl anion *were* aromatic, we would expect it to be unusually stable. This would mean that 1,3,5-cycloheptatriene should be unusually acidic. The fact that 1,3,5-cycloheptatriene is not unusually acidic (it is less acidic than 1,3,5-heptatriene) confirms the prediction made in the previous problem, that the 1,3,5-cycloheptatrienyl anion should not be aromatic.

14.7 (a) [structure with Br groups on seven-membered ring]   (b) [structure with Br groups on seven-membered ring] $\xrightarrow[\text{−HBr}]{\text{heat}}$ [cycloheptatriene ring] + Br⁻

Tropylium bromide

These results suggest that the bonding in tropylium bromide is ionic; that is, it consists of a positive tropylium ion and a negative bromide ion.

14.8 The fact that the cyclopentadienyl cation is antiaromatic means that the following hypothetical transformation would occur with an increase in π-electron energy.

[cyclopentadienyl cation structure] $\xrightarrow[\text{energy increases}]{\text{π-electron}}$ + [cyclopentadiene structure] + H₂

14.9 (a) The cyclopropenyl cation (below).

[triangle with + symbol]   or   [triangle with circle and + inside]   SbCl₆⁻

(b) Only one ¹³C NMR signal is predicted for this ion.

14.10 (a) 3   (b) 4   (c) 7   (d) 5

14.11 The upfield signal arises from the six methyl protons of *trans*-15,16-dimethyldihydropyrene, which by virtue of their location are strongly shielded by the magnetic field created by the aromatic ring current (see Figure 14.8).

14.12 Major contributors to the hybrid must be ones that involve separated charges. Contributors like the following would have separated charges, and would have aromatic five- and seven-membered rings.

14.13 (a) [purine structure with SH group]   (b) [purine structure with OH group]

14.14 Because of their symmetries, *p*-dibromobenzene would give two ^{13}C signals, *o*-dibromobenzene would give three, and *m*-dibromobenzene would give four.

Two signals   Three signals   Four signals

14.15

A   B   C   D

A. Strong absorption at 740 cm^{-1} is characteristic of ortho substitution.

B. A very strong absorption peak at 800 cm^{-1} is characteristic of para substitution.

C. Strong absorption peaks at 680 and 760 cm^{-1} are characteristic of meta substitution.

D. Very strong absorption peaks at 693 and 765 cm^{-1} are characteristic of a monosubstituted benzene ring.

Problems

Nomenclature

14.16 (a)   (b)   (c)

(d)   (e)   (f)

(g) [structure: benzene ring with OCH₂CH₃ (ethoxy) at top and Cl at bottom right]

(h) [structure: benzene ring with SO₃H at top and Cl at bottom]

(i) [structure: benzene ring with SO₂OCH₃ at top and CH₃ at bottom]

(j) [structure: benzene ring with CH₂Br substituent]

(k) [structure: benzene ring with NH₂ at top and NO₂ at bottom]

(l) [structure: benzene ring with two CH₃ groups (1,2-dimethyl)]

(m) [structure: benzene ring with tert-butyl group]

(n) [structure: benzene ring with OH at top and CH₃ at bottom]

(o) [structure: benzene ring with acetyl group (C=O) at top and Br at bottom]

(p) [structure: cyclohexane ring with OH and C₆H₅]

(q) [structure: HO–CH₂–CH(CH₃)–CH(CH₃)–C₆H₅]

(r) [structure: benzene ring with OCH₃ and Cl (ortho)]

14.17 (a)

1,2,3-Tribromo-
benzene

Br, Br, Br [structure: benzene with Br groups at 1,2,4 positions]

1,2,4-Tribromo-
benzene

Br, Br, Br [structure: benzene with Br groups at 1,3,5 positions]

1,3,5-Tribromo-
benzene

(b)

2,3-Dichloro-
phenol

2,4-Dichloro-
phenol

2,5-Dichloro-
phenol

2,6-Dichloro-
phenol

3,4-Dichloro-
phenol

3,5-Dichloro-
phenol

(c)

4-Nitroaniline
(*p*-nitroaniline)

3-Nitroaniline
(*m*-nitroaniline)

2-Nitroaniline
(*o*-nitroaniline)

(d)

4-Methylbenzene-
sulfonic acid
(*p*-toluenesulfonic
acid)

3-Methylbenzene-
sulfonic acid
(*m*-toluenesulfonic
acid)

2-Methylbenzene-
sulfonic acid
(*o*-toluenesulfonic
acid)

(e)

Butylbenzene

Isobutylbenzene

sec-Butylbenzene

tert-Butylbenzene

Aromaticity

14.18   (a) Antiaromatic

(b) Aromatic

(c) Aromatic

(d) Antiaromatic

(e) Aromatic

(f) Aromatic

(g) Antiaromatic

(h) Antiaromatic

(i) Aromatic

(j) Nonaromatic

(k) Aromatic

(l) Aromatic

14.19   (a) Energy

Antibonding

Bonding

(b) Energy

Antibonding

Bonding

14.20 (a)

(b)

14.21

The conjugate base of **A** is a substituted cyclopentadienyl anion and both rings are aromatic. In **B**, the five-membered ring does not contribute to anion stability. The additional stabilization provided by the cyclopentyldienyl moiety is the reason for the greater acidity of **A**.

14.22

The major resonance form leaves the electrons in the cyclopentadiene ring to make both rings aromatic. Therefore, the central bond has more single bond character and rotation around that bond is easily achieved.

14.23 Hückel's rule should apply to both pentalene and heptalene. Pentalene's antiaromaticity can be attributed to its having 8 π electrons. Heptalene's lack of aromaticity can be attributed to its having 12 π electrons. Neither 8 nor 12 is a Hückel number.

14.24 (a) The extra two electrons go into the two partly filled (nonbonding) molecular orbitals (Fig. 14.7), causing them to become filled. The dianion, therefore, is not a diradical. Moreover, the cyclooctatetraene dianion has 10 π electrons (a Hückel number), and this apparently gives it the stability of an aromatic compound. (The highest occupied molecular orbitals may become slightly lower in energy and become bonding molecular orbitals.) The stability gained by becoming aromatic is apparently large enough to overcome the extra strain involved in having the ring of the dianion become planar.

(b) The strong base (butyllithium) removes two protons from the compound on the left. This acid-base reaction leads to the formation of the 10 π electron pentalene dianion, an aromatic dianion.

Pentalene dianion

14.25 The bridging —CH$_2$— group causes the 10 π electron ring system (below) to become planar. This allows the ring to become aromatic.

14.26 (a) Resonance contributions that involve the carbonyl group of **I** resemble the *aromatic* cycloheptatrienyl cation and thus stabilize **I**. Similar contributors to the hybrid of **II** resemble the *antiaromatic* cyclopentadienyl cation (see Problem 14.8) and thus destabilize **II**.

(a)

I ⟷ **IA**

Contributors like **IA** are exceptionally stable because they resemble an aromatic compound. They therefore make large stabilizing contributions to the hybrid.

II ⟷ **IIA**

Contributors like **IIA** are exceptionally unstable because they resemble an anti-aromatic compound. Any contribution they make to the hybrid is destabilizing.

(b)

14.27 Ionization of 5-chloro-1,3-cyclopentadiene would produce a cyclopentadienyl cation, and the cyclopentadienyl cation (see Problem 14.8) would be highly unstable because it would be antiaromatic.

Antiaromatic ion
(highly unstable)

14.28 (a) The cyclononatetraenyl anion with 10 π electrons obeys Hückel's rule.

Cyclononatetraenyl anion
10 π electrons
Aromatic

Cyclohexadecaoctaenyl dianion
18 π electrons
Aromatic

(b) By adding 2 π electrons, [16] annulene becomes an 18 π electron system and therefore obeys Hückel's rule.

14.29 As noted in Problem 13.42, furan can serve as the diene component of Diels–Alder reactions, readily losing all aromatic character in the process. Benzene, on the other hand, is so unreactive in a Diels–Alder reaction that it can be used as a nonreactive solvent for Diels–Alder reactions.

Spectroscopy and Structure Elucidation

14.30 (a)

4 total signals
aromatic region, 2 doublets

6 total signals
aromatic region, 2 doublets
and 2 doublets of doublets

(b)

Signal 1 (~4 ppm, singlet)
Signal 2 (~2 ppm, singlet)

Signal 3 (~2 ppm, singlet)
Signal 4 (~2 ppm, singlet)

(c)

5 total signals
aromatic region: 1 singlet,
2 doublets, and 1 doublet
of doublets

3 total signals
aromatic region: 2 doublets

14.31 A

CH₃ (*a*)
H   (*b*)
CH₃ (*a*)
(*c*)

(*a*) doublet δ 1.25
(*b*) septet δ 2.9
(*c*) multiplet δ 7.3

B

CH₃ (*a*)
H   (*b*)
NH₂ (*c*)
(*d*)

(*a*) doublet δ 1.35
(*b*) quartet δ 4.1
(*c*) singlet δ 1.7
(*d*) multiplet δ 7.3

(a) triplet δ 2.9
(b) quintet δ 2.05
(c) multiplet δ 7.1

14.32 A ^{1}H NMR signal this far upfield indicates that cyclooctatetraene is a cyclic polyene and is not aromatic; its π electrons are not fully delocalized.

14.33 Compound F is *p*-isopropyltoluene. ^{1}H NMR assignments are shown in the following spectrum.

The IR absorptions from 3020 to 2870 cm^{-1} indicate *sp*^{2} and *sp*^{3} C-H stretching vibrations, as would be present in an alkyl substituted benzene ring. The absorptions at 1517 and 1463 cm^{-1} range are characteristic of benzene ring stretching vibrations. The absorption at 818 cm^{-1} suggests para substitution of the benzene ring.

We can make the following ^{1}H NMR assignments:

(a) doublet δ 1.25   (c) septet δ 2.85

(b) singlet δ 2.3   (d) multiplet δ 7.1

14.34 Compound **L** is allylbenzene,

(c) H $\quad\quad$ H *(a)*

$$C = C$$

CH_2 $\quad$ H *(b)*

(d)

(e)

(d) Doublet, δ 3.1 (2H)
(a) or *(b)* Multiplet, δ 4.8
(a) or *(b)* Multiplet, δ 5.1
(c) Multiplet, δ 5.8
(e) Multiplet, δ 7.1 (5H)

The following IR assignments can be made.

3035 cm^{-1}, C—H stretching of benzene ring
3020 cm^{-1}, C—H stretching of —CH=CH$_2$ group
2925 cm^{-1} and 2853 cm^{-1}, C—H stretching of —CH$_2$— group
1640 cm^{-1}, C=C stretching
990 cm^{-1} and 915 cm^{-1}, C—H bendings of —CH=CH$_2$ group
740 cm^{-1} and 695 cm^{-1}, C—H bendings of —C$_6$H$_5$ group

The UV absorbance maximum at 255 nm is indicative of a benzene ring that is not conjugated with a double bond.

14.35 Compound **M** is *m*-ethyltoluene. We can make the following assignments in the spectrum.

CH_3*(c)*

(d)

CH_2—CH_3
(b) $\quad$ *(a)*

(a) triplet δ 1.4
(b) quartet δ 2.6
(c) singlet δ 2.4
(d) multiplet δ 7.05

Meta substitution is indicated by the very strong peaks at 690 and 780 cm^{-1} in the IR spectrum.

14.36 Compound **X** is *m*-xylene. The upfield signal at δ 2.3 arises from the two equivalent methyl groups. The downfield signals at δ 6.9 and 7.1 arise from the protons of the benzene ring. Meta substitution is indicated by the strong IR peak at 680 cm^{-1} and very strong IR peak at 760 cm^{-1}.

14.37 The broad IR peak at 3400 cm^{-1} indicates a hydroxyl group, and the two bands at 720 and 770 cm^{-1} suggest a monosubstituted benzene ring. The presence of these groups is also indicated by the peaks at δ 4.4 and δ 7.2 in the ^{1}H NMR spectrum. The ^{1}H NMR spectrum also shows a triplet at δ 0.85 indicating a —CH$_3$ group coupled with an adjacent —CH$_2$— group. There is a complex multiplet at δ 1.7 and there is also a triplet at δ 4.5 (1H). Putting these pieces together in the only way possible gives us the following structure for **Y**.

14.38 (a) Four unsplit signals.

(b) Absorptions arising from: $=\!\overset{/}{\underset{\backslash H}{C}}$, —CH$_3$, and $\overset{\backslash}{\underset{/}{C}}\!=\!O$ groups.

Challenge Problems

14.39 The vinylic protons of *p*-chlorostyrene should give a spectrum approximately like the following:

14.40

A $\longrightarrow$ B

14.41

C $\longrightarrow$ D   (a "sandwich compound")

14.42 Ph——≡——≡——Ph
                    E

14.43

Third unoccupied π MO (five nodal planes)   ___

Second unoccupied π MO (four nodal planes)   ___

First unoccupied π MO (three nodal planes)   ___

Highest energy occupied π MO (two nodal planes)   ⇅

Second highest energy occupied π MO (one nodal plane)   ⇅

Lowest energy occupied π MO (no nodal planes)   ⇅

QUIZ

14.1 Which of the following reactions of benzene is inconsistent with the assertion that benzene is aromatic?

(a) Br_2/25 °C ⟶ no reaction

(b) H_2/Pt/25 °C ⟶ no reaction

(c) Br_2/FeBr$_3$ ⟶ C_6H_5Br + HBr

(d) $KMnO_4$/H_2O/25 °C ⟶ no reaction

(e) None of the above

14.2   Which is the correct name of the compound shown?

(a) 3-Chloro-5-nitrotoluene

(b) *m*-Chloro-*m*-nitrotoluene

(c) 1-Chloro-3-nitro-5-toluene

(d) *m*-Chloromethylnitrobenzene

(e) More than one of these

14.3   Which is the correct name of the compound shown?

(a) 2-Fluoro-1-hydroxyphenylbenzene

(b) 2-Fluoro-4-phenylphenol

(c) *m*-Fluoro-*p*-hydroxybiphenyl

(d) *o*-Fluoro-*p*-phenylphenol

(e) More than one of these

14.4   Which of the following molecules or ions is not aromatic according to Hückel's rule?

(a)   (b)   (c)   (d)

(e) All are aromatic.

14.5   Give the structure of a compound with the formula C_7H_7Cl that is capable of undergoing both S_N1 and S_N2 reactions.

14.6   Write the name of an aromatic compound that is isomeric with naphthalene.

15 REACTIONS OF AROMATIC COMPOUNDS

SOLUTIONS TO PROBLEMS

15.1

15.2 The rate is dependent on the concentration of $^+NO_2$ ion formed from protonated nitric acid.

$$H-\overset{+}{\underset{H}{O}}-NO_2 \ + \ HA \ \longrightarrow \ ^+NO_2 \ + \ H_3O^+ \ + \ A^-$$

(where $HA = HNO_3$ or $HOSO_3H$)

Because H_2SO_4 ($HOSO_3H$) is a stronger acid, a mixture of it and HNO_3 will contain a higher concentration of protonated nitric acid than will nitric acid alone.

That is, the reaction,

$$H-O-NO_2 \ + \ HOSO_3H \ \rightleftharpoons \ H-\overset{+}{\underset{H}{O}}-NO_2 \ + \ HSO_4^-$$

Protonated
nitric acid

produces more protonated nitric acid than the reaction,

$$H-O-NO_2 \ + \ H-O-NO_2 \ \rightleftharpoons \ H-\overset{+}{\underset{H}{O}}-NO_2 \ + \ NO_3^-$$

15.3 *Step 1*

Step 2

Step 3

15.4

Rearrangement to the secondary carbocation occurs as the leaving group departs.

Both carbocations are then attacked by the ring.

15.5 (a)

(b)

15.6 As the following structures show, attack at the ortho and para positions of phenol leads to arenium ions that are more stable (than the one resulting from meta attack) because they are hybrids of four resonance structures, one of which is relatively stable. Only three resonance structures are possible for the meta arenium ion, and none is relatively stable.

Ortho attack

Meta attack

Para attack

15.7 (a) The atom (an oxygen atom) attached to the benzene ring has an unshared electron pair that it can donate to the arenium ions formed from ortho and para attack, stabilizing them. (The arenium ions are analogous to the previous answer with a —COCH$_3$ group replacing the —H of the phenolic hydroxyl).

(b) Structures such as the following compete with the benzene ring for the oxygen electrons, making them less available to the benzene ring.

This effect makes the benzene ring of phenyl acetate less electron rich and, therefore, less reactive.

(c) Because the acetamido group has an unshared electron pair on the nitrogen atom that it can donate to the benzene ring, it is an ortho-para director.

(d) Structures such as the following compete with the benzene ring for the nitrogen electrons, making them less available to the benzene ring.

15.8 *Ortho attack*

Relatively stable

Para attack

Relatively stable

15.9 The phenyl group, as the following resonance structures show, can act as an electron-releasing group and can stabilize the arenium ions formed from ortho and para attack.

Ortho attack

In the case of the arenium ions above and following, the unsubstituted ring can also be shown in the alternative Kekulé structure.

Para attack

In the case of both ortho and para substitution, the phenyl group functions as an activating group, resulting in a faster reaction than in the case of benzene.

15.10   As the following resonance structures show, each mode of attack has three possible resonance forms. However, for both ortho and para attack, there is one resonance form which puts a positive charge next to the inductively withdrawing CF_3 group. That result is unfavorable, and renders meta attack the best possible option since such an unfavorable resonance from is avoided.

Ortho attack

Meta attack

Para attack

15.11   (a)

(b)

(c)

(d)   minor   +   major

15.12 (a) (b) (c)

15.13 The addition of hydrogen bromide to 1-phenylpropene proceeds through a benzylic radical in the presence of peroxides, and through a benzylic cation in their absence (cf., a and b as follows).

(a) Hydrogen bromide addition in the presence of peroxides.

Chain Initiation

Step 1 $R-O-O-R \longrightarrow 2\,R-O\cdot$

Step 2 $RO\cdot\; +\; H-Br \longrightarrow R-O-H\; +\; Br\cdot$

Step 3 $Br\cdot\; +\; \underset{C_6H_5}{\overset{}{\text{(alkene)}}} \longrightarrow C_6H_5\underset{Br}{\overset{\cdot}{\text{(radical)}}}$
A benzylic radical

Chain Propagation

Step 4 $C_6H_5\underset{Br}{\overset{\cdot}{\text{(radical)}}}\; +\; H-Br \longrightarrow C_6H_5\underset{Br}{\text{(product)}}\; +\; Br\cdot$
2-Bromo-1-phenylpropane

The mechanism for the addition of hydrogen bromide to 1-phenylpropene in the presence of peroxides is a chain mechanism analogous to the one we discussed when we described anti-Markovnikov addition in Section 10.9. The step that determines the orientation of the reaction is the first chain-propagating step. Bromine attacks the second carbon atom of the chain because by doing so the reaction produces a more stable benzylic radical. Had the bromine atom attacked the double bond in the opposite way, a less stable secondary radical would have been formed.

$$C_6H_5 \diagup\diagdown + Br\cdot \xrightarrow{\quad\times\quad} C_6H_5\underset{Br}{\overset{|}{CH}}\diagdown \cdot$$

A secondary radical

(b) Hydrogen bromide addition in the absence of peroxides.

$$C_6H_5\diagup\diagdown + HBr \longrightarrow C_6H_5\overset{+}{\diagup}\diagdown + Br^-$$

A benzylic cation

$$\downarrow$$

$$C_6H_5\underset{\overset{|}{Br}}{}\diagdown$$

1-Bromo-1-phenylpropane

In the absence of peroxides, hydrogen bromide adds through an ionic mechanism. The step that determines the orientation in the ionic mechanism is the first, where the proton attacks the double bond to give the more stable benzylic cation. Had the proton attacked the double bond in the opposite way, a less stable secondary cation would have been formed.

$$C_6H_5\diagup\diagdown + HBr \xrightarrow{\quad\times\quad} C_6H_5\overset{+}{\diagdown}\diagup + Br^-$$

A secondary cation

15.14 (a) [structure: phenyl with CH(Cl)CH₂CH₃] because the more stable carbocation intermediate is the benzylic

carbocation, [structure: phenyl with C⁺HCH₂CH₃] , which then reacts with a chloride ion.

(b) [structure: phenyl with CH(OH)CH₃] because the more stable intermediate is a mercurinium ion in which a partial positive charge resides on the benzylic carbon, which then reacts with H_2O.

15.15 (a) The first method would fail because introducing the chlorine substituent first would introduce an ortho-para directing group. Consequently, the subsequent Friedel-Crafts reaction would not then take place at the desired meta position.

The second method would fail for essentially the same reasons. Introducing the ethyl group first would introduce an ortho-para director, and subsequent ring chlorination would not take place at the desired meta position.

(b) If we introduce an acetyl group first, which we later convert to an ethyl group, we install a meta director. This allows us to put the chlorine atom in the desired position. Conversion of the acetyl group to an ethyl group is then carried out using the Clemmensen (or Wolff–Kishner) reduction.

15.16 (a)

15.17 That *o*-chlorotoluene leads to the formation of two products (*o*-cresol and *m*-cresol) when submitted to the conditions used in the Dow process suggests that an elimination-addition mechanism takes place.

Inasmuch as chlorobenzene and *o*-chlorotoluene should have similar reactivities under these conditions, it is reasonable to assume that chlorobenzene also reacts by an elimination-addition mechanism in the Dow process.

15.18 2-Bromo-1,3-dimethylbenzene, because it has no α-hydrogen, cannot undergo an elimination. Its lack of reactivity toward sodium amide in liquid ammonia suggests that those compounds (e.g., bromobenzene) that do react do so by a mechanism that begins with an elimination.

15.19

Reactions and Synthesis

15.20 (a)

(b)

(c)

(d)

(e)

(f)

(g)

(h)

15.21 (a) (b)

(c) (d)

(e)

(f)

15.22 (a)

(b)

(c)

15.23 (a) [structure: 1-chloroethylbenzene] (b) [structure: (E)-1-propenylbenzene] (c) [structure: 1-butenylbenzene]

(d) [structure: 2-bromobutylbenzene] (e) [structure: 1-hydroxybutylbenzene] (f) [structure: butylbenzene]

(g) [structure: benzoic acid] (h) [structure: 2,6-dibromo-4-methylphenol]

15.24 (a) [benzene] + [isopropyl chloride] $\xrightarrow{\text{AlCl}_3}$ [isopropylbenzene]

(b) [benzene] + [tert-butyl chloride] $\xrightarrow{\text{AlCl}_3}$ [tert-butylbenzene]

(c) [benzene] + [propanoyl chloride] $\xrightarrow{\text{AlCl}_3}$ [propiophenone] $\xrightarrow[\text{heat}]{\text{H}_2\text{NNH}_2, \text{KOH}}$ [propylbenzene]

(*Note:* The use of Cl⌒⌒ in a Friedel-Crafts synthesis gives mainly the rearranged product, isopropylbenzene.)

(d) [benzene] + [butanoyl chloride] $\xrightarrow{\text{AlCl}_3}$ [butyrophenone]

$\xleftarrow[\substack{\text{Zn(Hg)} \\ \text{HCl,} \\ \text{reflux}}]{\substack{\text{or H}_2\text{NNH}_2, \\ \text{KOH, heat}}}$ [butylbenzene]

(e) [tert-butylbenzene] + Cl$_2$ $\xrightarrow[\text{dark}]{\text{FeCl}_3}$ [4-chloro-tert-butylbenzene]

[from (b)]

(f)

(g) [from (f)]

$$(1)\ BH_3:THF$$
$$(2)\ H_2O_2,\ HO^-$$
(syn addition)

+ enantiomer

(h) + HNO_3 $\xrightarrow{H_2SO_4}$ ... $\xrightarrow[H_2SO_4]{HNO_3}$ ...

(i) + Br_2 $\xrightarrow{FeBr_3}$ ...

(j) + Br_2 $\xrightarrow{FeBr_3}$ ... $\xrightarrow[H_2SO_4]{HNO_3}$ ... + ortho isomer

(k) + Cl_2 $\xrightarrow{FeCl_3}$ ... $\xrightarrow[H_2SO_4]{SO_3}$ ... + ... (separate)

H_2O/H_2SO_4, heat

(l) [from (k)]

Cl —[SO_3 / H_2SO_4]→ Cl, SO_3H (+ ortho isomer) —[HNO_3 / H_2SO_4]→ Cl, NO_2, SO_3H —[H_2O/H_2SO_4 heat]→ Cl, NO_2

(m) [from (h)]

NO_2 —[SO_3 / H_2SO_4]→ NO_2, SO_3H

15.25 (a) C_6H_5—CH=CH$_2$ —[Cl_2]→ C_6H_5—CHCl—CH$_2$Cl

(b) C_6H_5—CH=CH$_2$ —[H_2 / Ni pressure]→ C_6H_5—CH$_2$—CH$_3$

(c) C_6H_5—CH=CH$_2$ —[1) OsO_4, pyridine / 2) $NaHSO_3$/H_2O]→ C_6H_5—CH(OH)—CH$_2$OH

(d) C_6H_5—CH=CH$_2$ —[(1) $KMnO_4$, HO^-, heat / (2) H_3O^+]→ C_6H_5—COOH

(e) C_6H_5—CH=CH$_2$ —[H_2O / H_2SO_4]→ C_6H_5—CH(OH)—CH$_3$

(f) C_6H_5—CH=CH$_2$ —[HBr / no peroxides]→ C_6H_5—CHBr—CH$_3$

(g) C_6H_5—CH=CH$_2$ —[(1) BH_3:THF / (2) H_2O_2, HO^-]→ C_6H_5—CH$_2$—CH$_2$OH

(h) C_6H_5 — $\xrightarrow[\text{(2)}]{\text{(1) BH}_3\text{:THF}}$ — C_6H_5 — D

(i) C_6H_5 — $\xrightarrow[\text{peroxides}]{\text{HBr}}$ — C_6H_5 — Br

(j) C_6H_5 — Br + NaI $\longrightarrow$ C_6H_5 — I
   [from (i)]

(k) C_6H_5 — Br + $^-$CN $\longrightarrow$ C_6H_5 — CN
   [from (i)]

(l) C_6H_5 — $\xrightarrow[\substack{\text{Ni} \\ \text{pressure}}]{\text{D}_2}$ C_6H_5 — D (D)

(m) [diene] + [styrene] $\xrightarrow{\text{heat}}$ [4-phenylcyclohexene] $\xrightarrow[\substack{\text{Ni} \\ \text{pressure}}]{\text{H}_2}$ [phenylcyclohexane]

(n) C_6H_5 — OH $\xrightarrow{\text{Na}}$ C_6H_5 — ONa $\xrightarrow{\text{CH}_3\text{I}}$ C_6H_5 — OCH_3
   [from (g)]

15.26 (a) [toluene] $\xrightarrow[\text{heat}]{\text{KMnO}_4, \text{HO}^-}$ $\xrightarrow{\text{H}_3\text{O}^+}$ [benzoic acid] $\xrightarrow[\text{FeCl}_3]{\text{Cl}_2}$ [3-chlorobenzoic acid]

(b) [toluene] + [acetyl chloride] $\xrightarrow{\text{AlCl}_3}$ [4-methylacetophenone] + ortho isomer

(c)

(+ ortho isomer)

(d)

(+ ortho isomer)

(e)

(f)

+ ortho isomer

(g)

+ ortho isomer

(h)

(i)

(j)

15.27 (a)

(b)

[from (a)]

(c)

excess
[from (b)]

(d)

[from (b)]

(e)

15.28 (a) Step (2) will fail because a Friedel–Crafts reaction will not take place on a ring that bears an —NO$_2$ group (or any meta director).

no Friedel–Crafts
reaction

(b) The last step will first brominate the double bond.

15.29 This problem serves as another illustration of the use of a sulfonic acid group as a blocking group in a synthetic sequence. Here we are able to bring about nitration between two meta substituents.

15.30

15.31 (a)

(b)

(c)

(d)

15.32 (a)

A   B   C   D

(b)

E

(c)

F

(d)

G

15.33 (a) 1. Generate the electrophile

2. Attack the electrophile with aromatic ring to form the sigma complex.

3. Elimination to regain aromaticity

(b) 1. Generate the electrophile

$$Br-Br \ + \ FeBr_3 \ \longrightarrow \ Br^+ \ + \ FeBr_4^-$$

2. The aromatic ring attacks the electrophile to generate a resonance-stabilized sigma complex.

3. Elimination to regain aromaticity

(c) 1. Generate the electrophile

2. Attack the electrophile with the aromatic ring to form the sigma complex.

3. Elimination to regain aromaticity

15.34

15.35 (a) Electrophilic aromatic substitution will take place as follows:

(b) The ring directly attached to the oxygen atom is activated toward electrophilic attack because the oxygen atom can donate an unshared electron pair to it and stabilize the intermediate arenium ion when attack occurs at the ortho or para position.

15.36

15.37 (a)

(b)

15.38 (a)

(b) 1,2-Addition.

(c) Yes. The carbocation given in (a) is a hybrid of *secondary allylic* and *benzylic* contributors and is therefore more stable than any other possibility; for example,

A hybrid of allylic contributors only

(d) Since the reaction produces only *the more stable isomer*—that is, the one in which the double bond is conjugated with the benzene ring—the reaction is likely to be under equilibrium control:

15.39

15.40

Benzyne and Nucleophilic Aromatic Substitution

15.41

15.42 Both *o*- and *m*-toluidine are formed in the following way:

Both *m*- and *p*-toluidine are formed from another benzyne-type intermediate.

15.43 (a)

(b)

15.44 One can draw a resonance structure for the 2,6-di-*tert*-butylphenoxide ion which shows the carbon para to the oxygen as a nucleophilic center; that is, the species is an ambident nucleophile.

Given the steric hindrance about the oxygen, the nucleophilic character of the para carbon is dominant and an S_NAr reaction occurs at that position to produce this biphenyl derivative:

15.45

General Problems

15.46 (a) C_6H_5—CH₂Br $\xrightarrow[\substack{\text{DMF} \\ (-\text{NaBr})}]{\text{NaCN}}$ C_6H_5—CH₂CN

(b) C_6H_5—CH₂Br $\xrightarrow[\substack{\text{CH}_3\text{OH} \\ (-\text{NaBr})}]{\text{CH}_3\text{ONa}}$ C_6H_5—CH₂OCH₃

(c) C_6H_5—CH₂Br $\xrightarrow{\qquad}$ C_6H_5—CH₂—O—C(=O)CH₃

CH₃C(=O)ONa
CH₃C(=O)OH
(−NaBr)

(d) C_6H_5—CH₂Br $\xrightarrow[\substack{\text{acetone} \\ (-\text{NaBr})}]{\text{NaI}}$ C_6H_5—CH₂I

(e) CH₂=CH—CH₂Br $\xrightarrow[\substack{\text{acetone} \\ (-\text{NaBr})}]{\text{NaN}_3}$ CH₂=CH—CH₂N₃

(f) CH₂=CH—CH₂Br $\xrightarrow{\qquad}$ CH₂=CH—CH₂—O—CH₂CH(CH₃)₂

(CH₃)₂CHCH₂ONa
(CH₃)₂CHCH₂OH
(−NaBr)

15.47 A = ⬡(cyclohexene)   B = (bromo-cyclohexadiene with Br)

15.48

(p-xylene) $\xrightarrow[\text{H}_2\text{SO}_4]{\text{HNO}_3}$ (2-nitro-p-xylene) (only possible mononitro product)

15.49

$C_8H_{13}OCl$

15.50 (a and b) The *tert*-butyl group is easily introduced by any of the variations of the Friedel–Crafts alkylation reaction, and, because of the stability of the *tert*-butyl cation, it is easily removed under acidic conditions.

(c) In contrast to the $-SO_3H$ group often used as a blocking group, $-C(CH_3)_3$ activates the ring to further electrophilic substitution.

15.51 At the lower temperature, the reaction is kinetically controlled, and the usual *o,p*-directive effects of the $-CH_3$ group are observed. At higher temperatures, the reaction is thermodynamically controlled. At reaction times long enough for equilibrium to be reached, the most stable isomer, *m*-toluenesulfonic acid, is the principal product.

15.52 (a)

BHA

(b)

BHT

Notice that both reactions are Friedel–Crafts alkylations.

Synthesizing the Material

15.53

Toluene + Succinic anhydride $\xrightarrow{\text{AlCl}_3}$ **A** ($C_{11}H_{12}O_3$) $\xrightarrow[\text{heat}]{\text{H}_2\text{NNH}_2, \text{KOH}}$

B ($C_{11}H_{14}O_2$) $\xrightarrow{\text{SOCl}_2}$ **C** ($C_{11}H_{13}ClO$) $\xrightarrow{\text{AlCl}_3}$

D ($C_{11}H_{12}O$) $\xrightarrow{\text{NaBH}_4}$ **E** ($C_{11}H_{14}O$) $\xrightarrow[\text{heat}]{\text{H}_2\text{SO}_4}$

F ($C_{11}H_{12}$) $\xrightarrow[\text{light}]{\text{NBS}}$ **G** ($C_{11}H_{11}Br$) $\xrightarrow[\substack{\text{EtOH} \\ \text{heat}}]{\text{NaOEt}}$

2-Methylnaphthalene

15.54 (a)

(major product)

(b)

(c)

Note: another solution would be F-C acylation,
Wolff-Kishner reduction, then a second F-C
acylation and Wolff-Kishner reduction

15.55 (a)

(major product)

3) O₃
4) Me₂S

5) NaBH₄, H₃O⁺

A

(b)

1) KMnO₄, HO⁻
(H₃O⁺ workup)

2) LiAlH₄ in Et₂O
3) H₂O/H₂SO₄
4) KOt-Bu, t-BuOH

5) NaH, CH₃I

B

Note: step 4 does
no effective chemistry
and can be omitted
from the sequence

(c)

1) ⟋⟍⟋ , Δ

2) H₂, Pd/C

3) Br₂, FeBr₃

(±)-**C** (major product)

15.56 (a)

Cl—(C₆H₄)—CH₂OEt
$\xrightarrow{\text{4) NaH, EtI}}$
Cl—(C₆H₄)—CH₂OH
$\xrightarrow{\text{3) NaBH}_4, \text{ MeOH}}$
Cl—(C₆H₄)—CHO

$\Big\Downarrow$ 2) Cl₂, AlCl₃

benzene
$\xleftarrow[\text{AlCl}_3]{\text{1) HCOCl,}}$
benzaldehyde

A

(b)

NO₂-substituted benzene with Cl and CH₂OH (HO)
$\xrightarrow[\text{5) H}_2\text{O/H}_2\text{SO}_4]{\text{4) LiAlH}_4 \text{ in Et}_2\text{O}}$
NO₂-substituted benzene with Cl and COOH
$\xrightarrow[\text{HO}^-, \Delta]{\text{3) KMnO}_4,}$
NO₂-substituted benzene with Cl and CH₃

$\Big\Downarrow$ 2) HNO₃, H₂SO₄

chlorobenzene
$\xleftarrow[\text{AlCl}_3]{\text{1) CH}_3\text{Cl,}}$
Cl and CH₃ substituted benzene

B

Note: Alkyl is the stronger directing group for step 2, and step 1 will produce *para*-methyl substituted material as well that must be separated when conducted in forward direction. Precursor for step 3 must be methyl based on step 1.

Challenge Problems

15.57

2,4,5-Trichlorophenol (2,4,5-TCP)

15.58 Examination of the resonance structures that can be drawn for the S_NAr intermediate in the first case reveals that the five-membered ring has cyclopentadienyl anion character. This is a stabilizing feature due to aromaticity.

In the latter case, electron-delocalization into the other (seven-membered) ring results in eight electrons in that ring, which does not provide aromatic stabilization.

15.59 Resonance structures for adding the E^+ to the 2 position

Resonance structures for adding the E^+ to the 3 position

There are more resonance structures for substitution at the 2 position of furan. The pathway for EAS should be lower in energy for substitution at the 2 position over the 3 position.

15.60 The evidence indicates that the mechanistic step in which the C—H bond is broken is *not* rate determining. (In the case cited, it makes no difference kinetically if a C—H or C—D bond is broken in electrophilic aromatic substitution.) This evidence is consistent with the two-step mechanism given in Section 15.2. The step in which the aromatic compound reacts with the electrophile (NO_2^+) is the slow rate-determining step. Proton (or deuteron) loss from the arenium ion to return to an aromatic system is a rapid step and has no effect on the overall rate.

15.61 The final product is *o*-nitroaniline. (The reactions are given in Section 15.12B.) The presence of six signals in the ^{13}C NMR spectrum confirms that the substitution in the final product is ortho and not para. A final product with para substitution (i.e., *p*-nitroaniline) would have given only four signals in the ^{13}C NMR spectrum.

15.62 (a) Large ortho substituents prevent the two rings from becoming coplanar and prevent rotation about the single bond that connects them. If the correct substitution patterns are present, the molecule as a whole will be chiral. Thus, enantiomeric forms are possible even though the molecules do not have a chirality center. The compound with 2-NO_2, 6-CO_2H, 2'-NO_2, 6'-CO_2H is an example.

The molecules are atropisomers (see Section 5.18). Furthermore, they are nonsuperposable mirror images and, thus, are enantiomers.

(b) Yes

and

These molecules are enantiomeric atropisomers (Section 5.18).

(c) This molecule has a plane of symmetry; hence, its mirror image forms are equivalent.

The plane of the page is a plane of symmetry.

15.63 The proximity of the two —OH groups results in the two naphthalene moieties not being coplanar. As a result, the two enantiomeric forms are nonequivalent and can be separated by a resolution technique.

$\neq$

15.64

(Glycerol). After sodium borohydride reduction of the aldehyde, ozonolysis oxidatively degrades the aromatic ring, leaving only the polyhydroxy side chain. Water is an alternative to HOAc sometimes used to work up ozonolysis reactions of benzene rings.

15.65

D
(Threonic acid)

Ozonolysis oxidatively degrades the aromatic rings, leaving only the carboxyl carbon as a remnant of the alkyl-substituted benzene ring. Water is an alternative to HOAc sometimes used to work up ozonolysis reactions of benzene rings.

15.66 (a) HOMO-1 best represents the region where electrons of the additional π bond would be found because it has a large lobe extending toward the periphery of the carbon ring, signifying the high probability of finding the additional π electrons in this region. HOMO-2 also has a small lobe inside the ring, corresponding to the second lobe expected for a typical π bond. Overall, the molecular orbital bears less resemblance to that of a simple alkene π bond, however, because of the complexity of the molecule.

(b) Only a vacant orbital can accept an electron pair from a Lewis base, hence it is the LUMO+1 (next to the lowest *unoccupied* molecular orbital) that is involved in the nucleophilic addition step.

QUIZ

15.1 Which of the following compounds would be most reactive toward ring bromination?

(a)   (b)   (c)   (d)   (e)

15.2 Which of the following is *not* a meta-directing substituent when present on a benzene ring?

(a) $-C_6H_5$ (b) $-NO_2$ (c) $-\overset{+}{N}(CH_3)_3$ (d) $-C\equiv N$ (e) $-CO_2H$

15.3 The major product(s), **C**, of the reaction,

$\xrightarrow[\text{FeBr}_3]{\text{Br}_2}$ **C**, would be

(a)   (b)   (c)

(d) Equal amounts of (a) and (b)   (e) Equal amounts of (a) and (c)

15.4 Complete the following syntheses.

16

ALDEHYDES AND KETONES. NUCLEOPHILIC ADDITION TO THE CARBONYL GROUP

SOLUTIONS TO PROBLEMS

16.1 (a)

Pentanal

2-Methylbutanal

3-Methylbutanal

2,2-Dimethylpropanal

2-Pentanone

3-Pentanone

3-Methyl-2-butanone

(b)

Acetophenone
(methyl phenyl ketone)

Phenylethanal
(phenylacetaldehyde)

2-Methylbenzaldehyde
(*o*-tolualdehyde)

3-Methylbenzaldehyde
(*m*-tolualdehyde)

4-Methylbenzaldehyde
(*p*-tolualdehyde)

16.2 (a) 1-Pentanol, because its molecules form hydrogen bonds to each other.

(b) 2-Pentanol, because its molecules form hydrogen bonds to each other.

(c) Pentanal, because its molecules are more polar.

(d) 2-Phenylethanol, because its molecules form hydrogen bonds to each other.

(e) Benzyl alcohol because its molecules form hydrogen bonds to each other.

16.3 (a)

(b)

16.4 (a)

(b)

(c)

(d)

(e)

(f)

16.5 (a) The nucleophile is the negatively charged carbon of the Grignard reagent *acting as a carbanion.*

(b) The magnesium portion of the Grignard reagent acts as a Lewis acid and accepts an electron pair of the carbonyl oxygen. This acid-base interaction makes the carbonyl carbon even more positive and, therefore, even more susceptible to nucleophilic attack.

(c) The product that forms initially (above) is a magnesium alkoxide salt.

(d) On addition of water, the organic product that forms is an alcohol.

16.6 The nucleophile is a hydride ion.

16.7

16.8 *Acid-Catalyzed Reaction*

Base-Catalyzed Reaction

$$OH^- \;+\; H_2{}^{18}O \;\rightleftharpoons\; H_2O \;+\; {}^{18}OH^-$$

16.9

(hemiacetal)

(acetal)

16.10

Sucrose

Sucrose

16.11

16.12

16.13 (a)

(b) Addition would take place at the ketone group as well as at the ester group. The product (after hydrolysis) would be

16.14 (a)

(b) Tetrahydropyranyl ethers are acetals; thus, they are stable in aqueous base and hydrolyze readily in aqueous acid.

5-Hydroxypentanal

(c)

16.15 (a)

(b)

16.16 (a)

Lactic acid

(b) A racemic form

16.17 (a) CH_3I $\xrightarrow[\text{(2) RLi}]{\text{(1) } (C_6H_5)_3P:}$ $H_2\ddot{C} - \overset{+}{P}(C_6H_5)_3$ →

(b) ∼Br $\xrightarrow[\text{(2) RLi}]{\text{(1) } (C_6H_5)_3P:}$ $\overset{+}{P}(C_6H_5)_3$ →

(c) $H_2\ddot{C} - \overset{+}{P}(C_6H_5)_3$ →
[from part (a)]

(d) $H_2\ddot{C} - \overset{+}{P}(C_6H_5)_3$ →
[from part (a)]

(e) ∼∼Br $\xrightarrow[\text{(2) RLi}]{\text{(1) } (C_6H_5)_3P:}$ $\overset{+}{P}(C_6H_5)_3$ →

(f) ∼∼Br $\xrightarrow[\text{(2) RLi}]{\text{(1) } (C_6H_5)_3P:}$ $\overset{+}{P}(C_6H_5)_3$ →

(g) $\xrightarrow[\text{(2) RLi}]{\text{(1) } (C_6H_5)_3P:}$ $\overset{+}{P}(C_6H_5)_3$ →

16.18

16.19

16.20

16.21

Problems

16.22 (a)   Methanal

(b)   Ethanal

(c) C_6H_5 Phenylethanal

(d)   Propanone

(e)   Butanone

(f)   1-Phenylethanone or methyl phenyl ketone

(g)   Diphenylmethanone or diphenyl ketone

(h)   2-Hydroxybenzaldehyde or o-hydroxybenzaldehyde

(i)   4-Hydroxy-3-methoxybenzaldehyde

(j)   3-Pentanone

(k)   2-Methyl-3-pentanone

(l)   2,4-Dimethyl-3-pentanone

(m) 5-Nonanone

(n) 4-Heptanone

(o) (*E*)-3-Phenylpropenal

16.23 (a) OH   (h)

(b)   (i) O$^-$ $^+$NH$_4$ + Ag ↓

(c) OH   (j)

(d)   (k)

(e)   (l)

(f) OH   (m)

(g)   (n) + CH$_3$CH$_3$ + NiS

(o) OH

16.24 (a)

(b)

(c)

(d) No reaction

(e)

(f)

(g)

(h)

(i) No reaction

(j)

(k)

(l) No reaction

(m)

(n) $CH_2=CH_2$ (or alkene) + CH_3CH_3 + NiS

16.25 (a)

(b)

(c)

(d)

(e)

(f)

(g)

16.26 (a)

(b)

(c)

(d)

(e)

(f)

(g)

16.27 (a) cyclopentanecarbaldehyde + CH_3NH_2, cat. HA → N-methyl imine

(b) cyclohexanone + pyrrolidine (N—H), cat. HA → enamine

(c) 2-(2-aminoethyl)cyclohexanone, cat. HA → bicyclic imine

(d) cyclohexanone + phosphorane (PPh$_3$) → alkene

(e) ketone, (1) HS—CH$_2$CH$_2$—SH, (2) Raney Ni, H$_2$ → alkane

(f) chromanol + :$\bar{C}H_2$—$\overset{+}{P}Ph_3$ (excess) → phenol with butenyl chain

(g) bicyclic diacetal, H$_2$O, HA → HO—...—CHO—...—OH

(h) methyl-substituted lactone

(i) 1-ethylindane

16.28 (a)

(b)

H$_2$SO$_4$ (cat.)

(c)

HA (cat.)

(d)

(1) HCN
(2) LiAlH$_4$
(3) H$_3$O$^+$

(e)

16.29 (a)

PCC
or Swern oxidation

(b)

(1) CH$_3$MgBr
(2) H$_3$O$^+$
(3) PCC or Swern oxidation
(4) CH$_3$CH$_2$MgBr
(5) H$_3$O$^+$

(c)

(1) OsO$_4$
(2) HA (cat.)

(d)

(1) O$_3$
(2) CH$_3$SCH$_3$
(3) HA (cat.)
HO$\diagdown\diagup$OH

(e)

HA (cat.)

(f)

HS$\diagup\diagdown$SH , HA (cat.)

(g) $\dfrac{\text{H}_2\text{NNH}_2,\ \text{KOH}}{\text{heat}}$

(h) mCPBA

16.30 (a)

(b)

(d)

16.31

(1) Swern oxidation or PCC

(2) ⌒⌒MgBr

(3) H_3O^+

(1) Swern oxidation or PCC

(2) ⌒MgBr

(3) H_3O^+

Synthesis

16.32 (a)

or

(b)

(Other methods are possible, e.g., NaBH$_4$ reduction of the starting ketone followed by dehydration and hydrogenation.)

16.33 (a)

(b)

(c)

[from (b)]

(d)

(e)

(f)

(g)

[from (a)]

(h)

[from (g)]

or

or Wolff–Kishner reduction of benzaldehyde

(i) $C_6H_5CHO \xrightarrow[\text{cat. HA}]{CH_3OH}$ $C_6H_5CH(OCH_3)_2$

(j) $C_6H_5CHO \xrightarrow[H_3{}^{18}O^+]{H_2{}^{18}O}$ $C_6H_5CH{}^{18}O$ (See Problem 16.8 for the mechanism)

(k) $C_6H_5CHO \xrightarrow[(2)\ H_3O^+]{(1)\ NaBD_4}$ $C_6H_5CHD(OH)$ + enantiomer

(l) $C_6H_5CHO \xrightarrow{HCN}$ $C_6H_5CH(OH)CN$ + enantiomer

(a cyanohydrin)

(m) $C_6H_5CHO \xrightarrow[HA]{NH_2OH}$ $C_6H_5CH=N-OH$ + stereoisomer

(an oxime)

(n) C_6H_5CHO + $H_2N-N(H)C_6H_5$ $\xrightarrow{H_3O^+}$ $C_6H_5CH=N-N(H)-C_6H_5$ + stereoisomer

(a phenylhydrazone)

(o) C_6H_5CHO + $(C_6H_5)_3\overset{+}{P}-\overset{..}{CH}-CH=CH_2$ (a Wittig reagent) $\longrightarrow$ $C_6H_5CH=CH-CH=CH_2$

16.34 (a) C_6H_6 + CH_3CH_2COCl $\xrightarrow{AlCl_3}$ $C_6H_5COCH_2CH_3$

(b)

$$+ \quad \xrightarrow{\text{(2) } H_3O^+}$$

(or ethyl magnesium bromide)

(c)

$$+ \quad \text{MgBr} \quad \xrightarrow{\text{(2) } H_3O^+} \quad \xrightarrow[\text{acetone}]{H_2CrO_4}$$

(or ethyl lithium)

16.35 (a)

$$\xrightarrow{\text{Swern oxidation or PCC}}$$

(b)

$$\xrightarrow{SOCl_2} \quad \xrightarrow[\text{Et}_2O, -78\ °C]{\text{LiAlH(O-}t\text{-Bu)}_3}$$

(c)

$$\xrightarrow[\text{(2) } H_3O^+]{\text{(1) } KMnO_4,\ HO^-} \quad \xrightarrow{\text{[as in (b)]}}$$

(d)

$$\xrightarrow[\text{(2) } H_3O^+]{\text{(1) } KMnO_4,\ HO^-} \quad \xrightarrow{\text{[as in (b)]}}$$

(e)

$$\xrightarrow[\text{(2) } H_2O]{\text{(1) DIBAL-H, } -78\ °C}$$

(f)

$$\xrightarrow[\text{(2) } H_2O]{\text{(1) DIBAL-H, } -78\ °C}$$

16.36

16.37

The compound $C_7H_6O_3$ is 3,4-dihydroxybenzaldehyde. The reaction involves hydrolysis of the acetal of formaldehyde.

16.38 (a)

(b) [from (a)]

(c)

(a Wittig reagent)

(d)

[from (a)]

+ (2) H₃O⁺

racemic

$\dfrac{H_2CrO_4}{acetone}$

16.39 HO $\overbrace{\hspace{2cm}}$ OH , cat. HA

A

$\dfrac{Mg}{Et_2O}$

B

(1)

(2) H₃O⁺, H₂O

C

⇌

(1) CH₃OH
(2) cat. HA

(a hemiacetal)   **D** (an acetal)

16.40

(7S)   (3S)   (7R)   (3S)

3,7-Diethyl-9-phenylnonan-2-one (dianeackerone) stereoisomers

16.41

$\dfrac{PCC}{CH_2Cl_2}$

A

$\dfrac{CH_3OH}{cat.\ HA}$

B

$\dfrac{KMnO_4,\ HO^-}{cold,\ dilute}$

C

$\dfrac{H_3O^+}{H_2O}$

Glyceraldehyde

The product would be racemic since no chiral reagents were used.

16.42

(R)-3-Phenyl-2-pentanone → NaBH$_4$ → (S) (R) + (R) (R)

Diastereomers

16.43

$$Br \underset{7}{(\ )} Br \xrightarrow[\text{(2) 2 RLi}]{\text{(1) 2 (C}_6\text{H}_5)_3\text{P}} (C_6H_5)_3\overset{+}{P} \underset{7}{(\ )} \overset{+}{P}(C_6H_5)_3$$

A

B → $\xrightarrow[\text{pressure}]{\text{H}_2 \text{ Pt}}$ → **C**

16.44 (a) *Retrosynthetic Analysis*

Synthesis

(b) **Retrosynthetic Analysis**

Synthesis

16.45

16.46 The two nitrogen atoms of semicarbazide that are adjacent to the C=O group bear partial positive charges because of resonance contributions made by the second and third structures below.

This nitrogen is the most nucleophilic.

16.47 Hydrolysis of the acetal linkage of multistriatin produces the ketodiol below.

16.48

16.49 **Compound W is**

16.50 Each ^{1}H NMR spectrum (Figs. 16.4 and 16.5) has a five-hydrogen peak near δ 7.2, suggesting the **Y** and **Z** each has a C_6H_5- group. The IR spectrum of each compound shows a strong peak near 1710 cm^{-1}. This absorption indicates that each compound has a $C=O$ group not adjacent to the phenyl group. We have, therefore, the following pieces,

If we subtract the atoms of these pieces from the molecular formula,

$$
\begin{array}{l}
C_{10}H_{12}O \\
\underline{-\ C_7H_5O\ \ (C_6H_5 + C{=}O)} \\
C_3H_7
\end{array}
$$

we are left with

In the ^{1}H NMR spectrum of **Y**, we see an ethyl group [triplet, δ 1.0 (3H) and quartet, δ 2.45 (2H)] and an unsplit —CH$_2$— group signal [singlet, δ 3.7 (2H)]. This means that **Y** must be

1-Phenyl-2-butanone

In the ^{1}H NMR spectrum of **Z**, we see an unsplit —CH$_3$ group signal [singlet, δ 2.1 (3H)] and two triplets at δ 2.7 and 2.9. This means **Z** must be

4-Phenyl-2-butanone

16.51   That compound **A** forms a phenylhydrazone, gives a negative Tollens' test, and gives an IR band near 1710 cm^{-1} indicates that **A** is a ketone. The ^{13}C spectrum of **A** contains only four signals indicating that **A** has a high degree of symmetry. The information from the DEPT ^{13}C NMR spectra enables us to conclude that **A** is diisobutyl ketone:

Assignments:

(a) δ 22.6

(b) δ 24.4

(c) δ 52.3

(d) δ 210.0

16.52   That the ^{13}C spectrum of **B** contains only three signals indicates that **B** has a highly symmetrical structure. The information from DEPT spectra indicates the presence of equivalent methyl groups (CH$_3$ at δ 19), equivalent —$\overset{|}{\underset{|}{C}}$— groups (at δ 71), and equivalent $\overset{\backslash}{\underset{/}{C}}$=O groups (at δ 216). These features allow only one possible structure for **B**:

Assignments:

(a) δ 19

(b) δ 71

(c) δ 216

Synthesizing the Material

16.53 (a)

(b)

(c)

(d)

16.54 (a)

A

(b)

B

(c)

C

(d)

1) NaH, CH₃I →

2) ClCOCH₃, AlCl₃ →

and

3) EtMgBr; H₃O⁺ ↓

4) PBr₃, pyr ←

[Note: reaction likely goes through carbocation intermediate stabilized by aromatic ring]

and

and

5) t-BuOK, t-BuOH ↓

and

D

16.55 (a)

4) NaH, CH₃I ⟹

3) MeMgBr; H₃O⁺ ⟹

or

2) mCPBA ⟹

mCPBA step is most critical for determining the correct structure of **A**   X⁻ ⁺PPh₃

1) (a) NaH; (b) HCOCH₃ ←

A

(b)

4) H_2NNH_2,
cat. HA
5) KOH, Δ

3) MeMgBr;
H_3O^+

Must be nitrile since only a single methyl addition and ketone product

2) KCN, DMF

OMs

1) MsCl, pyr

OH

B

Challenge Problems

16.56 (a) O—H stretch at about 3300 cm^{-1}; C=O stretch at about 1710 cm^{-1}

(b) OH (Intramolecular hemiacetal from C)

16.57 H_3CO H_3CO OH

E

QUIZ

16.1 Which Wittig reagent could be used to synthesize ? (Assume any other needed reagents are available.)

(a) $P(C_6H_5)_3$
$+$

(d) More than one of these

(b) $P(C_6H_5)_3$
$+$

(e) None of these

(c) $P(C_6H_5)_3$
$+$

16.2 Which compound is an acetal?

(a)

(b)

(c)

(d) More than one of these

(e) None of these

16.3 Which reaction sequence could be used to convert to ?

(a) O_3, then Me_2S, then $AlCl_3$, then

(b) (1) H_2, P-2 cat.; (2) $Hg(OAc)_2$/THF-H_2O; (3) $NaBH_4$, HO^-; (4) PCC or Swern oxidation or H_2CrO_4

(c) HCl, then

(d) O_3, then Me_2S, then H_2SO_4, $HgSO_4$, H_2O, heat

(e) OH, then H_2O_2, HO^-/H_2O

16.4 Complete the following syntheses. If more than one step is required for a transformation, list them as (1), (2), (3), and so on.

(a) C₆H₅—CH₃ $\xrightarrow[hv]{NBS}$ **A** [ ] **B** [ ] $\xrightarrow{THF}$

C₆H₅CH₂CN **C** [ ] → C₆H₅CH₂COCH₃

(b) Cl—C₆H₄—CH₂OH **A** [ ] → Cl—C₆H₄—CHO

B [ ] → Cl—C₆H₄—CH(OCH₂CH₂O) **C** [ ] → ClMg—C₆H₄—CH(OCH₂CH₂O)

(c) CH₃CH₂CH₂Br **A** [ ] → **B** [ ] $\xrightarrow{C_6H_5Li}$

CH₃CH₂CH=P(C₆H₅)₃ **C** [ ] → (cyclopentylidene propene)

(d) C₆H₅CHO **A** [ ] → C₆H₅COCH₃ **B** [ ] →

C₆H₅C(CH₃)(OH)CN (racemic) **C** [ ] → C₆H₅C(CH₃)(OH)CH₂NH₂ (racemic)

16.5 An industrial synthesis of benzaldehyde makes use of toluene and molecular chlorine as starting materials to produce C₆H₅CHCl₂. This compound is then converted to benzaldehyde. Suggest what steps are involved in the process.

16.6 In the case of aldehydes and unsymmetrical ketones, two isomeric oximes are possible. What is the origin of this isomerism?

17

CARBOXYLIC ACIDS AND THEIR DERIVATIVES: NUCLEOPHILIC ADDITION-ELIMINATION AT THE ACYL CARBON

SOLUTIONS TO PROBLEMS

17.1 (a) 2-Methylbutanoic acid

(b) (Z)-3-Pentenoic acid or (Z)-pent-3-enoic acid

(c) Sodium 4-bromobutanoate

(d) 5-Phenylpentanoic acid

(e) (E)-3-Ethyl-3-pentenoic acid or (E)-3-Ethylpent-3-enoic acid

17.2 (a) (—F is more electronegative than —H)

(b) (—F is more electronegative than —Cl)

(c) (—F is closer to —CO$_2$H)

(d) [—$\overset{+}{N}Me_3$ is more electronegative than —H]

(e) (—CF$_3$ is more electronegative than —CH$_3$)

17.3 (a)

(b)

(c)

(d)

(e)

(f)

(g)

(h)

(i)

(j)

17.4 (a) $\xrightarrow[\text{(2) H}_3\text{O}^+]{\text{(1) KMnO}_4,\ \text{HO}^-,\ \text{heat}}$ + CO_2

(b)

Mg / Et₂O → PhMgBr → CO₂ → PhC(O)OMgBr → H₃O⁺ → PhCOOH

(c)

(1) Cl₂/NaOH
(2) H₃O⁺

→ benzoic acid + CHCl₃

(d)

(1) KMnO₄, HO⁻, heat
(2) H₃O⁺

→ benzoic acid

(e)

(1) KMnO₄, HO⁻, heat
(2) H₃O⁺

→ benzoic acid

(f)

(1) KMnO₄, HO⁻
(2) H₃O⁺

→ benzoic acid

17.5 These syntheses are easy to see if we work backward using a retrosynthetic analysis before writing the synthesis.

(a) *Retrosynthetic analysis*

Synthesis

(b) *Retrosynthetic analysis*

Synthesis

(c) *Retrosynthetic analysis*

Synthesis

(d) *Retrosynthetic analysis*

Synthesis

(e) *Retrosynthetic analysis*

Synthesis

17.6 (a)

Retrosynthetic analysis

Synthesis

Retrosynthetic analysis

Synthesis

Retrosynthetic analysis

Synthesis

(b) A nitrile synthesis. Preparation of a Grignard reagent from HO⌒⌒⌒Br would not be possible because of the presence of the weakly acidic hydroxyl group.

17.7 Since maleic acid is a cis dicarboxylic acid, dehydration occurs readily:

Maleic acid   Maleic anhydride

Being a trans dicarboxylic acid, fumaric acid must undergo isomerization to maleic acid first. This isomerization requires a higher temperature.

Fumaric acid

17.8 The labeled oxygen atom should appear in the carboxyl group of the acid. (Follow the reverse steps of the mechanism in Section 17.7A of the text using $H_2^{18}O$.)

17.9

Intermolecular proton transfer

17.10 (a) (1)

(2)

(3)

E

F

(4)

F

(b) Method (3) should give a higher yield of **F** than method (4). Since the hydroxide ion is a strong base and since the alkyl halide is secondary, method (4) is likely to be accompanied by considerable elimination. Method (3), on the other hand, employs a weaker base, acetate ion, in the S_N2 step and is less likely to be complicated by elimination. Hydrolysis of the ester **E** that results should also proceed in high yield.

17.11 (a) Steric hindrance presented by the di-ortho methyl groups of methyl mesitoate prevents formation of the tetrahedral intermediate that must accompany attack at the acyl carbon.

(b) Carry out hydrolysis with labeled $H^{18}O^-$ in labeled $H_2^{18}O$. The label should appear in the methanol.

17.12 (a) C_6H_5

(b)

(c)

17.13 **(a)**

(b) An elimination reaction would take place because ⁻CN is a strong base, and the substrate is a 3° alkyl halide.

17.14 **(a)**

(b)

(c)

(d)

(e)

(f)

17.15 (a) By decarboxylation of a β-keto acid:

(b) By decarboxylation of a β-keto acid:

(c) By decarboxylation of a β-keto acid:

(d) By decarboxylation of a β-keto acid:

17.16 (a) The oxygen-oxygen bond of the diacyl peroxide has a low homolytic bond dissociation energy ($DH° \sim 139$ kJ mol^{-1}). This allows the following reaction to occur at a moderate temperature.

$$
\underset{R}{\overset{O}{\|}}\underset{O-O}{}\underset{R}{\overset{O}{\|}} \longrightarrow 2\ \underset{R}{\overset{O}{\|}}\underset{O\cdot}{}
$$

(b) By decarboxylation of the carboxyl radical produced in part (a).

$$
\underset{R}{\overset{O}{\|}}\underset{O\cdot}{} \longrightarrow R\cdot + CO_2
$$

(c) **Chain Initiation**

Step 1 $\quad \underset{R}{\overset{O}{\|}}\underset{O-O}{}\underset{R}{\overset{O}{\|}} \xrightarrow{\text{heat}} 2\ \underset{R}{\overset{O}{\|}}\underset{O\cdot}{}$

Step 2 $\quad \underset{R}{\overset{O}{\|}}\underset{O\cdot}{} \longrightarrow R\cdot + CO_2$

Chain Propagation

Step 3 $\quad R\cdot\ +\ = \longrightarrow R\diagup\diagdown\cdot$

Step 4 $\quad R\diagup\diagdown\cdot\ +\ = \longrightarrow R\diagup\diagdown\diagup\cdot$

Steps 3, 4, 3, 4, and so on.

Problems

Structure and Nomenclature

17.17 (a)

$$\text{CH}_3\text{CH}_2\text{CH}_2\text{CH}_2\text{CH}_2\text{CH}_2\underset{\text{OH}}{\overset{O}{\|}}$$

(b)

$$\text{CH}_3\text{CH}_2\underset{\text{NH}_2}{\overset{O}{\|}}$$

(c)

$$\text{CH}_3\text{CH}_2\text{CH}_2\text{CH}_2\underset{N}{\overset{O}{\|}}\diagup\diagdown$$

(d)

$$\underset{\text{OH}}{\overset{O}{\|}}$$

[the (Z) isomer]

(e)

(f)

(g)

(h)

(i)

(j)

(k)

(l)

(m)

(n)

17.18 (a) Benzoic acid

(b) Benzoyl chloride

(c) Acetamide or ethanamide

(d) Acetic anhydride or ethanoic anhydride

(e) Benzyl benzoate

(f) Phenyl propanoate or phenyl propionate

(g) Isopropyl acetate or 1-methylethyl ethanoate

(h) Acetonitrile or ethanenitrile

17.19 Alkyl groups are electron releasing; they help disperse the positive charge of an alkyl-ammonium salt and thereby help to stabilize it.

$$R\ddot{N}H_2 \;+\; H_3O^+ \longrightarrow R{\rightarrow}\overset{+}{N}H_3 \;+\; H_2O$$

*Stabilized by
electron-releasing
alkyl group*

Consequently, alkylamines are somewhat stronger bases than ammonia.

Amides, on the other hand, have acyl groups, $R\overset{\text{O}}{\underset{}{\diagup\!\!\diagdown}}$, attached to nitrogen, and acyl groups are electron withdrawing. They are especially electron withdrawing because of resonance contributions of the kind shown here,

This kind of resonance also *stabilizes* the amide. The tendency of the acyl group to be electron withdrawing, however, *destabilizes* the conjugate acid of an amide, and reactions such as the following do not take place to an appreciable extent.

*Stabilized
by
resonance*   *Destabilized by
electron-withdrawing
acyl group*

17.20 (a) The conjugate base of an amide is stabilized by resonance.

*This structure is especially
stable because the negative
charge is on oxygen.*

There is no resonance stabilization for the conjugate base of an amine, $R\bar{\ddot{N}}H$.

(b) The conjugate base of an imide is stabilized by additional resonance structures,

An imide

Functional Group Transformations

17.21 (a)

CH_3COOH + HCl

(b)

(c)

(d)

(e)

(f)

(g)

(h)

(i)

(j)

(k)

(l)

(m)

17.22 (a) $CH_3CONH_2 + CH_3CO_2{}^- \overset{+}{N}H_4$

(b) $2\ CH_3CO_2H$

(c) $CH_3CO_2CH_2CH_2CH_3 + CH_3CO_2H$

(d) $C_6H_5COCH_3 + CH_3CO_2H$

(e) $CH_3CONHCH_2CH_3 + CH_3CO_2{}^- \ CH_3CH_2\overset{+}{N}H_3$

(f) $CH_3CON(CH_2CH_3)_2 + CH_3CO_2{}^-(CH_3CH_2)_2\overset{+}{N}H_2$

(g) $+$ $CH_3CO_2{}^-$

17.23 (a)   (b)

(c)   (d)

(e)   (f)

17.24 (a) $+$

(b) $+$

(c) $+$

(d) $+$

(e) $+$

(f) $+$

17.25 (a)

(b)

(c)

17.26 (a)

(b)

Acid-catalyzed dehydration is favored due to conjugation and because formation of the 4-membered ring lactone is disfavored geometrically.

(c)

and

(d)

(e)

(f)

General Problems

17.27 (a)

(b)

(c)

(d)

(e)
$$\underset{\text{Br}}{\bigcirc} \xrightarrow[\text{(3) H}_3\text{O}^+]{\substack{\text{(1) Mg}^\circ \\ \text{(2) CO}_2}} \text{cyclopentanecarboxylic acid (OH)}$$

(f) cyclohexyl acetate $\xrightarrow[\text{(2) H}_3\text{O}^+]{\text{(1) NaOH, H}_2\text{O}}$ cyclohexanol + acetic acid

(g) $\xrightarrow{\text{H}_2\text{O, H}_2\text{SO}_4 \text{ (cat.)}}$ carboxylic acid + cyclopropyl ammonium

(h) benzyl alcohol (OH) $\xrightarrow[\text{(2) SOCl}_2]{\text{(1) H}_2\text{CrO}_4}$ benzoyl chloride

(i) Phenol + phthalic anhydride $\longrightarrow$ phenyl hydrogen phthalate

17.28 (a) benzyl alcohol (OH) $\xrightarrow[\text{(2) CH}_3\text{OH (excess), H}_2\text{SO}_4 \text{ (cat.)}]{\text{(1) H}_2\text{CrO}_4}$ methyl benzoate (OCH$_3$)

(b) δ-valerolactone $\xrightarrow[\text{(2) H}_3\text{O}^+]{\text{(1) NaOH, H}_2\text{O}}$ hydroxy acid (OH, OH)

(c) cyclobutanecarbonitrile ($\equiv$N) $\xrightarrow[\text{(2) SOCl}_2]{\text{(1) H}_2\text{O, H}_2\text{SO}_4 \text{ (cat.)}}$ cyclobutanecarbonyl chloride (Cl)

(d) cyclopentanecarbonyl chloride (Cl) $\xrightarrow{\text{LiAlH(O-}t\text{-Bu)}_3}$ cyclopentanecarbaldehyde (H)

(e) methyl benzoate (OCH$_3$) $\xrightarrow{\text{LiAlH(O-}t\text{-Bu)}_3}$ benzaldehyde (H)

(f)

17.29 (a)

(b)

(c)

(d)

(e)

(f)

17.30 (a)

(b)

(c)

(d)

(e)

Mechanisms

17.31 See the mechanisms in Section 17.8F, where R = CH$_2$CH$_3$ for propanamide.

17.32 (a)

(b)

(c)

Tautomerization

Intermolecular proton transfer

17.33 *cis*-4-Hydroxycyclohexanecarboxylic acid can assume a boat conformation that permits lactone formation.

Neither of the chair conformations nor the boat form of *trans*-4-hydroxycyclohexanecarboxylic acid places the —OH group and the —CO$_2$H group close enough together to permit lactonization.

Synthesis

17.34 **(a)**

$$\underset{\text{(2) } H_3O^+}{\overset{\text{(1) } KMnO_4, HO^-, \text{ heat}}{\longrightarrow}}$$

(b)

[from (a)]

$$\underset{\substack{\text{(2) } LiAlH(O\text{-}t\text{-}Bu)_3 \\ Et_2O, -78\ °C}}{\overset{\text{(1) } SOCl_2}{\longrightarrow}}$$

HCN

$$\xrightarrow[\text{heat}]{H_3O^+, H_2O}$$

(c)

$$\xrightarrow[hv \text{ or } ROOR]{NBS}$$

NaCN

$$\xrightarrow[\text{heat}]{H_3O^+, H_2O}$$

(d)

[from (c)]

$$\underset{\substack{\text{(2) } LiAlH(O\text{-}t\text{-}Bu)_3 \\ Et_2O, -78\ °C}}{\overset{\text{(1) } SOCl_2}{\longrightarrow}}$$

HCN

$$\xrightarrow[\text{heat}]{H_3O^+, H_2O}$$

$[(E) + (Z)]$

17.35 (a) HO-(CH₂)₄-OH $\xrightarrow[\text{(2) H}_3\text{O}^+]{\text{(1) KMnO}_4,\ \text{HO}^-,\ \text{heat}}$ HO-C(=O)-CH₂-CH₂-C(=O)-OH

(b) cyclohexanol with OH $\xrightarrow[\text{(2) H}_3\text{O}^+]{\text{(1) KMnO}_4,\ \text{HO}^-,\ \text{heat}}$ HO-C(=O)-(CH₂)₄-C(=O)-OH

17.36 (a) pentanol chain —OH $\xrightarrow[\text{(2) H}_3\text{O}^+]{\text{(1) KMnO}_4,\ \text{HO}^-,\ \text{heat}}$ carboxylic acid —C(=O)—OH

(b) —Br $\xrightarrow[\text{(2) CO}_2]{\text{(1) Mg, Et}_2\text{O}}$ —C(=O)—$OMgBr$ $\xrightarrow{\text{H}_3\text{O}^+}$ —C(=O)—OH

—Br $\xrightarrow{^-CN}$ —CN $\xrightarrow[\text{heat}]{\text{H}_3\text{O}^+,\ \text{H}_2\text{O}}$ —C(=O)—OH

(c) alkene chain $[(E)\ \text{or}\ (Z)]$ $\xrightarrow[\text{(2) H}_3\text{O}^+]{\text{(1) KMnO}_4,\ \text{HO}^-,\ \text{heat}}$ 2 —C(=O)—OH

(d) —C(=O)—H $\xrightarrow[\text{(2) H}_3\text{O}^+]{\text{(1) Ag(NH}_3)_2\text{OH}}$ —C(=O)—OH

17.37 3-methylbenzoic acid (C(=O)OH, CH₃) $+$ $SOCl_2$ $\longrightarrow$ acid chloride (C(=O)Cl, CH₃) $\xrightarrow[\text{(excess)}]{\text{Et}_2\text{NH}}$ amide (C(=O)N(Et)₂, CH₃)

17.38

17.39 (a)

(b)

(c)

17.40 (a)

(b)

(R)-(−)-2-butanol $\xrightarrow[\substack{\text{pyridine} \\ \text{(inversion)}}]{\text{PBr}_3}$ **E** $\xrightarrow[\text{(inversion)}]{^-\text{CN}}$ **F**

$\xrightarrow[\text{(retention)}]{\text{H}_2\text{SO}_4,\ \text{H}_2\text{O}}$ (−)-**C** $\xrightarrow[\substack{\text{(2) H}_2\text{O} \\ \text{(retention)}}]{\text{(1) LiAlH}_4}$ (+)-**D**

(c)

A $\xrightarrow[\text{(inversion)}]{}$ **G** $\xrightarrow[\text{(retention)}]{\text{HO}^-}$ (+)-**H** +

(S)-(+)-2-butanol

(d)

(−)-**D** $\xrightarrow[\text{(retention)}]{\text{PBr}_3}$ **J** $\xrightarrow[\substack{\text{Et}_2\text{O} \\ \text{(retention)}}]{\text{Mg}}$

K $\xrightarrow[\substack{\text{(2) H}_3\text{O}^+ \\ \text{(retention)}}]{\text{(1) CO}_2}$ **L**

(e)

(R)-(+)-Glyceraldehyde $\xrightarrow{\text{HCN}}$ **M** + **N**

(f) **M** $\xrightarrow[\text{heat}]{\text{H}_2\text{SO}_4,\ \text{H}_2\text{O}}$ **P** $\xrightarrow[\text{HNO}_3]{[\text{O}]}$ *meso*-Tartaric acid

(g) N $\xrightarrow[\text{heat}]{\text{H}_2\text{SO}_4, \text{H}_2\text{O}}$ **Q** $\xrightarrow[\text{HNO}_3]{[\text{O}]}$ (−)-Tartaric acid

17.41

(R)-(+)-
Glyceraldehyde $\xrightarrow[{[\text{O}]}]{\text{Br}_2, \text{H}_2\text{O}}$ (R)-(−)-
Glyceric acid $\xrightarrow{\text{PBr}_3}$ (S)-(−)-3-Bromo-
2-hydroxypropanoic
acid

(R)-(+)-Malic acid $\xleftarrow[\text{heat}]{\text{H}_3\text{O}^+}$ (R)-(C$_4$H$_5$NO$_3$) $\xleftarrow{\text{NaCN}}$

17.42 (a) (R)-(+)-
Glyceraldehyde $\xrightarrow{\text{HCN}}$ **M** + **N** [cf. Problem
17.40(e)]

N $\xrightarrow[\text{H}_2\text{O}]{\text{H}_2\text{SO}_4}$ $\xrightarrow[\text{HNO}_3]{[\text{O}]}$ (−)-Tartaric acid

[cf. Problem 17.40(g)]

(S)-(−)-Malic acid $\xleftarrow[\text{H}_3\text{O}^+]{\text{Zn}}$ $\xleftarrow{\text{PBr}_3}$

(b) Replacement of either alcoholic —OH by a reaction that proceeds with inversion produces the same stereoisomer.

(c) Two. The stereoisomer given in (b) and the one given next, below.

(d) It would have made no difference because treating either isomer (or both together) with zinc and acid produces (−)-malic acid.

(−)-Malic acid

Spectroscopy

17.43

Phenacetin → Phenetidine

An interpretation of the ^{1}H NMR spectral data for phenacetin is as follows:

(a) triplet δ 1.4

(b) singlet δ 2.1

(c) quartet δ 3.95

(d) doublet δ 6.8; doublet, δ 7.4

(e) broad singlet δ 9.0

17.44 (a)

Interpretation:

(a) Triplet δ 1.2 (6H)

(b) Singlet δ 2.5 (4H)

(c) Quartet δ 4.1 (4H)

, 1740 cm^{-1} (ester carbonyl)

(b)

Interpretation:

(a) Doublet δ 1.0 (6H)

(b) Multiplet δ 2.1 (1H)

(c) Doublet δ 4.1 (2H)

(d) Multiplet δ 7.8 (5H)

, 1720 cm^{-1} (ester carbonyl)

(c)
(d)
(b)
(a)
(c)

Interpretation:

(a) Triplet δ 1.2 (3H)

(b) Singlet δ 3.5 (2H)

(c) Quartet δ 4.1 (2H)

(d) Multiplet δ 7.3 (5H)

, 1740 cm^{-1} (ester carbonyl)

(d)
(b)
H   O
Cl
Cl   OH
(a)

Interpretation:

(a) Singlet δ 6.0

(b) Singlet δ 11.70

—OH, 2500–2700 cm^{-1}

, 1705 cm^{-1} (carboxylic acid carbonyl)

(e) (b)
Cl   O
H
H   O
(b)   (a)
(c)

Interpretation:

(a) Triplet δ 1.3

(b) Singlet δ 4.0

(c) Quartet δ 4.2

, 1745 cm^{-1} (ester carbonyl)

17.45 The very low hydrogen content of the molecular formula of **Y** ($C_8H_4O_3$) indicates that **Y** is highly unsaturated. That **Y** dissolves slowly in warm aqueous $NaHCO_3$ suggests that **Y** is a carboxylic acid anhydride that hydrolyzes and dissolves because it forms a carboxylate salt:

$$\underset{\text{(insoluble)}}{\underset{R}{\overset{O}{\parallel}}\text{-C-O-C-}\underset{R}{\overset{O}{\parallel}}} \xrightarrow[\text{H}_2\text{O, heat}]{\text{NaHCO}_3} \underset{\text{(soluble)}}{R\text{-C-O}^-\text{Na}^+ + R\text{-C-O}^-\text{Na}^+}$$

The infrared absorption peaks at 1779 and 1854 cm^{-1} are consistent with those of an aromatic carboxylic anhydride (Figure 17.2).

That only four signals appear in the ^{13}C spectrum of **Y** indicates a high degree of symmetry for **Y**. Three of the signals occur in the aromatic region (δ 120-140) and one signal is downfield (δ 162)

These signals and the information from the DEPT ^{13}C NMR spectra lead us to conclude that **Y** is phthalic anhydride. The assignments are

(a) δ 125
(b) δ 130
(c) δ 136
(d) δ 162

Z is phthalic acid and **AA** is ethyl hydrogen phthalate.

Synthesizing the Material

17.46 (a) $CH_3O_2C-C\equiv C-CO_2CH_3$. This is a Diels-Alder reaction.

(b) H_2, Pd. The disubstituted double bond is less hindered than the tetrasubstituted double bond and hence is more reactive.

(c) $CH_2=CH-CH=CH_2$. Another Diels-Alder reaction.

(d) $LiAlH_4$

(e)
$$CH_3-\overset{\displaystyle O}{\underset{\displaystyle O}{\overset{\|}{\underset{\|}{S}}}}-Cl \text{ and pyridine}$$

(f) $CH_3CH_2S^-$

(g) OsO_4, then $NaHSO_3$

(h) Raney Ni

(i) O_3, followed by oxidation

(j) Heat

17.47 (a)

(b)

(c)

(d)

17.48 (a)

(b)

(c)

(1) CH₃MgBr
(excess);
H₃O⁺

(2) NaH, CH₃I

(3) ⬡ , Δ

(4) H₂, Pd/C

C

(d)

(1) Cl₂, AlCl₃

(2) NaOH,
H₂O, Δ

(3) H₃O⁺

(4) SOCl₂, pyr

D

17.49 (a)

R = alkyl, aryl

A

(b)

B

Challenge Problems

17.50 (a) Ethyl acetate   (b) Acetic anhydride   (c) *N*-Ethylacetamide.

17.51 In the first instance, nucleophilic attack by the amine occurs preferentially at the less hindered carbon of the formyl group. (Recall that aldehydes are more reactive than ketones toward nucleophiles for the same reason.) In the second case,

is a better leaving group than

since the former is the conjugate base of the stronger acid.

17.52

17.53 C_6H_6 +

AlCl$_3$ →

Clemmensen reduction →

H$_2$C=O (H—C(=O)—H)
HCl, ZnCl$_2$ →

KCN →

NaNH$_2$
CH$_3$I →

H$_3$O$^+$
heat →

17.54 **A** =

C =

B =

D =

(racemic)

In the last step, HI/red P accomplishes both the reduction of —OH to —H and the hydrolysis of the nitrile function.

17.55 (a) The signal at δ 193.8 is consistent with the carbonyl carbon of an aldehyde and shows that the PCC reaction produced cinnamaldehyde.

(b) The signal at δ 164.5 is consistent with the carbonyl carbon of a carboxylic acid, and suggests that the oxidation with $K_2Cr_2O_7$ in sulfuric acid produced cinnamic acid.

QUIZ

17.1 Which of the following would be the strongest acid?

(a) Benzoic acid   (b) 4-Nitrobenzoic acid   (c) 4-Methylbenzoic acid

(d) 4-Methoxybenzoic acid   (e) 4-Ethylbenzoic acid

17.2 Which of the following would yield (*S*)-2-butanol?

(a) (*R*)-2-Bromobutane + [structure: CH₃C(=O)O⁻ Na⁺] $\longrightarrow$ product $\xrightarrow[\text{heat}]{\text{HO}^-,\ \text{H}_2\text{O}}$

(b) (*R*)-2-Bromobutane $\xrightarrow[\text{heat}]{\text{HO}^-,\ \text{H}_2\text{O}}$

(c) (*S*)-2-Butyl acetate $\xrightarrow[\text{heat}]{\text{HO}^-,\ \text{H}_2\text{O}}$

(d) All of the above

(e) None of the above

17.3 Which reagent would serve as the basis for a simple chemical test to distinguish between hexanoic acid and hexanamide?

(a) Cold dilute NaOH   (b) Cold dilute NaHCO₃

(c) Cold concd H₂SO₄   (d) More than one of these

(e) None of these

17.4 Give an acceptable name for:

A   **B**   **C**

17.5 Complete the following synthesis.

(a)

C₆H₅—CH₃ →[**A** ⬚] benzoic acid (C₆H₅COOH) →[**B** ⬚]

benzoyl chloride (C₆H₅COCl) →[(CH₃)₂NH] **C** ⬚ + (CH₃)₂NH₂⁺Cl⁻

↓ **D** ⬚

benzoic anhydride (C₆H₅CO–O–COC₆H₅) →[CH₃CH₂OH] **E** ⬚ + C₆H₅COOH

↓ C₆H₅—NH₂

F ⬚ + C₆H₅N⁺H₃   C₆H₅CO–O⁻

(b)

C₆H₅COCl + CH₃—¹⁸OH ⟶ **A** ⬚ + HCl

B ⬚ + **C** ⬚ ←[NaOH, H₂O heat] (from A)

(c)

cyclopentanecarbonyl chloride →[NH₃ (excess)] **A** ⬚ + N⁺H₄Cl⁻

↓ P₄O₁₀, heat

C ⬚ ←[(1) DIBAL-H hexane, −78 °C (2) H₂O] **B** ⬚

17.6 Which of these acids would undergo decarboxylation most readily?

18

REACTIONS AT THE α CARBON OF CARBONYL COMPOUNDS: ENOLS AND ENOLATES

SOLUTIONS TO PROBLEMS

18.1

Cyclohexa-2,4-dien-1-one   Phenol
(keto form)   (enol form)

The enol form is aromatic, and it is therefore stabilized by the resonance energy of the benzene ring.

18.2 No.

does not have a hydrogen attached to its α-carbon atom

(which is a chirality center) and thus enol formation involving the chirality center is not

possible. With

the chirality center is a β carbon and thus enol

formation does not affect it.

18.3

cis-Decalone   Protonation on the bottom face   trans-Decalone   + EtO⁻

A large group is axial in cis-decalone.

No bulky groups are axial in trans-decalone.

18.4   The reaction is said to be "base promoted" because base is consumed as the reaction takes place. A catalyst is, by definition, not consumed, and remains unchanged by the end of the reaction.

18.5   (a) The slow step in base-catalyzed racemization is the same as that in base-promoted halogenation—*the formation of an enolate anion.* (Formation of an enolate anion from 2-methyl-1-phenylbutan-1-one leads to racemization because the enolate anion is achiral. When it accepts a proton, it yields a racemic form.) The slow step in acid-catalyzed racemization is also the same as that in acid-catalyzed halogenation—*the formation of an enol.* (The enol, like the enolate anion, is achiral and tautomerizes to yield a racemic form of the ketone.)

(b) According to the mechanism given, the slow step for acid-catalyzed iodination (formation of the enol) is the same as that for acid-catalyzed bromination. Thus, we would expect both reactions to occur at the same rate.

(c) Again, the slow step for both reactions (formation of the enolate anion) is the same, and consequently, both reactions take place at the same rate.

18.6 (a)

(b)

18.7 Working backward,

(a)

(b)

18.8   (a) Reactivity is the same as with any S_N2 reaction. With primary halides substitution is highly favored, with secondary halides elimination competes with substitution, and with tertiary halides elimination is the exclusive course of reaction.

(b) Acetoacetic ester and 2-methylpropene (isobutylene)

(c) Bromobenzene is unreactive to nucleophilic substitution because the halogen is bonded to an sp^2 carbon.

18.9   The carboxyl group that is lost more readily is the one that is β to the keto group (cf. Section 17.10 of the text).

18.10   Working backward

18.11   Working backward

18.12   Working backward,

(b)

heat
$-CO_2$

(1) HO^-, H_2O, heat
(2) H_3O^+

MeI
t-BuOK

(1) NaOEt
(2) ⌇⌇Br

(c)

heat
$-CO_2$

(1) HO^-, H_2O, heat
(2) H_3O^+

(1) NaOEt
(2) ⌇⌇Br

18.13   2 ⌇⌇Br + CN⌇⌇OEt   base
-2 HBr

HO^-, heat
$-CO_2$, $-EtOH$

(1) HO^-, heat
$-NH_3$
(2) H_3O^+

Valproic acid

18.14 These syntheses are easier to see if we work backward.

18.15 (a)

1   2   4   3

(b)

4   1   3   2

(c)

4   3   1   2

18.16 Abstraction of an α hydrogen at the ring junction yields an enolate anion that can then accept a proton to form either *trans*-1-decalone or *cis*-1-decalone. Since *trans*-1-decalone is more stable, it predominates at equilibrium.

$$A:^-$$
$$(-H^+)$$

$$HA$$

(95%)   (5%)
trans-1-Decalone   *cis*-1-Decalone
(more stable)   (less stable)

18.17 In a polar solvent, such as water, the keto form is stabilized by solvation. When the interaction with the solvent becomes minimal, the enol form achieves stability by internal hydrogen bonding.

18.18 (a)

(1) LDA
(2)   I

(b)   Br₂, CH₃COOH

(c)   NaH

(d)   Br₂ (excess), NaOH

(e)

(1) LDA
(2) Cl
(3) H₂O

(f)

(1) I₂, NaOH
(2) H₃O⁺

18.19   (a)

(b)

(c)

(d)

18.20 (a)

(resonance
stabilized by
carbonyl)

(more stable
isomer due to
conjugation)

(b)

(c)

(d)

SUMMARY OF ACETOACETIC ESTER AND MALONIC ESTER SYNTHESES

A. Acetoacetic Ester Synthesis

B. Malonic Ester Synthesis

Acetoacetic Ester and Malonic Ester Syntheses

18.21   (a)

(b)

(c)

(d)   (1) NaOEt
      (2) Br–CH2–CO–OEt

      (1) HO⁻, H₂O, heat
      (2) H₃O⁺

heat
−CO₂

NaBH₄

(e)   (1) NaOEt
      (2) CH₂=CH–CH2Br

      (1) LiAlH₄
      (2) H₃O⁺

(f) Continuing from problem 18.11

      NaBH₄

18.22 (a)   (1) NaOEt
                 (2) Br

      (1) t-BuOK
      (2) MeI

      (1) HO⁻, H₂O, heat
      (2) H₃O⁺

      heat
      −CO₂

(b)   (1) LiAlH₄
      (2) H₃O⁺

[from Problem 18.12 (c)]

(c)   (1) LiAlH₄
      (2) H₃O⁺

[from (a) above]

(d)

18.23   (a)

(b)

18.24   The following reaction took place:

18.25 (a)

(b)

D
Racemic form

E
Meso compound

(c)

18.26 (a)

(1) NaOEt

(2)

(1) NaOH
(2) H_3O^+ or H_3O^+

(ester hydrolysis
under either
basic or acidic
conditions)

heat

(b)

(1) NaOEt

(2) Cl⎓⎓⎓Cl

(1) NaOH
(2) H_3O^+ or H_3O^+

$\xrightarrow{\text{LiAlH}_4 \; / \; \text{Et}_2\text{O}}$

heat

(c)

(1) NaOEt

(2) Br⎓

(1) NaOH
(2) H_3O^+ or H_3O^+

heat

(f)

(1) NaOEt

(2) ⌒Br

(1) NaOEt

(2) BrCH₃

(1) NaOH

(2) H₃O⁺   or H₃O⁺

heat

General Problems

18.27 (a)

(1) NaOEt

(2) ⌒Br

(3) H₃O⁺, heat

(b)

(1) NaOEt

(2) ⌒Br

(3) NaOEt

(4) ⌒Br

(5) H₃O⁺, heat

(c)

(1) LDA

(2) CH₃I

(d)

(1) ⌒⌒OCH₃

NaOCH₃ (excess)

(2) H₃O⁺, heat

(e)

(f)

18.28

F

G

H

I

linalool

18.29

$(C_{10}H_{17}BrO_4)$

$(C_{10}H_{16}O_4)$

$(C_6H_{12}O_2)$

$$\xrightarrow[\text{(2) } H_3O^+]{\text{(1) } HO^-,\ H_2O}$$

(C$_9$H$_{12}$O$_4$) $\xrightarrow{\text{heat}}$ J   (C$_8$H$_{12}$O$_2$)   +   CO$_2$

18.30 (a) H$_3$CO Br + (C$_6$H$_5$)$_3$P $\xrightarrow{\text{(2) RLi}}$ H$_3$CO P(C$_6$H$_5$)$_3$

(b) Hydrolysis of the ether yields a hemiacetal that then goes on to form an aldehyde.

[also (Z)] $\xrightleftharpoons[\text{:A}^-]{\text{HA}}$

$\xrightleftharpoons[-H_2O]{+H_2O}$ $\xrightleftharpoons[\text{HA}]{\text{:A}^-}$ (hemiacetal)

$\xrightleftharpoons[+CH_3OH]{-CH_3OH}$

(c)   +   H$_3$CO P(C$_6$H$_5$)$_3$ $\longrightarrow$ $\xrightleftharpoons[\text{:A}^-]{\text{HA}}$

$\xrightleftharpoons[-H_2O]{+H_2O}$ $\xrightleftharpoons[\text{HA}]{\text{:A}^-}$

$\xrightleftharpoons[+CH_3OH]{-CH_3OH}$

18.31 (a) The hydrogen atom that is added to the aldehyde carbon atom in the reduction must come from the other aldehyde rather than from the solvent. It must be transferred as a hydride ion and directly from molecule to molecule, since if it were ever a free species it would react immediately with the solvent. A possible mechanism is the following:

(b) Although an aldol reaction occurs initially, the aldol reaction is reversible. The Cannizzaro reaction, though slower, is irreversible. Eventually, all the product is in the form of the alcohol and the carboxylate ion.

18.32

Spectroscopy

18.33 (a) Compound **U** is ethyl phenyl ketone. (b) Compound **V** is benzyl methyl ketone:

δ 7.7   **U**

δ 2.0 — 1705 cm^{-1}

δ 3.5

δ 7.1   **V**

18.34 **A** is

δ 2.1 δ 2.6 δ 4.7 δ 3.2

$$\xrightarrow[\text{NaOH}]{\text{I}_2} \text{CHI}_3 \downarrow$$

$$\xrightarrow{\text{Ag(NH}_3)_2{}^+\text{OH}^-} \text{no reaction}$$

$\big\downarrow \text{H}_3\text{O}^+$

$$\xrightarrow{\text{Ag(NH}_3)_2{}^+\text{OH}^-} \text{Ag} \downarrow +$$

Synthesizing the Material

18.35 (a)

$$\xrightarrow{\substack{\text{Jones} \\ \text{oxidation}}}$$

$$\xrightarrow{\substack{\text{NaH, CH}_3\text{I} \\ \text{or CH}_3\text{OH, HA}}}$$

$$\xrightarrow[\substack{\text{Br} \diagup\diagdown\diagup \text{Br}}]{\text{NaOCH}_3 \text{ (2 equiv),}}$$

$$\xleftarrow[\Delta]{\text{H}_3\text{O}^+}$$

(b)

Note: could also use (1) LDA and (2) CH$_3$I starting from acetone.

(c)

(d)

18.36 (a)

(b)

(1) pyrrolidine, HA (cat.)
(2) MeI
(3) H₃O⁺
(4) (a) LiAlH₄; (b) H₃O⁺

B

(c)

(1) PBr₃, pyr
(2) NaOEt
(3) NaH, Br₂
(4) NaOEt, EtOH

C

(d)

(1) Δ
(2) EtMgBr (excess); H₃O⁺
(3) PBr₃, pyr
(4) KCN
(5) (a) DIBAL-H; (b) H₃O⁺

D

18.37   (a)

Note: all choices working
backwards are shown to help
consider various possibilities

(5) H₂, Pd/C

(4) H₂C=PPh₃

(2) AcCl, AlCl₃

(3) LDA; MeI

[no longer a
possibility
due to step 2]

(1) Br₂, FeBr

A

(b)

(6) DCC,
H₂NCH₃

(5) Raney Ni, H₂

Note: key here is that the Raney Ni step be
considered in conjuction with other operations.

(2) NaOH, H₂O
(3) H₃O⁺

(4) HS‑‑‑‑SH,
acid

R = alkyl or aryl

(1) NaOEt;
CH₃I

R = alkyl or aryl

B

Challenge Problem

18.38

QUIZ

18.1 What would be the major product of the following reaction?

(a)

(b)

(c)

(d)

(e) None of these

18.2 Supply formulas for the missing reagents and intermediates in the following synthesis.

18.3 What alkyl halide would be used in the malonic ester synthesis of 4-methylpentanoic acid?

(a) ⌒⌒Br

(b) isopropyl–Br

(c) isobutyl–Br

(d) isoamyl–Br

(e) ⌄⌄Br

18.4 Which is the preferred base for the alkylation of a monoalkylacetoacetic ester?

(a) NaOH

(b) KOH

(c) EtONa

(d) (CH₃)₂CH–ONa

(e) *t*-BuOK

18.5 To which of these can the haloform reaction be applied for the synthesis of a carboxylic acid?

(a)

(b)

(c)

(d)

(e)

19

CONDENSATION AND CONJUGATE ADDITION REACTIONS OF CARBONYL COMPOUNDS: MORE CHEMISTRY OF ENOLATES

SOLUTIONS TO PROBLEMS

19.1 (a) *Step 1*

Step 2

Step 3

(b)

19.2

19.3 (a)

(b) To undergo a Dieckmann condensation, diethyl 1,5-pentanedioate would have to form a highly strained four-membered ring.

19.4

19.5 (a)

(b)

19.6 (a)

(b)

(c)

19.7

Intermolecular proton transfer

2,6-Dimethylhepta-2,5-dien-4-one

19.8 Drawing the molecules as they will appear in the final product helps to visualize the necessary steps:

Mesitylene

The two molecules that lead to mesitylene are shown as follows:

(4-Methylpent-3-en-2-one is formed by an acid-catalyzed condensation between two molecules of acetone as shown in the text.)

The mechanism is the following.

19.9 (a)

(b) To form [structure], a hydroxide ion would have to remove a β proton

in the first step. This does not happen because the anion that would be produced,

[structure], cannot be stabilized by resonance.

(c)

[both (E) and (Z)]

19.10 (a) 2 [structure] $\xrightarrow[\text{H}_2\text{O}]{\text{HO}^-}$ [structure]

(b) Product of (a) $\xrightarrow[-\text{H}_2\text{O}]{\text{HA}}$ [structure]

[$(E) + (Z)$]

$\xrightarrow[\text{(2) H}_2\text{O}]{\text{(1) LiAlH}_4, \text{Et}_2\text{O}}$ [structure]

[$(E) + (Z)$]

(c) Product of (b) $\xrightarrow[\substack{\text{Pt} \\ \text{pressure}}]{\text{H}_2}$ [structure]

(d) Product of (a) $\xrightarrow{\text{NaBH}_4}$ [structure]

19.11

$C_{11}H_{14}O$

$C_{14}H_{18}O$

Lily aldehyde ($C_{14}H_{20}O$)

19.12 Three successive aldol additions occur.

First
Aldol
Addition

Second
Aldol
Addition

Third
Aldol
Addition

19.13 (a)

Pseudoionone

α-Ionone

β-Ionone

(b) In β-ionone both double bonds and the carbonyl group are conjugated; thus, it is more stable.

(c) β-Ionone, because it is a fully conjugated unsaturated system.

19.14 (a)

(b)

19.15 (a)

(b)

(c)

19.16

19.17 (a) (b) (c)

Notice that starting compounds are drawn so as to indicate which atoms are involved in the cyclization reaction.

19.18 It is necessary for conditions to favor the intramolecular reaction rather than the intermolecular one. One way to create these conditions is to use very dilute solutions when we carry out the reaction. When the concentration of the compound to be cyclized is very low (i.e., when we use what we call a "high dilution technique"), the probability is greater that one end of a molecule will react with the other end of that same molecule rather than with a different molecule.

19.19 (a)

(b)

19.20

19.21 (a)

(b) 2-Methylcyclohexane-1,3-dione is more acidic than cyclohexanone because its enolate ion is stabilized by an additional resonance structure.

19.22 (a)

(b)

(c)

Problems

Claisen Condensation Reactions

19.23 (a)

(b)

(c)

(d)

(e)

(f)

19.24 Working backward in each case:

(a)

(b)

(c)

19.25

(a)

(b)

(c)

or

(or other pairings of reactants)

(or other pairings of reactants)

19.26

NaOEt, EtOH

(major)

(minor)

(The minor product has no alpha hydrogen, so a retro-Dieckmann reaction takes place, favoring the starting material instead.)

Claisen and Dieckmann condensations are reversible. The driving force is the deprotonation of the alpha hydrogen in the product between the two carbonyls. This deprotonation prevents the reaction from reversing, pulling the product out of the equilbrium.

Molecule can not undergo retro-Dieckmann in deprotonated form.

19.27 Intramolecular cyclization (which would give a product of formula $C_6H_8O_3$) is not favored because of ring strain. The formula of the product actually obtained suggests a 1:1 intermolecular reaction:

$C_{12}H_{16}O_6$

19.28 The synthesis actually involves two sequential Claisen condensations, with ethyl acetate serving as the source of the carbanionic species. A final decarboxylation finishes the product.

not isolated

19.29

$R \overset{O}{\underset{}{\parallel}} \text{CH}_2\text{CO}^-$ + EtOH

19.30 A gamma hydrogen is removed by base (as is an alpha hydrogen in the usual Claisen) to give a resonance-stabilized enolate:

Ethyl crotonate differs from ethyl acetate by −CH=CH−, a vinyl group. The transmission of the stabilizing effect of the −COOC$_2$H$_5$ group is an example of vinylogous reactivity (reactivity through conjugation with a double bond).

19.31

The two-stage reaction involves a reverse Dieckmann condensation followed by a forward Dieckmann using a different enolate. The original enolate cannot lead back to the starting material since the crucial α-hydrogen is lacking.

19.32

Aldol Reactions

19.33 (a)

(b)

(c)

(d)

(e)

19.34

1   2   3   4

19.35 (a)

(b)

(c)

(d)

(e)

(f)

(g)

19.36 (a)   (b)

(c)   (d)

(e) +

(f) +

(g)

(h)

19.37 (a)

(b)

(c)

(d)

19.38 (a) The conjugate base is a hybrid of the following structures (cf. Rev. Prob. 19.30):

This structure is especially stable because the negative charge is on the oxygen atom.

(b)

19.39

19.40 This difference in behavior indicates that, for acetaldehyde, the capture of a proton from the solvent (the reverse of the reaction by which the enolate anion is formed) occurs much more slowly than the attack by the enolate anion on another molecule.

When acetone is used, the equilibrium for the formation of the enolate anion is unfavorable, but more importantly, enolate attack on another acetone molecule is disfavored due to steric hindrance. Here proton capture (actually deuteron capture) competes very well with the aldol reaction.

Conjugate Addition Reactions

19.41 (a)

(b)

(c)

The Michael reaction is reversible, and the reaction just given is an example of a reverse Michael reaction.

19.42 Two reactions take place. The first is a reaction called the Knoevenagel condensation, initiated by attack of the enolate from the dicarbonyl compound on the ketone,

Then the α,β-unsaturated diketone reacts with a second mole of the active methylene compound in a Michael addition.

19.43

19.44 There are two stages to the reaction. The initial Michael addition is followed by the reaction of an enolate with an ester carbonyl to form a diketone (in equilibrium with an enol form).

Another (less likely) possibility is the formation of a seven-membered ring.

General Problems

19.45 (a)

(b)

(c)

(1) NaOEt

(2) [isobutenyl bromide]

(3) NaOEt

(4) [ethyl acrylate]

(5) H₃O⁺, heat

19.46

Repeat aldol reaction on other side;
same steps as above

19.47 (a)

(b)

(c)

(d)

19.48 (a)

(b)

(c)

(d)
$$\xrightarrow[\substack{(1)\ LDA\ (1.1\ equiv) \\ (2)\ \\ (3)\ H_3O^+}]{}$$

19.49
$$\xrightarrow[\substack{(1)\ CH_3MgI \\ (2)\ ^+NH_4/H_2O}]{}$$ **A** $$\xrightarrow[\text{heat}]{HA}$$ **B** $$\xrightarrow[\substack{(1)\ O_3 \\ (2)\ Me_2S}]{}$$

C $$\xrightarrow[\text{pyridine}]{}$$ Queen substance

$$\xrightarrow[Pd]{H_2}$$ **D** $$\xrightarrow[\substack{(1)\ I_2/NaOH \\ (2)\ H_3O^+}]{}$$

E

19.50 C_6H_5 $+$ H H $+$ $HN(CH_3)_2$ $\xrightarrow[(-H_2O)]{\text{Mannich reaction}}$ C_6H_5 $\underset{CH_3}{\overset{CH_3}{N}}$

$$\xrightarrow[\substack{(1)\ C_6H_5 \\ (2)\ H_3O^+}]{MgBr}$$ C_6H_5 $\underset{C_6H_5}{\overset{OH}{\vert}}$ $\underset{CH_3}{\overset{CH_3}{N}}$ $$\xrightarrow{}$$

C_6H_5 C_6H_5 $\underset{CH_3}{\overset{CH_3}{N}}$

Darvon

19.51

19.52

(1) Aromatization by proton removal using A:⁻

(2) HA (Protonation of the ether.)

19.53 (a)

(b) Decarboxylation of the epoxy acid gives an enolate anion which, on protonation, gives an aldehyde.

(c)

β-Ionone

19.54 (a)

(b)

Spectroscopy

19.55 (a) In simple addition, the carbonyl peak (1665–1780 cm^{-1} region) does not appear in the product; in conjugate addition, it does.

(b) As the reaction takes place, the long-wavelength absorption arising from the conjugated system should disappear, One could follow the rate of the reaction by following the rate at which this absorption peak disappears.

19.56

X

Formed by the aldol condensation of acetone with two molecules of benzaldehyde.

Synthesizing the Material

19.57 (a)

(b) KMnO$_4$, HO$^-$, then H$_3$O$^+$

(c) CH$_3$OH, HA

(d) CH$_3$ONa, then H$_3$O$^+$

(e) and (f)

and

(g) HO$^-$, H$_2$O, then H$_3$O$^+$

(h) heat (−CO$_2$)

(i) CH$_3$OH, HA

(j) CH$_3$ONa, then H$_3$O$^+$ and heat (−CO$_2$)

(k) 2 NaNH$_2$ + 2 CH$_3$I

19.58 (a)

Note: In α,β-unsaturated systems, LiAlH₄ gives 1,2 addition at the carbonyl, leading to an alcohol, whereas NaBH₄ can lead to both 1,2 or 1,4 addition, reducing the carbonyl or the alkene. Hence, we choose LiAlH₄ here.

(b)

(c)

Note: the amide could also result from the ester through (1) saponification, (2) conversion to an acid chloride, and (3) reaction with excess NH₃ without the need for using high heat to convert an ester to an amide directly.

19.59 (a)

(b)

19.60 *Brief retrosynthesis*

Forward synthesis

(1) Br$_2$,
  FeBr$_3$
(2) Mg°
(3) CO$_2$;
  H$_3$O$^+$

A

(1) LiAlH$_4$ in Et$_2$O
(2) H$_2$O/H$_2$SO$_4$
(3) Swern oxidation

B

CH$_3$l

(1) Ph$_3$P
(2) *n*-BuLi

$=$PPh$_3$
C

(1) Br$_2$,
  AcOH
(2) *t*-BuOK,
  t-BuOH

$=$PPh$_3$
C

Note: propanal could also be converted
into 1,3-butadiene by a Wittig reaction
followed by allylic bromination and
elimination.

B

NaOEt/
EtOH

$+$

Δ

A

HO

(1) NaBH$_4$,
  MeOH
(2) H$_2$, Pd/C

HA $\quad$ $-$[H$_2$O]

Challenge Problems

19.61 (a)

Acid is added after formation of the anion is complete at each β-dicarbonyl group.

(b)

19.62 (a,b)

19.63 (a)

Same concept as above, but here just showing key intermediates with a retro-aldol condensation followed by forward aldol condensation

(b) In both cases, these reactions go to the indicated product because they are the thermodynamically fovored product. In the first case, a better ring size is produced in terms of strain (5 versus 7). In both cases, the more substituted enolate attaks the sterically least encumbered carbonyl, which is the favored pathway overall for reaching the thermodynamic product.

QUIZ

19.1 Which hydrogen atoms in the following ester are most acidic?

$$
\underset{a}{CH_3}-\underset{b}{CH_2}-\overset{O}{\underset{\|}{C}}-\underset{c}{CH_2}-\overset{O}{\underset{\|}{C}}-\underset{d}{OCH_2}-\underset{e}{CH_3}
$$

(a) *a*   (b) *b*   (c) *c*   (d) *d*   (e) *e*

19.2 What would be the product of the following reaction?

19.3 What starting materials could be used in a crossed Claisen condensation to prepare the following compound?

(e) More than one of the above

19.4 Supply the missing reagents, intermediates, and products.

(a)

A

B

t-BuOK / *t*-BuOH

C

(1) HO⁻, H₂O (2) H₃O⁺

D

E

heat −CO₂

(b)

A

Et₂NH / EtOH

B

NaOEt, EtOH

C

(1) HO⁻, H₂O (2) H₃O⁺

heat −2 CO₂

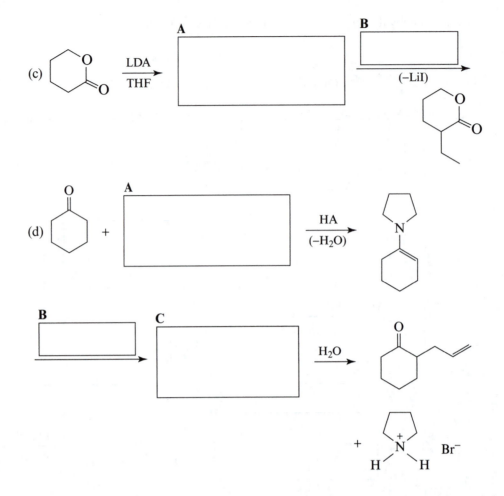

19.5 Supply formulas for the missing reagents and intermediates in the following synthesis.

+ enantiomer and
simple addition product

19.6   Which would be formed in the following reaction?

(a)   (b)   (c)

(d)   (e) All of these will be formed.

19.7   Supply formulas for the missing reagents and intermediates in the following synthesis.

19.8   Supply formulas for the missing reagents and intermediates in the following synthesis.

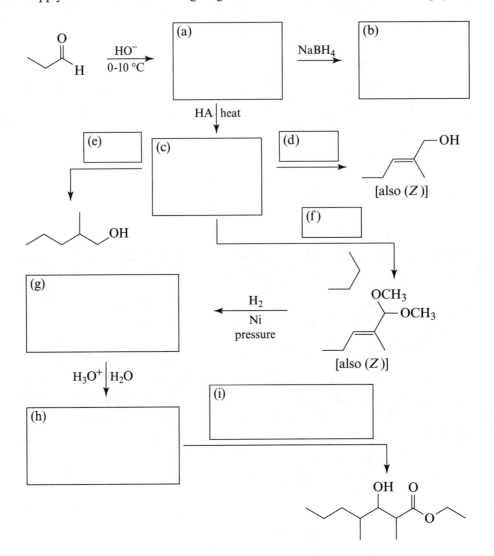

20 AMINES

PREPARATION AND REACTIONS OF AMINES

A. Preparation (Section 20.4)

1. Preparation via nucleophilic substitution reactions.

Substitution by azide and reduction

Gabriel synthesis

S_N2 with NH_3 (multiple products, hence a poor method)

2. Preparation through reduction of nitro compounds.

3. Preparation via reductive amination.

Aldehyde or ketone

4. Preparation of amines through reduction of amides, oximes, and nitriles.

5. Preparation through the Hofmann rearrangement of amides.

B. Reaction of Amines (Section 20.5)

1. As a base or a nucleophile

As a base

As a nucleophile in alkylation

As a nucleophile in acylation

2. With nitrous acid

1° aliphatic   (unstable)   alkenes,
alcohols,
and alkyl halides

$$\text{Ar}-\text{NH}_2 \xrightarrow[\text{HX (0–5 °C)}]{\text{HONO}} \text{Ar}-\text{N}_2^+ \text{X}^-$$

1° aromatic

$\xrightarrow{\text{CuCl}} \text{ArCl} + \text{N}_2$

$\xrightarrow{\text{CuBr}} \text{ArBr} + \text{N}_2$

$\xrightarrow{\text{CuCN}} \text{ArCN} + \text{N}_2$

$\xrightarrow{\text{KI}} \text{ArI} + \text{N}_2$

$\xrightarrow{\text{HBF}_4} \text{ArN}_2^+ \text{BF}_4^- \xrightarrow{\Delta} \begin{cases} \text{ArF} + \\ \text{N}_2 + \\ \text{BF}_3 \end{cases}$

$\xrightarrow[\text{Cu}^{2+},\ \text{H}_2\text{O}]{\text{Cu}_2\text{O}} \text{ArOH} + \text{N}_2$

$\xrightarrow{\text{H}_3\text{PO}_2} \text{ArH} + \text{N}_2$

(phenol–OH) $\longrightarrow \text{Ar}-\text{N}=\text{N}-\!\!\!\!\bigcirc\!\!\!\!-\text{OH}$

(phenyl–NR$_2$) $\longrightarrow \text{Ar}-\text{N}=\text{N}-\!\!\!\!\bigcirc\!\!\!\!-\text{NR}_2$

$$\text{R}_2\text{NH} \xrightarrow{\text{HONO}} \text{R}_2\text{N}-\text{N}=\text{O}$$

2° aliphatic

$$\text{ArNHR} \xrightarrow{\text{HONO}} \text{ArN}-\text{R} \ (\overset{\displaystyle \text{N}=\text{O}}{|})$$

2° aromatic

$$\text{R}_3\text{N} \xrightarrow[\text{HX}]{\text{NaNO}_2} \text{R}_3\overset{+}{\text{N}}\text{H}\ \text{X}^- + \text{R}_3\overset{+}{\text{N}}-\text{N}=\text{O}\ \text{X}^-$$

3° aliphatic

$$\text{R}_2\text{N}-\!\!\!\!\bigcirc\!\!\!\! \xrightarrow{\text{HONO}} \text{R}_2\text{N}-\!\!\!\!\bigcirc\!\!\!\!-\text{N}=\text{O}$$

3° aromatic

3. With sulfonyl chlorides

$$\text{R}-\text{NH}_2 + \text{ArSO}_2\text{Cl} \xrightarrow[(-\text{HCl})]{} \text{RNHSO}_2\text{Ar} \underset{\text{H}_3\text{O}^+}{\overset{\text{HO}^-}{\rightleftharpoons}} [\text{RNSO}_2\text{Ar}]^- + \text{H}_2\text{O}$$

1° amine

$$\text{R}_2\text{NH} + \text{ArSO}_2\text{Cl} \xrightarrow[(-\text{HCl})]{} \text{R}_2\text{NSO}_2\text{Ar}$$

2° amine

4. The Hofmann elimination

$$HO^- + \underset{H}{\overset{}{\diagup}}\hspace{-0.5em}\bigvee\hspace{-0.5em}\overset{+}{N}(CH_3)_3 \xrightarrow{\text{heat}} \hspace{-0.5em}\diagup\hspace{-0.3em}=\hspace{-0.3em}\diagdown + (CH_3)_3N + H_2O$$

SOLUTIONS TO PROBLEMS

20.1 Dissolve both compounds in diethyl ether and extract with aqueous HCl. This procedure gives an ether layer that contains cyclohexane and an aqueous layer that contains hexylaminium chloride. Cyclohexane may then be recovered from the ether layer by distillation. Hexylamine may be recovered from the aqueous layer by adding aqueous NaOH (to convert hexylaminium chloride to hexylamine) and then by ether extraction and distillation.

20.2 We begin by dissolving the mixture in a water-immiscible organic solvent such as CH_2Cl_2 or Et_2O. Then, extractions with aqueous acids and bases allow us to separate the components. [We separate 4-methylphenol (*p*-cresol) from benzoic acid by taking advantage of benzoic acid's solubility in the more weakly basic aqueous $NaHCO_3$, whereas *p*-cresol requires the more strongly basic, aqueous NaOH.]

20.3

When the first two molecules react, they lead as shown to a new amine (ethylamine), a material that is also a nucleophile and can compete with the NH_3 molecules for the ethyl bromide. In addition, an additional molecule of NH_3 is consumed in the first reaction as a base, meaning less of it is available as a nucleophile as the reaction progresses. As a result, there are double and triple alkylation products with ethyl bromide and a mixture of products is formed.

20.4   (a) Neglecting Kekulé forms of the ring, we can write the following resonance structures for the phthalimide anion.

(b) Phthalimide is more acidic than benzamide because its anion is stabilized by resonance to a greater extent than the anion of benzamide. (Benzamide has only one carbonyl group attached to the nitrogen atom, and thus fewer resonance contributors are possible.)

(c)

Then,

20.5

KOH

C_6H_5 Br
$(-KBr)$

$\dfrac{H_2NNH_2}{\text{EtOH reflux}}$ $C_6H_5\!\!-\!\!NH_2$ + Benzylamine

20.6 (a) + NH_3 $\dfrac{H_2}{\text{Ni pressure}}$ NH$_2$

(b) + $C_6H_5NH_2$ $\dfrac{LiBH_3CN}{MeOH}$ N$-$C$_6$H$_5$ H

(c) C_6H_5 H + $(CH_3)_2NH$ $\dfrac{LiBH_3CN}{MeOH}$ C_6H_5 N CH$_3$ CH$_3$

20.7 (a) C_6H_5 OH $\xrightarrow{SOCl_2}$ C_6H_5 Cl NH$_2$

C_6H_5 N H $\xrightarrow{LiAlH_4}$ C_6H_5 N H

(b) Br $\xrightarrow{NaCN}$ N

$\xrightarrow{LiAlH_4}$ NH$_2$

(c) OH $\xrightarrow{SOCl_2}$ Cl N H

N $\xrightarrow{LiAlH_4}$ N

(d) $\xrightarrow{NH_2OH}$ HO N $\xrightarrow{Na/EtOH}$ NH$_2$

20.8 (a)

(+ ortho; separate)

(b)

(c)

(excess)

(d)

(e)

20.9 An amine acting as a base.

An amine acting as a nucleophile in an alkylation reaction.

An amine acting as a nucleophile in an acylation reaction.

$$(CH_3)_2\ddot{N}H \;+\; \text{(acyl chloride)} \longrightarrow (CH_3)_2N\text{—C(=O)} \;+\; (CH_3)_2\overset{+}{N}H_2\ Cl^-$$
(excess)

An amino group acting as an activating group and as an ortho-para director in electrophilic aromatic substitution.

20.10 (a, b)

$$^-O{-}\ddot{N}{=}O \;+\; H_3O^+ \;\rightleftharpoons\; HO{-}\ddot{N}{=}O \;+\; H_2O$$

$$HO{-}\ddot{N}{=}O \;+\; H_3O^+ \;\rightleftharpoons\; H\overset{+}{O}{-}\ddot{N}{=}O \;+\; H_2O$$
$$\quad \big|\ H$$

$$H\overset{+}{O}{-}\ddot{N}{=}O \;\rightleftharpoons\; H_2O \;+\; :\overset{+}{N}{=}O$$
$$\big|\ H$$

(c) The $\overset{+}{N}O$ ion is a weak electrophile. For it to react with an aromatic ring, the ring must have a powerful activating group such as $-OH$ or $-NR_2$.

20.11 (a)

(b)

[from part (a)]

(c) [as in (a)]

$\xrightarrow[\text{(2) HO}^-]{\text{(1) Fe, HCl, heat}}$

[aniline, NH$_2$]

$\xrightarrow{\text{(acetic anhydride)}}$

[acetanilide, HN-C(=O)CH$_3$]

$\xrightarrow[\text{H}_2\text{SO}_4]{\text{concd}}$

[p-acetamidobenzenesulfonic acid, SO$_3$H]

$\xrightarrow{\text{HNO}_3}$

[2-nitro-4-sulfo acetanilide, NO$_2$, SO$_3$H]

$\xrightarrow[\text{(2) HO}^-]{\text{(1) dil. H}_2\text{SO}_4, \text{ heat}}$

[o-nitroaniline, NH$_2$, NO$_2$]

(d) [from part (c)]

[acetanilide]

$\xrightarrow[\text{H}_2\text{SO}_4]{\text{HNO}_3}$

[p-nitroacetanilide, NO$_2$]
(plus a trace of ortho)

$\xrightarrow[\text{(2) HO}^-]{\text{(1) H}_3\text{O}^+}$

[p-nitroaniline, NH$_2$, NO$_2$]

20.12 (a) Toluene $\xrightarrow[\text{H}_2\text{SO}_4]{\text{HNO}_3}$ *p*-Nitrotoluene (+ *o*-nitrotoluene) $\xrightarrow[\text{(2) HO}^-]{\text{(1) Fe, HCl}}$ [*p*-toluidine, CH$_3$, NH$_2$] $\xrightarrow{\text{(acetic anhydride)}}$

[p-methylacetanilide, CH$_3$, HN-C(=O)CH$_3$] $\xrightarrow[\text{H}_2\text{SO}_4]{\text{HNO}_3}$ [CH$_3$, NO$_2$, HN-C(=O)CH$_3$] $\xrightarrow[\text{(2) HO}^-]{\text{(1) H}_3\text{O}^+}$ [CH$_3$, NO$_2$, NH$_2$] $\xrightarrow[\text{(0–5°C)}]{\text{HONO}}$

[CH$_3$, NO$_2$, N$_2^+$] $\xrightarrow[\text{(−N}_2)]{\text{H}_3\text{PO}_2}$ [CH$_3$, NO$_2$] $\xrightarrow[\text{(2) HO}^-]{\text{(1) Fe, HCl}}$ [CH$_3$, NH$_2$]
m-toluidine

20.13

20.14

Orange II

20.15

Butter yellow

20.16

A

B

C   Phenacetin

20.17

Use of the same sequence starting from benzylamine and using methyl bromide instead of benzyl bromide would also yield *N*-methylbenzylamine.

20.18 (a)

Aniline

Sulfathiazole

(b)

Succinylsulfathiazole

20.19

(a) Benzyl-N-methylamine structure with CH₃ on N-H

(b) $\left(\right)_3 N$ — triisopropylamine structure

(c) N-methyl-N-ethylaniline structure with CH₃ and ethyl on N

(d) 3-methylaniline structure with CH₃ and NH₂

(e) 2-methylpyrrole structure with N-H and CH₃

(f) N-ethylpiperidine structure

(g) N-ethylpyridinium bromide structure, ^+N Br$^-$

(h) nicotinic acid structure with $\overset{O}{\overset{\|}{C}}$ OH

(i) indole structure with N-H

(j) acetanilide structure

(k) $CH_3-\overset{\overset{\displaystyle H}{|}}{\underset{\underset{\displaystyle CH_3}{|}}{N^{\pm}}}-H$ Cl$^-$

(l) 2-methylimidazole structure with CH₃ and N-H

(m) H_2N ‒‒‒ OH

(n) $\left(\right)_4 N^+$ Cl$^-$ tetrabutylammonium chloride

(o) pyrrolidine structure with N-H

(p) N,N,4-trimethylaniline structure with CH₃, N, CH₃ and H_3C

(q)

NH$_2$ / CH$_3$O—

(r) (CH$_3$)$_4$N$^+$ HO$^-$

(s)

O / OH / H$_2$N—

(t)

H / N / CH$_3$

20.20

(a) Propylamine (or 1-propanamine)

(b) *N*-Methylaniline

(c) Isopropyltrimethylammonium iodide

(d) 2-Methylaniline (*o*-toluidine)

(e) 2-Methoxyaniline (or *o*-methoxyaniline)

(f) Pyrazole

(g) 2-Aminopyrimidine

(h) Benzylaminium chloride

(i) *N*,*N*-Dipropylaniline

(j) Benzenesulfonamide

(k) Methylaminium acetate

(l) 3-Amino-1-propanol
(or 3-aminopropan-1-ol)

(m) Purine

(n) *N*-Methylpyrrole

Amine Synthesis and Reactivity

20.21

sp^3 hybridization stabilizes the corresponding conjugate acid.

Alkyl groups are electron donating, stabilizing the corresponding conjugate acid.

Lone pair is not involved in aromaticity.

20.22 (a)

N + LiAlH$_4$ $\xrightarrow{\text{(2) H}_2\text{O}}$ NH$_2$

(b)

O / NH$_2$ + LiAlH$_4$ $\xrightarrow{\text{(2) H}_2\text{O}}$ NH$_2$

(c)

(d)

(e)

(f)

(g)

20.23 (a)

(b)

Hofmann rearrangement

(c)

20.24 (a)

(b) [from part (a)]

(c)

(d)

20.25 (a)

(b)

(c) [from part (a)] $\xrightarrow[\text{H}_2\text{SO}_4]{\text{HNO}_3}$ (p-nitroacetanilide) (+ trace of ortho) $\xrightarrow[\text{(2) HO}^-]{\text{(1) H}_3\text{O}^+, \text{H}_2\text{O}}$ (p-nitroaniline)

(d) [from part (a)] $\xrightarrow{\text{HOSO}_2\text{Cl}}$ (SO$_2$Cl product) $\xrightarrow[\substack{\text{(2) H}_3\text{O}^+, \text{ heat} \\ \text{(3) HO}^-}]{\text{(1) NH}_3}$ (SO$_2$NH$_2$ product)

(e) (aniline, NH$_2$) $\xrightarrow[\text{base}]{2\ \text{CH}_3\text{I}}$ (N,N-dimethylaniline, H$_3$C–N–CH$_3$)

(f) (aniline, NH$_2$) $\xrightarrow[(0\text{–}5\ °\text{C})]{\text{HONO}}$ (N$_2^+$X$^-$) $\xrightarrow{\text{HBF}_4}$ (N$_2^+$BF$_4^-$) $\xrightarrow{\text{heat}}$ (F)

(g) [from part (f)] (N$_2^+$X$^-$) $\xrightarrow{\text{CuCl}}$ (Cl)

(h) [from part (f)] (N$_2^+$X$^-$) $\xrightarrow{\text{CuBr}}$ (Br)

(i) [from part (f)] (N$_2^+$X$^-$) $\xrightarrow{\text{KI}}$ (I)

(j) [from part (f)] (N$_2^+$X$^-$) $\xrightarrow{\text{CuCN}}$ (N≡C)

(k) [from part (j)] (C≡N) $\xrightarrow[\text{(2) HO}^-]{\text{(1) H}_3\text{O}^+, \text{H}_2\text{O}}$ (benzoic acid, COOH)

(l)

$N_2^+X^-$

[from part (f)]

$\xrightarrow[\text{Cu}^{2+}, \text{H}_2\text{O}]{\text{Cu}_2\text{O}}$

OH

(m)

$N_2^+X^-$

[from part (f)]

$\xrightarrow[\text{H}_2\text{O}]{\text{H}_3\text{PO}_2}$

(n)

$N_2^+X^-$

[from part (f)]

+

OH

[from part (l)]

$\xrightarrow[\text{pH 8-10}]{\text{HO}^-}$

N=N—⟨ ⟩—OH

(o)

$N_2^+X^-$

[from part (f)]

+

N(CH₃)₂

[from part (e)]

$\xrightarrow[\text{pH 5-7}]{\text{H}_3\text{O}^+}$

N=N—⟨ ⟩—N(CH₃)₂

20.26 (a)

NH₂ on cyclohexanone

$\xrightarrow{\substack{\text{(1) cat. HA} \\ \text{(2) NaBH}_3\text{CN}}}$

bicyclic amine (octahydroindole)

(b)

cyclohexyl-CH₂Br

$\xrightarrow{\substack{\text{(1) NaCN} \\ \text{(2) LiAlH}_4 \\ \text{(3) H}_3\text{O}^+}}$

cyclohexyl-CH₂CH₂NH₂

(c)

cyclohexyl-CH₂Br

$\xrightarrow{\substack{\text{(1) NaN}_3 \\ \text{(2) LiAlH}_4 \\ \text{(3) H}_3\text{O}^+}}$

cyclohexyl-CH₂NH₂

(d)

cyclopentanone

$\xrightarrow{\substack{\text{(1) NH}_2\text{OH, cat. HA} \\ \text{(2) NaBH}_3\text{CN}}}$

cyclopentyl-NH₂

(e)

cyclohexyl-NH₂

$\xrightarrow{\substack{\text{(1) acetic anhydride} \\ \text{(2) LiAlH}_4 \\ \text{(3) H}_3\text{O}^+}}$

cyclohexyl-NH-CH₂CH₃

20.27 (a)

(b)

(c)

(d)

(e)

20.28 (a)

(b)

(c)

(d)

This product is not stable and would revert to starting material upon reaction work-up (i.e. exposure to water) so in essence no reaction would be observed

(e)

20.29 (a)

(b)

20.30 (a) $2 \text{ } \diagup\diagdown \text{NH}_2 +$

(b) $2 \text{ CH}_3\text{NH}_2 +$

(c)

(d) [product of (c)] $\xrightarrow{\text{heat}}$

(e)

(f)

(g) 2

(h)

(i)

20.31 (a)

Separated

(b)

[from part (a)]

(c)

[from part (a)]

(d)

[from Problem 20.11(a)]

(e)

(f)

[from Problem 20.11(b)]

(g)

[from Problem 20.11(d)]

$$\xrightarrow[0-5\ ^\circ C]{H_2SO_4/NaNO_2}$$

$$\xrightarrow{H_3PO_2}$$

(h)

[from part (g)]

(1) Fe, HCl, heat
(2) HO⁻

(i)

[from part (g)]

(1) HBr/NaNO₂, 0–5 °C
(2) CuBr

(1) Fe, HCl, heat
(2) HO⁻

(1) H₂SO₄/NaNO₂, 0–5 °C
(2) Cu₂O, Cu²⁺, H₂O

(j)

[from part (i)]

(1) H₂SO₄/NaNO₂, 0–5 °C
(2) CuCN

(k)

(l)

(m)

(n)

[from part (c)]

$\xrightarrow{\text{H}_2\text{SO}_4/\text{NaNO}_2,\ 0\text{–}5\ ^\circ\text{C}}$

, pH 8-10

(o)

[from part (n)]

[from part (c)]
pH 8-10

20.32 (a) Benzylamine dissolves in dilute HCl at room temperature and would exhibit alkyl ^{1}H NMR signals.

Benzamide would not dissolve in dilute HCl, and would not exhibit alkyl ^{1}H NMR signals but would have a carbonyl absorption in its IR spectrum.

(b) Allylamine reacts with (and decolorizes) bromine instantly and would exhibit alkene ^{1}H NMR signals.

Propylamine would not react with bromine and would not show alkene ^{1}H NMR signals.

(c) H_3C—⬡—$\ddot{N}H_2$, as a primary amine, would show two peaks in the N-H region of its IR spectrum due to asymmetric and symmetric H_2N- stretching.

⬡—$\ddot{N}HCH_3$, as a secondary amine, would show only one peak in the N-H region of its IR spectrum.

(d) ⬡—$\ddot{N}H_2$, as a primary amine, would show two peaks in the N-H region of its IR spectrum due to asymmetric and symmetric H_2N- stretching.

⬡ N—H, as a secondary amine, would show only one peak in the N-H region of its IR spectrum.

(e) Pyridine dissolves in dilute HCl,

⬡N + H_3O^+ + Cl^- $\longrightarrow$ ⬡$\overset{+}{N}$—H Cl^-

benzene does not:

⬡ + H_3O^+ + Cl^- $\longrightarrow$ No reaction

(f) Aniline reacts with nitrous acid at 0–5 °C to give a stable diazonium salt that couples with 2-naphthol, yielding an intensely colored azo compound.

⬡—NH_2 $\xrightarrow[\text{0–5 °C}]{H_2SO_4/NaNO_2}$ ⬡—$\overset{+}{N_2}$ $\xrightarrow{\text{2-naphthol}}$

Cyclohexylamine reacts with nitrous acid at 0–5 °C to yield a highly unstable diazonium salt—one that decomposes so rapidly that the addition of 2-naphthol gives no azo compound.

⬡—NH_2 $\xrightarrow[\text{0–5 °C}]{H_2SO_4/NaNO_2}$ $\left[\text{⬡—}\overset{+}{N_2}\right]$ $\xrightarrow{-N_2}$ $\left[\text{⬡}+\right]$ $\longrightarrow$

alkenes, alcohols, and alkyl halides $\xrightarrow{\text{2-naphthol}}$ No reaction

(g) N would dissolve in dilute HCl, but as a tertiary amine, would not exhibit an absorption in the N-H region of its IR spectrum.

NH would dissolve in dilute HCl, but as a secondary amine, would exhibit a single peak in the N-H region of its IR spectrum.

(h) Tripropylaminium chloride reacts with aqueous NaOH to give a water insoluble tertiary amine.

Water soluble   Water insoluble

Tetrapropylammonium chloride does not react with aqueous NaOH (at room temperature), and the tetrapropylammonium ion remains in solution.

Water soluble   Water soluble

(i) Tetrapropylammonium chloride dissolves in water to give a neutral solution. Tetrapropylammonium hydroxide dissolves in water to give a strongly basic solution.

20.33 Follow the procedure outlined in the answer to Problem 20.2. Toluene will show the same solubility behavior as benzene.

Mechanisms

20.34

20.35 (a)

(b)

20.36 Carry out the Hofmann reaction using a mixture of ^{15}N labeled benzamide, $C_6H_5CO^*NH_2$, and p-chlorobenzamide. If the process is intramolecular, only labeled aniline, $C_6H_5^*NH_2$, and p-chloroaniline, will be produced.

If the process is one in which the migrating moiety truly separates from the remainder of the molecule, then, in addition to the two products mentioned above, there should be produced both the unlabeled aniline and labeled p-chloroaniline, p-$ClC_6H_4^*NH_2$.

Note: When this experiment is actually carried out, analysis of the reaction mixture by mass spectrometry shows that the process is intramolecular.

General Synthesis

20.37

20.38 (a)

(b)

20.39

20.40

$$\text{heat} \longrightarrow \overset{\text{W}}{\diagup\!\!\diagdown\!\!\diagup\!\!\diagdown} + H_2O + (CH_3)_3N$$

20.41 $(CH_3)_2NH$ +

Acetylcholine iodide

20.42

20.43

Diethylpropion

20.44

Spectroscopy

20.45 The solubility characteristics, molecular formula, and IR evidence indicate that compound **W** is a tertiary amine. The ^{1}H NMR provides evidence for the following:

(1) The signals at δ 7.2 and 6.7 suggest hydrogens bonded to a benzene ring. Their integrals relative to those for other signals suggests two $-$ C$_6$H$_5$ groups.
(2) A CH$_3$CH$_2-$ group (the quartet at δ 3.5 and the triplet at δ 1.2).
(3) An unsplit $-$CH$_2-$ group (the singlet at δ 4.5).

There is only one reasonable way to pull all of this together.

Thus **W** is *N*-benzyl-*N*-ethylaniline.

20.46 Compound **X** is benzyl bromide, $C_6H_5CH_2Br$. This is the only structure consistent with the ^1H NMR and IR data. (The monosubstituted benzene ring is strongly indicated by the 5H, δ 7.3 ^1H NMR absorption and is confirmed by the peaks at 690 and 770 cm^{-1} in the IR spectrum.)

Compound **Y**, therefore, must be phenylacetonitrile, $C_6H_5CH_2CN$, and **Z** must be 2-phenylethylamine, $C_6H_5CH_2CH_2NH_2$.

X	**Y**	**Z**
C_7H_7Br	C_8H_7N	$C_8H_{11}N$

Interpretations of the IR and ^1H NMR spectra of **Z** are as follows.

(a) singlet δ 1.0
(b) triplet δ 2.7
(c) triplet δ 2.9
(d) multiplet δ 7.25

20.47 That **A** contains nitrogen and is soluble in dilute HCl suggests that **A** is an amine. The two IR absorption bands in the 3300–3500 cm^{-1} region suggest that **A** is a primary amine. The ^{13}C spectrum shows only two signals in the upfield aliphatic region. There are four signals downfield in the aromatic region. The information from the DEPT spectra suggests an ethyl group *or two equivalent ethyl groups.* Assuming the latter, and assuming that **A** is a primary amine, we can conclude from the molecular formula that **A** is 2,6-diethylaniline. The assignments are

(*a*) δ 12.9
(*b*) δ 24.2
(*c*) δ 118.1
(*d*) δ 125.9
(*e*) δ 127.4
(*f*) δ 141.5

(An equally plausible answer would be that **A** is 3,5-diethylaniline.)

20.48   That **B** dissolves in dilute HCl suggests that **B** is an amine. That the IR spectrum of **B** lacks bands in the 3300–3500 cm^{-1} region suggests that **B** is a tertiary amine. The upfield signals in the ^{13}C spectrum and the DEPT information suggest two equivalent ethyl groups (as was also true of **A** in the preceding problem). The DEPT information for the downfield peaks (in the aromatic region) is consistent with a monosubstituted benzene ring. Putting all of these observations together with the molecular formula leads us to conclude that **B** is *N,N*-diethylaniline. The assignments are

(*a*) δ 12.5
(*b*) δ 44.2
(*c*) δ 112.0
(*d*) δ 115.5
(*e*) δ 128.1
(*f*) δ 147.8

Synthesizing the Material

20.49 (a)

(c)

Note: if concerned that the decarboxylation step would also impact the nitrile, the order of operations could instead be to add ethyl group first, decarboxylate, and then do Michael addition with the thermodynamic enolate.

(d)

(e)

(f)

20.50 (a)

(b)

(c)

20.51

(1) TBAF
(2) Jones oxidation

Repeat
same 4
steps

Challenge Problems

20.52 (a) $C_6H_5\overset{\cdot\cdot}{\underset{H}{N}}\overset{CH_3}{}$

I

(b) **I** $\xrightarrow[\substack{\text{(during} \\ \text{electron} \\ \text{impact} \\ \text{mass} \\ \text{spectrometry)}}]{-e^-}$ $C_6H_5\overset{+\cdot}{\underset{H}{N}}$ *m/z* 107

$\Big\downarrow$ −H·

$C_6H_5\overset{}{\underset{H}{\overset{+}{N}}}=CH_2$ *m/z* 106

20.53 (a, b)

A

B Diphenylcarbodiimide

20.54

20.55 An abbreviated version of one of several possible routes:

"triacetoneamine"

20.56

O-protonated acetamide cation   *N*-protonated acetamide cation

The *N*-protonated acetamide cation shows significant localization of positive charge near its NH_3^+ group, as indicated by the strong blue color at that location in its electrostatic potential map. The *N*-protonated acetamide model also shows noticeable negative charge (or less positive charge) near its oxygen atom, as indicated by yellow in that region. These observations suggest a lack of delocalization of the positive charge.

On the other hand, the *O*-protonated acetamide cation has some blue as well as green mapped to both its NH_2 group and O—H (protonated carbonyl) group, suggesting delocalization of the positive charge between these two locations.

An attempt to draw resonance structures for these two protonated forms shows that two resonance contributors are possible for *O*-protonated acetamide, whereas only one proper Lewis structure is possible for *N*-protonated acetamide. The resonance contributors for *O*-protonated acetamide distribute the formal positive charge over both the oxygen and nitrogen atoms.

O-protonated acetamide cation   *N*-protonated acetamide cation

QUIZ

20.1 Which of the following would be soluble in dilute aqueous HCl?

(a) [structure: benzene ring with NH_2]

(b) [structure: benzene ring with CH_2NH_2]

(c) [structure: benzene ring with $C(=O)NH_2$]

(d) Two of the above

(e) All of the above

20.2 Which would yield propylamine?

(a) [structure: CH group with Br] $\xrightarrow[\text{(2) LiAlH}_4]{\text{(1) NaCN}}$

(b) [structure: aldehyde with H] $\xrightarrow[\text{H}_2/\text{Ni}]{\text{NH}_3}$

(c) [structure: amide with NH_2] $\xrightarrow[\text{HO}^-]{\text{Br}_2}$

(d) Two of the above

(e) All of the above

20.3 Select the reagent or spectroscopic technique from the list below that could be the basis of a simple chemical test that would distinguish between each of the following:

(a) [structure: benzamide with $C(=O)NH_2$] and [structure: benzene ring with CH_2NH_2]

(b) [structure: benzene ring with CH_2NH_2] and [structure: benzene ring with $N(H)CH_3$]

1. Cold dilute $NaHCO_3$
2. Cold dilute HCl
3. IR spectroscopy
4. ^{1}H NMR spectrometry
5. Cold dilute NaOH

(c) [structure: benzene ring with NH_2] and [structure: cyclohexane ring with NH_2]

20.4 Complete the following syntheses:

(a)

(b)

20.5 Select the stronger base from each pair (in aqueous solution):

(a)

or

(1)   (2)

(b)

or

(1)   (2)

(c)

or

(1)   (2)

(d)

or

(1)   (2)

(e)

or

(1)   (2)

(f)

or

(1)   (2)

21

TRANSITION METAL COMPLEXES: PROMOTERS OF KEY BOND-FORMING REACTIONS

SOLUTIONS TO PROBLEMS

21.1 (a) +1 charge on metal, 16 valence electrons (8 from Rh, 8 from the ligands)

(b) +2 charge on metal, 22 valence electrons (note that there can be exceptions to the 18 electron rule; 10 from Hg, 12 from ligands)

(c) 0 charge on metal, 20 valence electrons (note that there can be exceptions to the 18 electron rule; 10 from Ni, 10 from ligands)

21.2 A syn addition of D_2 to the *trans* alkene would produce the following racemic form.

21.3 (a)

(b)

21.4 (a)

$X = Cl, Br, or I$

(b)

21.5

21.6 B(OH)₂ +

21.7 (a) (b) CF₃ / t-Bu

(c) t-Bu (d) O₂N / CO₂Et

21.8 (a) t-Bu—I + Bu₃Sn

(b) —I + Bu₃Sn

21.9 (a) OH (b) H₃CO / CO₂CH₃

21.10 (a) H + Cl₂C=CH₂

(b) EtO / Br + H≡Si(CH₃)₃

21.11

(a)

(b)

21.12 (a)

(b) OTDBMS

(c)

(d)

21.13 (a) +2 charge on metal, 16 valence electrons (8 from Pd, 8 from the ligands), coordinately unsaturated.

(b) +2 charge on metal, 18 valence electrons (6 from Ru, 12 from the ligands), coordinately saturated.

(c) 0 charge on metal, 18 valence electrons (8 from Fe, 10 from the ligands), coordinately saturated.

21.14

14 valence electrons
(+2 charge on metal)

16 valence electrons
(0 charge on metal)

18 valence electrons
(+2 charge on metal)

21.15 (a)

(b)

[Note: many other possibilities exist here]

(c)

21.16 There are numerous possibilities to answer this question. The solutions below focus on Stille and Suzuki reactions, but Heck couplings, Gilman couplings, other C–C bond formation events with different partners could also be used.

(a)

(b)

(c)

21.17 (a)

+ enantiomer

(b)

H₃CO...

+

(c)

+

+

(d)

OH

21.18 (a)

(b)

HO_2C CO_2H

(c)

21.19 As part of the mechanism of ring-closing metathesis, the metal carbene promoter forms a four-membered ring with one of the olefins through a [2+2] reaction, with a retro-[2+2] reaction then making a new carbene that can accomplish the final ring closure through a similar process intramolecularly. Key from an entropy perspective is the first retro-[2+2] reaction that produces a molecule of ethylene as a side-product. As a result, in the final balanced chemical equation it means that the full reaction process generates two molecules (i.e. ethylene and the ring-closed product) from one molecule of starting material. That positive contributor to the ΔS term will offset at least some of the penalty of the negative entropy associated with making a more rigid system from a more flexible precursor.

21.20 (a)

(b)

(c)

21.21 (a)

(b)

21.22

21.23 The mechanism for the process is shown below, with Step 4 regenerating the coordinatively unsaturated complex that serves as the catalyst for the process. In addition, there is one important but potentially subtle point in Step 3. There are two hydrogen atoms that could be lost during the 1,2-*syn* elimination step. The transition state shown here removes a proton leading to a *trans* alkene product, while the other transition state that is not drawn would afford a *cis* alkene. Between these two transition states, the one leading to the *trans* product has a lower energy barrier because it limits the potential for steric interactions between the phenyl and ethyl groups; similarly, the *trans* product is more stable also because it limits steric interactions. Thus, from both a kinetic and thermodynamic perspective, the *trans* product is favored to result from the 1,2-*syn* elimination step.

Synthesizing the Material

21.24 **(a)**

(b)

(c)

Note: there are many answers possible for this question based on different couplings. Here we are assuming product separation after the first step.

(d)

1) OsO$_4$
2) NaHSO$_3$

Swern
oxidation

21.25 (a)

1) HNO$_3$,
   H$_2$SO$_4$

2) Fe, HCl;
   HO$^-$

3) DCC, AcCl

4)   SnBu$_3$

Pd(Ph$_3$P)$_4$
base

A

(b)

1) LDA; NBS
2) t-BuOK,
   t-BuOH

3) Grubbs II

4) EtMgBr
   (excess);
   H$_3$O$^+$

Note: in the final step
both 1,2- and 1,4-addition
can occur.

B

(c)

1) O_3
2) Me_2S

3) $H_2C\!\!=\!\!PPh_3$
(excess)

4) $Pd(Ph_3P)_4$
Bu_3Sn ⌇ Ph | base
(excess)

Ph ⌇ Ph

C

(d)

1) Br_2
2) $NaNH_2$
(2 equiv.), Δ

3) ⌇ Br

$CuI, Pd(Ph_3P)_4,$
Et_2NH

D

Challenge Problems

21.26

$Ru\!\!=$ +

$Ru\!\!=$
(catalyst
regenerated)

[retro
2+2]

21.27 One of these products, the one above the starting material as drawn in the question, is the result of an intramolecular metathesis reaction (i.e. ring closing metathesis) while the other product, the one below the starting material as drawn in the question, is the result of an intermolecular metathesis reaction (i.e. cross metathesis). When the starting material is dilute, i.e. present in low concentration, it is more likely to react with itself and generate the ring-closed metathesis product since interactions with other molecules of starting material should be less frequent. When the starting material is present in a more concentrated manner, then the frequency of intermolecular collisions should be higher and intermolecular reactions could potentially predominate, as was the case here.

In general terms, intramolecular reactions tend to be faster than intermolecular ones. However, in this case, the large ring size that would result if an intramolecular reaction happened makes this process fairly slow, with the flexible nature of the atom chain making it hard to get the two ends to connect. As a result, intermolecular events could reasonable occur, especially if the concentration is high enough to help promote them.

21.28 The reaction conditions of the Stille coupling lead the initial intermediate shown in brackets, a compound in which a diene has now been formed. With heat, and the presence of an α,β-unsaturated carbonyl in close proximity, an intramolecular Diels–Alder reaction is then promoted which leads to the indicated ring system.

QUIZ

21.1 What is the oxidation state of the metal in the following complexes?

(a) $RhCl_2(PhP_3)_2$

(b) $Ru(CO)_5$

(c) $Pd(Ph_3P)_4$

21.2 Which of the two metathesis reactions shown below is more likely to lead to a mixture of products?

21.3 The key step which determines the stereochemistry of the Heck coupling reaction is:

(a) Oxidative addition

(b) Alkene insertion

(c) Reductive elimination

21.4 Dialkyl copper-based nucleophiles/coupling partners are known as:

(a) Heck reagents

(b) Suzuki reagents

(c) Gilman reagents

(d) Grignard reagents

21.5 Which of the following are inventors of important catalysts for the olefin metathesis reaction? More than one answer may be right.

(a) Akira Suzuki

(b) Robert Grubbs

(c) Herbert Brown

(d) Eiichi Negishi

(e) Richard Schrock

21.6 Which compound would be the most difficult to prepare using Pd-base coupling reactions?

ANSWERS TO SECOND REVIEW PROBLEM SET

These problems review concepts from Chapters 13–21.

1. Increasing acidity →

(a)

(b)

(c)

(d)

(e)

2. Increasing basicity →

(a)

(b)

(c) O_2N—〈 〉—NH_2 < 〈 〉—NH_2 < CH_3—〈 〉—NH_2

(d)

3. (a)

(b)

(c)

(d)

4.

5. (a)

(b)

6. (a) [structure: butanol] $\xrightarrow[\text{(or HBr)}]{\text{PBr}_3}$ [structure: butyl bromide]

(b) [structure: butyl bromide] + [potassium phthalimide, NK] $\longrightarrow$ [N-butylphthalimide]

[from part (a)]

$\xrightarrow[\text{heat}]{\text{H}_2\text{NNH}_2}$ [butylamine, NH$_2$] + [phthalhydrazide]

(c) [butyl bromide, Br] $\xrightarrow{\text{NaCN}}$ [pentanenitrile, CN] $\xrightarrow[\text{Et}_2\text{O}]{\text{LiAlH}_4}$ [pentylamine, NH$_2$]

[from part (a)]

(d) [butanol, OH] $\xrightarrow[\text{(2) H}_3\text{O}^+]{\text{(1) KMnO}_4,\ \text{HO}^-,\ \text{heat}}$ [butanoic acid, O OH]

(e) [nitrile, CN] $\xrightarrow[\text{heat}]{\text{H}_3\text{O}^+,\ \text{H}_2\text{O}}$ [carboxylic acid, O OH] + NH$_4^+$

[from part (c)]

(f) [butanoic acid, O OH] $\xrightarrow{\text{SOCl}_2}$ [butanoyl chloride, O Cl]

[from part (d)]

(g) [butanoyl chloride, O Cl] $\xrightarrow{\text{NH}_3}$ [butanamide, O NH$_2$]

[from part (f)]

(h) [butanoyl chloride, O Cl] $\xrightarrow[\text{base}]{\text{OH (butanol)}}$ [butyl butanoate, O O]

[from part (f)]

(i) [butanamide, O NH$_2$] $\xrightarrow[\text{(2) H}_3\text{O}^+]{\text{(1) Br}_2,\ \text{HO}^-}$ [propylamine, NH$_2$] + CO$_3^{2-}$

[from part (g)]

(j) [from part (f)]

(k) [from part (f)]

(l) [from part (a)]

7. (a)

(separate from ortho isomer)

(1) epoxide
(2) H_3O^+

Swern oxidation
or
PCC

HO^-
(aldol condensation)

$[(E)$ or $(Z)]$

(d)

CH_3

$\xrightarrow[\substack{\text{(separate from} \\ \text{ortho isomer)}}]{HNO_3, H_2SO_4}$

CH_3 / O_2N

$\xrightarrow[\text{(2) } H_3O^+]{\text{(1) } KMnO_4, HO^-, \text{ heat}}$

O_2N — benzoic acid with OH

$\xrightarrow{SOCl_2}$

O_2N — acyl chloride with Cl

$\xrightarrow[\substack{\text{diethyl ether} \\ -78°C}]{LiAlH(t\text{-BuO})_3}$

O_2N — aldehyde with H

$\xrightarrow{HO^-}$ (with acetophenone)

O_2N — chalcone product

[(E) or (Z)]

(e)

CH_3

$\xrightarrow[(hv)]{NBS}$

benzyl bromide Br

$\xrightarrow{Na^+ \ ^-\!:\!\equiv\!H}$

phenylpropyne

(1) H_2, Lindlar's catalyst
(2) (i) $Hg(OAc)_2$, THF-H_2O
    (ii) $NaBH_4$
(3) H_2CrO_4

$\longrightarrow$ phenylacetone (O)

$\xrightarrow{HCN}$ cyanohydrin CN / OH

$\xrightarrow[\text{heat}]{H_3O^+}$

[carboxylic acid OH / OH] $\xrightarrow{-H_2O}$ unsaturated acid OH

[(E) or (Z)]

8.

cyclopentadiene + ethylene $\longrightarrow$ norbornene

9.

Isodrin

10.

2-Methyl-1,3-
butadiene   Diethyl fumarate

A

+ enantiomer

B

+ enantiomer

C

+ enantiomer

D

+ enantiomer

11. (a) A is

C is

(b) **A** is an allylic alcohol and thus forms a carbocation readily. **B** is a conjugated enyne and is therefore more stable than **A**.

12.

D

E

F

Vitamin A acetate

13.

Bisphenol A

14.

(1) KMnO$_4$, HO$^-$, heat
(2) H$_3$O$^+$

A

SOCl$_2$

B

C

H$_2$
cat.

Procaine

15.

(1) ≡ :$^-$Na$^+$
(2) NH$_4$Cl/H$_2$O

A

B

NH$_3$

Ethinamate

16. (a)

(b) The last step probably takes place by an S_N1 mechanism. Diphenylmethyl bromide, **B**, ionizes readily because it forms the resonance-stabilized benzylic carbocation.

17. (a) For this synthesis we need to prepare the benzylic halide,

and then allow it to react with $(CH_3)_2NCH_2CH_2OH$ as in Problem 16.

This benzylic halide can be made as follows:

(b) For this synthesis we can prepare the requisite benzylic halide in two ways:

or

We shall then allow the benzylic halide to react with $(CH_3)_2NCH_2CH_2OH$ as in Problem 16.

18.

19.

20.

21.

22.

23.

Notice that the second step involves the oxidation of a secondary alcohol in the presence of a tertiary alcohol. This selectivity is possible because tertiary alcohols do not undergo oxidation readily (Section 12.4).

24. Working backward, we notice that methyl *trans*-4-isopropylcyclohexanecarboxylate has both large groups equatorial and is, therefore, more stable than the corresponding cis isomer. This stability of the trans isomer means that, if we were to synthesize the cis isomer or a mixture of both the cis and trans isomers, we could obtain the desired trans isomer by a base-catalyzed isomerization (epimerization):

We could synthesize a mixture of the desired isomers from phenol in the following way:

25. The phenoxide ion has nucleophilic character both at the oxygen and at the carbon para to it; it is ambident.

The benzyne produced by the elimination phase of the reaction can then undergo addition by attack by either nucleophilic center: reaction at oxygen producing **1** (diphenyl ether), and reaction at carbon producing **2** (4-hydroxybiphenyl).

26.

27. Compounds and reagents **A—M**, regarding the synthesis of dianeackerone:

(a)

G CH₃I
H H₂CrO₄

J LAH
K PCC

M H₂, Pd

(b)

(c) (3*S*, 7*R*)-Dianeackerone

28. The molecular formula indicates that the compound is saturated and is an alcohol or an ether. The IR data establish the compound as an alcohol. That **Y** gives a green opaque solution when treated with CrO_3 in aqueous H_2SO_4 indicates that **Y** is a primary or secondary alcohol. That **Y** gives a negative iodoform test indicates that **Y** does not contain the grouping $-\underset{\underset{OH}{|}}{C}HCH_3$. The ^{13}C spectrum of **Y** contains only four signals indicating that some of the carbons in **Y** are equivalent. The information from DEPT spectra helps us conclude that **Y** is 2-ethyl-1-butanol.

(a) δ 11.1
(b) δ 23.0
(c) δ 43.6
(d) δ 64.6

Notice that the most downfield signal is a CH_2 group. This indicates that this carbon atom bears the —OH group and that **Y** is a primary alcohol. The most upfield signals indicate the presence of the ethyl groups.

29. That **Z** decolorizes bromine indicates that **Z** is an alkene. We are told that **Z** is the more stable isomer of a pair of stereoisomers. This fact suggests that **Z** is a trans alkene. That the ^{13}C spectrum contains only three signals, even though **Z** contains eight carbon atoms, indicates that **Z** is highly symmetric. The information from DEPT spectra indicates that the upfield signals of the alkyl groups arise from equivalent isopropyl groups. We conclude, therefore, that **Z** is *trans*-2,5-dimethyl-3-hexene.

(a) δ 22.8
(b) δ 31.0
(c) δ 134.5

30. (a, b)

31. Compound **N** is $C_6H_5CH=CHOCH_3$. The absence of absorption peaks due to O—H or C=O stretching in the IR spectrum of **N** suggests that the oxygen atom is present as part of an ether linkage. The (5H) ^1H NMR multiplet between δ 7.1–7.6 strongly suggests the presence of a monosubstituted benzene ring; this is confirmed by the strong peaks at ~690 and ~770 cm^{-1} in the IR spectrum.

We can make the following assignments in the ^1H NMR spectrum:

$$\underset{C_6H_5}{(a)}—\underset{CH}{(b)}=\underset{CH}{(c)}—\underset{OCH_3}{(d)}$$

(*a*) Multiplet δ 7.1–7.6
(*b*) Doublet δ 6.1
(*c*) Doublet δ 5.2
(*d*) Singlet δ 3.7

32. That compound **X** does not dissolve in aqueous sodium bicarbonate indicates that **X** is not a carboxylic acid. That **X** has an IR absorption peak at 1740 cm^{-1} indicates the presence of a carbonyl group, probably that of an ester. That the molecular formula of **X** ($C_7H_{12}O_4$) contains four oxygen atoms suggests that **X** is a diester.

The ^{13}C spectrum shows only four signals indicating a high degree of symmetry for **X**. The single signal at δ 166.7 is that of an ester carbonyl carbon, indicating that both ester groups of **X** are equivalent.

Putting these observations together with the information gathered from DEPT ^{13}C spectra and the molecular formula leads us to the conclusion that **X** is diethyl malonate. The assignments are

(*a*) δ 14.2
(*b*) δ 41.6
(*c*) δ 61.3
(*d*) δ 166.7

33.

34. That **C** gives a positive Tollens' test indicates the presence of an aldehyde group; the solubility of **C** in aqueous HCl suggests that **C** is also an amine. The absence of bands in the 3300–3500-cm^{-1} region of the IR spectrum of **C** suggests that **C** is a tertiary amine. The signal at δ 189.7 in the ^{13}C spectrum can be assigned to the aldehyde group. The signal at δ 39.7 is the only one in the aliphatic region and is consistent with a methyl group or with two equivalent methyl groups. The remaining signals are in the aromatic region.

If we assume that **C** has a benzene ring containing a $\overset{\overset{\displaystyle O}{\|}}{\diagdown}$H group and a —N(CH$_3$)$_2$ group, then the aromatic signals and their DEPT spectra are consistent with **C** being *p*-(*N*,*N*-dimethylamino)benzaldehyde. The assignments are

(a) δ 39.7
(b) δ 110.8
(c) δ 124.9
(d) δ 131.6
(e) δ 154.1
(f) δ 189.7

35. **X** is a phenol because it dissolves in aqueous NaOH but not in aqueous NaHCO$_3$. It gives a dibromo derivative and must therefore be substituted in the ortho or para position. The broad IR peak at 3250 cm^{-1} also suggests a phenol. The peak at 830 cm^{-1} indicates para substitution. The ^{1}H NMR singlet at δ 1.3 (9H) suggests nine equivalent methyl hydrogen atoms, which must be a *tert*-butyl group. The structure of **X** is

36. The broad IR peak at 3200–3600 cm^{-1} suggests a hydroxyl group. The two ^{1}H NMR peaks at δ 1.67 and δ 1.74 are not a doublet because their separation is not equal to other splittings; therefore, these peaks are singlets. Reaction with Br$_2$ suggests an alkene. If we put these bits of information together, we conclude that **Z** is 3-methyl-2-buten-1-ol (3-methylbut-2-en-1-ol).

(a) and (b) singlets δ 1.67 and δ 1.74
(c) multiplet δ 5.40
(d) doublet δ 4.10
(e) singlet δ 2.3

37.

38. (a, b)

22 CARBOHYDRATES

SUMMARY OF SOME REACTIONS OF MONOSACCHARIDES

$$
\begin{array}{c}
\text{CO}_2\text{H} \\
|\\
(\text{CHOH})_n \\
|\\
\text{CH}_2\text{OH}
\end{array}
\quad \text{Aldonic acid}
$$

via $\dfrac{\text{Br}_2}{\text{H}_2\text{O}}$

$$
\begin{array}{c}
\text{CO}_2\text{H} \\
|\\
(\text{CHOH})_n \\
|\\
\text{CO}_2\text{H}
\end{array}
\quad \text{Aldaric acid}
$$

via HNO_3

$$
\begin{array}{c}
\text{CH}=\text{NNHC}_6\text{H}_5 \\
|\\
\text{C}=\text{NNHC}_6\text{H}_5 \\
|\\
(\text{CHOH})_{n-1} \\
|\\
\text{CH}_2\text{OH}
\end{array}
\quad \text{Osazone}
$$

via $\text{C}_6\text{H}_5\text{NHNH}_2$

$$
\begin{array}{c}
\text{CN} \\
|\\
(\text{CHOH})_{n+1} \\
|\\
\text{CH}_2\text{OH}
\end{array}
\quad \text{Cyanohydrin}
\xrightarrow[\text{steps}]{\text{several}}
\begin{array}{c}
\text{CHO} \\
|\\
(\text{CHOH})_{n+1} \\
|\\
\text{CH}_2\text{OH}
\end{array}
$$

via HCN (Kiliani-Fischer synthesis)

Aldose with one more carbon

$$
\begin{array}{c}
\text{CH}_2\text{OH} \\
|\\
(\text{CHOH})_n \\
|\\
\text{CH}_2\text{OH}
\end{array}
\quad \text{Alditol}
$$

via NaBH_4

$$
\begin{array}{c}
\text{CHO} \\
|\\
(\text{CHOH})_{n-1} \\
|\\
\text{CH}_2\text{OH}
\end{array}
\quad \text{Aldose with one less carbon atom}
$$

via (1) $\text{Br}_2/\text{H}_2\text{O}$ (2) $\text{H}_2\text{O}_2/\text{Fe}^{3+}$ (Ruff degradation)

$$
\begin{array}{c}
\text{O}=\text{C} \!-\! \text{H} \\
|\\
(\text{CHOH})_n \\
|\\
\text{CH}_2\text{OH}
\end{array}
$$

Open-chain form of aldose

Cyclic form of D-glucose
$\xrightarrow[\text{HA}]{\text{CH}_3\text{OH}}$
Methyl glucoside
$\xrightarrow[\text{HO}^-]{(\text{CH}_3)_2\text{SO}_4}$

SOLUTIONS TO PROBLEMS

22.1 (a) Two,
$$
\begin{array}{c}
\text{CHO} \\
| \\
{}^*\text{CHOH} \\
| \\
{}^*\text{CHOH} \\
| \\
\text{CH}_2\text{OH}
\end{array}
$$

(b) Two,
$$
\begin{array}{c}
\text{CH}_2\text{OH} \\
| \\
\text{C}{=}\text{O} \\
| \\
{}^*\text{CHOH} \\
| \\
{}^*\text{CHOH} \\
| \\
\text{CH}_2\text{OH}
\end{array}
$$

(c) There would be four stereoisomers (two sets of enantiomers) for each general structure: $2^2 = 4$.

22.2

22.3 (a)

(b)

$$
\begin{array}{c}
\text{---OH} \\
\text{---O} \\
\text{H---OH} \\
\text{H---OH} \\
\text{HO---H} \\
\text{H---OH} \\
CH_2OH
\end{array}
$$

22.4 (a)

and

D-(+)-Glucose   2-Hydroxybenzyl alcohol

(b)

Salicin

22.5 Dissolve D-glucose in ethanol and then bubble in gaseous HCl.

22.6 α-D-Glucopyranose will give a positive test with Benedict's or Tollens' solution because it is a hemiacetal in equilibrium with an aldehyde. Methyl α-D-glucopyranoside, because it is an acetal, will not.

22.7 (a) Yes   (b)

D-Mannaric acid

(c) Yes   (d)

(e) No   (f)

D-Threose $\xrightarrow{\text{HNO}_3}$ D-Tartaric acid

(g) The aldaric acid obtained from D-erythrose is *meso*-tartaric acid; the aldaric acid obtained from D-threose is D-tartaric acid.

22.8

or   and   or

22.9 One way of predicting the products from a periodate oxidation is to place an —OH group on each carbon atom at the point where C—C bond cleavage has occurred:

$\xrightarrow{\text{IO}_4^-}$

Then if we recall (Section 16.7A) that *gem*-diols are usually unstable and lose water to produce carbonyl compounds, we get the following results:

Let us apply this procedure to several examples here while we remember that for every C—C bond that is broken 1 mol of HIO_4 is consumed.

(d) chemical reaction scheme

(e) chemical reaction scheme

(f) chemical reaction scheme

(g) chemical reaction scheme

(h)

D-Erythrose

22.10 Oxidation of an aldohexose and a ketohexose would each require 5 mol of HIO_4 but would give different results.

Aldohexose

Ketohexose

22.11 (a) Yes, D-glucitol would be optically active; only those alditols whose molecules possess a plane of symmetry would be optically inactive.

(b)

Optically inactive

Optically inactive

22.12 (a)

(b) This experiment shows that D-glucose and D-fructose have the same configurations at C3, C4, and C5.

22.13 (a)

L-(+)-Erythrose   L-(+)-Threose

(b) L-Glyceraldehyde

22.14 (a)

D-(–)-Erythrose

HCN

CN

Epimeric
cyanohydrins
(separated)

(1) Ba(OH)$_2$
(2) H$_3$O$^+$

Epimeric
aldonic acids

–H$_2$O

Epimeric
γ-aldonolactones

Na-Hg, H$_2$O
pH 3-5

(b)

D-(−)-Ribose   Optically inactive

D-(−)-Arabinose   Optically active

22.15   A Kiliani-Fischer synthesis starting with D-(−)-threose would yield **I** and **II**.

I	**II**
D-(+)-Xylose	D-(−)-Lyxose

I must be D-(+)-xylose because, when oxidized by nitric acid, it yields an optically inactive aldaric acid:

Optically inactive

II must be D-(−)-lyxose because, when oxidized by nitric acid, it yields an optically active aldaric acid:

Optically active

22.16

L-(+)-Ribose   L-(+)-Arabinose   L-(–)-Xylose   L-(+)-Lyxose

22.17 Since D-(+)-galactose yields an optically inactive aldaric acid, it must have either structure **III** or structure **IV**.

III $\xrightarrow{HNO_3}$ Optically inactive   **IV** $\xrightarrow{HNO_3}$ Optically inactive

A Ruff degradation beginning with **III** would yield D-(–)-ribose

III $\xrightarrow[H_2O]{Br_2}$ $\xrightarrow[Fe_2(SO_4)_3]{H_2O_2}$

D-(–)-Ribose

A Ruff degradation beginning with **IV** would yield D-(–)-lyxose: thus, D-(+)-galactose must have structure **IV**.

IV $\xrightarrow[H_2O]{Br_2}$ $\xrightarrow[Fe_2(SO_4)_3]{H_2O_2}$

D-(–)-Lyxose

22.18 D-(+)-glucose, as shown here.

The other γ-lactone of D-glucaric acid $\xrightarrow{Na-Hg}$ $\rightleftharpoons$ $\xrightarrow[pH\ 3-5]{Na-Hg}$ D-(+)-Glucose

22.19

D-Galactose

KMnO₄ / HO⁻

D-Galacturonic acid   + 2 (acetone)

Problems

Carbohydrate Structure and Reactions

22.20 (a)

(b)

(c)

or

(d)

(e)

(f)

(g)

(h) or

(i) or

(j) Any sugar that has a free aldehyde or ketone group or one that exists as a cyclic hemiacetal. The following are examples:

(k) (l)

(m) Any two diastereomers that differ in configuration at only one chirality center. (See Section 22.8.) D-Erythrose and D-threose are examples.

D-Erythrose   D-Threose

(n) Cyclic sugars that differ in configuration only at the hemiacetal or acetal carbon. The following are examples:

and

(o)

$$CH{=}NNHC_6H_5$$
$$C{=}NNHC_6H_5$$
$$(CHOH)_n$$
$$OH$$

(p) Maltose is an example:

(q) Amylose is an example:

(r) Any sugar in which all potential carbonyl groups are present as acetals (i.e., as glycosides). Sucrose (Section 22.12A) is an example of a nonreducing disaccharide; the methyl D-glucopyranosides (Section 22.4) are examples of nonreducing monosaccharides.

22.21 (a)

(b)

(c)

22.22

(chemical structures of furanose and pyranose forms)

and

(structure) + HIO$_4$ → (structure)

(structure) + 2 HIO$_4$ → (structure) + H–C(=O)–OH

The above would apply in the same way to the β anomers. A methyl ribofuranoside would consume only 1 mol of HIO$_4$; a methyl ribopyranoside would consume 2 mol of HIO$_4$ and would also produce 1 mol of formic acid.

22.23 One anomer of D-mannose is dextrorotatory ($[\alpha]_D^{25} = +29.3$); the other is levorotatory ($[\alpha]_D^{25} = -17.0$).

22.24 The microorganism selectively oxidizes the –CHOH group of D-glucitol that corresponds to C5 of D-glucose.

(Fischer projections)

D-Glucose $\xrightarrow{H_2}{Ni}$ D-Glucitol $\xrightarrow[\text{suboxydans}]{O_2 \\ Acetobacter}$ (structure) ≡ L-Sorbose

22.25 L-Gulose and L-idose would yield the same phenylosazone as L-sorbose.

CH=NNHC$_6$H$_5$
C=NNHC$_6$H$_5$
(Fischer projection)
Same phenylosazone

$\xleftarrow{C_6H_5NHNH_2}$

(Fischer projections)

L-Sorbose   L-Gulose   L-Idose

22.26

D-Psicose $\xrightarrow{C_6H_5NHNH_2}$ osazone $\xleftarrow{C_6H_5NHNH_2}$ D-Allose

D-Tagatose $\xrightarrow{C_6H_5NHNH_2}$ osazone $\xleftarrow{C_6H_5NHNH_2}$ D-Galactose

22.27 **A** is D-altrose, **B** is D-talose, **C** is D-galactose.

D-Altrose $\xrightarrow{H_2/Ni}$ Same alditol $\xleftarrow{H_2/Ni}$ D-Talose

A   **B**

A $\xrightarrow{C_6H_5NHNH_2}$ $\xleftarrow{C_6H_5NHNH_2}$ B

Different phenylosazones

D-Galactose **C** + C₆H₅NHNH₂ → Same phenylosazone ← C₆H₅NHNH₂ + D-Talose **B**

D-Galactose
C

Same phenylosazone

D-Talose
B

H₂, Ni (for C) → alditol; H₂, Ni (for B) → alditol

Different alditols

(*Note*: If we had designated D-talose as **A**, and D-altrose as **B**, then **C** is D-allose).

22.28

D-Xylose → (NaBH₄ or H₂/Pt) → D-Xylitol

22.29

D-Glucose → (Br₂ / H₂O) → (pyridine, epimerization) →

−H₂O ⇌ (Na-Hg, pH 3-5) → D-Mannose

22.30 The conformation of D-idopyranose with four equatorial —OH groups and an axial —CH$_2$OH group is more stable than the one with four axial —OH groups and an equatorial —CH$_2$OH group.

More stable   Less stable

4 Equatorial —OH groups   4 Axial —OH groups
1 Axial —CH$_2$OH group   1 Equatorial —CH$_2$OH group

Structure Elucidation

22.31 (a) The anhydro sugar is formed when the axial —CH$_2$OH group reacts with C1 to form a cyclic acetal.

β-D-Altropyranose

Anhydro sugar

Because the anhydro sugar is an acetal (i.e., an internal glycoside), it is a nonreducing sugar.

Methylation followed by acid hydrolysis converts the anhydro sugar to 2,3,4-tri-*O*-methyl-D-altrose:

Anhydro β-D-altropyranose   2,3,4-Tri-*O*-methyl-D-altrose

(b) Formation of an anhydro sugar requires that the monosaccharide adopt a chair conformation with the —CH$_2$OH group axial. With β-D-altropyranose this requires that two —OH groups be axial as well. With β-D-glucopyranose, however, it requires that all four —OH groups become axial and thus that the molecule adopt a very unstable conformation:

β-D-Glucopyranose

Highly unstable conformation

HA (−H$_2$O)

Anhydro-β-D-glucopyranose

22.32 1. The molecular formula and the results of acid hydrolysis show that lactose is a disaccharide composed of D-glucose and D-galactose. The fact that lactose is hydrolyzed by a *β-galactosidase* indicates that galactose is present as a glycoside and that the glycosidic linkage is beta to the galactose ring.

2. That lactose is a reducing sugar, forms a phenylosazone, and undergoes mutarotation indicates that one ring (presumably that of D-glucose) is present as a hemiacetal and thus is capable of existing to a limited extent as an aldehyde.

3. This experiment confirms that the D-glucose unit is present as a cyclic hemiacetal and that the D-galactose unit is present as a cyclic glycoside.

4. That 2,3,4,6-tetra-*O*-methyl-D-galactose is obtained in this experiment indicates (by virtue of the free —OH at C5) that the galactose ring of lactose is present as a pyranoside. That the methylated gluconic acid obtained from this experiment has a free —OH group at C4 indicates that the C4 oxygen atom of the glucose unit is connected in a glycosidic linkage to the galactose unit.

Now only the size of the glucose ring remains in question, and the answer to this is provided by experiment 5.

5. That methylation of lactose and subsequent hydrolysis gives 2,3,6-tri-*O*-methyl-D-glucose—that it gives a methylated glucose derivative with a free —OH at C4 and C5 demonstrates that the glucose ring is present as a pyranose. (We know already that the oxygen at C4 is connected in a glycosidic linkage to the galactose unit; thus, a free —OH at C5 indicates that the C5 oxygen atom is a part of the hemiacetal group of the glucose unit and that the ring is six membered.)

22.33

6-*O*-(α-D-Galactopyranosyl)-D-glucopyranose

We arrive at this conclusion from the data given:

1. That melibiose is a reducing sugar and that it undergoes mutarotation and forms a phenylosazone indicate that one monosaccharide is present as a cyclic hemiacetal.

2. That acid hydrolysis gives D-galactose and D-glucose indicates that melibiose is a disaccharide composed of one D-galactose unit and one D-glucose unit. That melibiose is hydrolyzed by an α-galactosidase suggests that melibiose is an α-D-galactosyl-D-glucose.

3. Oxidation of melibiose to melibionic acid and subsequent hydrolysis to give D-galactose and D-gluconic acid confirms that the glucose unit is present as a cyclic hemiacetal and that the galactose unit is present as a glycoside. (Had the reverse been true, this experiment would have yielded D-glucose and D-galactonic acid.)

Methylation and hydrolysis of melibionic acid produces 2,3,4,6-tetra-*O*-methyl-D-galactose and 2,3,4,5-tetra-*O*-methyl-D-gluconic acid. Formation of the first product—a galactose derivative with a free —OH at C5—demonstrates that the galactose ring is six membered; formation of the second product—a gluconic acid derivative with a free —OH at C6—demonstrates that the oxygen at C6 of the glucose unit is joined in a glycosidic linkage to the galactose unit.

4. That methylation and hydrolysis of melibiose gives a glucose derivative (2,3,4-tri-*O*-methyl-D-glucose) with free —OH groups at C5 and C6 shows that the glucose ring is also six membered. Melibiose is, therefore, 6-*O*-(α-D-galactopyranosyl-D-glucopyranose.

22.34 Trehalose has the following structure:

α-D-Glucopyranosyl-α-D-glucopyranoside

We arrive at this structure in the following way:

1. Acid hydrolysis shows that trehalose is a disaccharide consisting only of D-glucose units.

2. Hydrolysis by α-glucosidases and not by β-glucosidases shows that the glycosidic linkages are alpha.

3. That trehalose is a nonreducing sugar, that it does not form a phenylosazone, and that it does not react with bromine water indicate that no hemiacetal groups are present. This means that C1 of one glucose unit and C1 of the other must be joined in a glycosidic linkage. Fact 2 (just cited) indicates that this linkage is alpha to each ring.

4. That methylation of trehalose followed by hydrolysis yields only 2,3,4,6-tetra-*O*-methyl-D-glucose demonstrates that both rings are six membered.

22.35 (a) Tollens' reagent or Benedict's reagent will give a positive test with D-glucose but will give no reaction with D-glucitol.

(b) D-Glucaric acid will give an acidic aqueous solution that can be detected with blue litmus paper. D-Glucitol will give a neutral aqueous solution.

(c) D-Glucose will be oxidized by bromine water and the red brown color of bromine will disappear. D-Fructose will not be oxidized by bromine water since it does not contain an aldehyde group.

(d) Nitric acid oxidation will produce an *optically active* aldaric acid from D-glucose but an *optically inactive* aldaric acid will result from D-galactose.

(e) Maltose is a reducing sugar and will give a positive test with Tollens' or Benedict's solution. Sucrose is a nonreducing sugar and will not react.

(f) Maltose will give a positive Tollens' or Benedict's test; maltonic acid will not.

(g) 2,3,4,6-Tetra-*O*-methyl-β-D-glucopyranose will give a positive test with Tollens' or Benedict's solution; methyl β-D-glucopyranoside will not.

(h) Periodic acid will react with methyl α-D-ribofuranoside because it has hydroxyl groups on adjacent carbons. Methyl 2-deoxy-α-D-ribofuranoside will not react.

22.36 That the Schardinger dextrins are nonreducing shows that they have no free aldehyde or hemiacetal groups. This lack of reaction strongly suggests the presence of a *cyclic* structure. That methylation and subsequent hydrolysis yields only 2,3,6-tri-*O*-methyl-D-glucose indicates that the glycosidic linkages all involve C1 of one glucose unit and C4 of the next. That α-glucosidases cause hydrolysis of the glycosidic linkages indicates that they are α-glycosidic linkages. Thus, we are led to the following general structure.

$n = 3, 4,$ or 5

Note: Schardinger dextrins are extremely interesting compounds. They are able to form complexes with a wide variety of compounds by incorporating these compounds in the cavity in the middle of the cyclic dextrin structure. Complex formation takes place, however, only when the cyclic dextrin and the guest molecule are the right size. Anthracene molecules, for example, will fit into the cavity of a cyclic dextrin with eight glucose units but will not fit into one with seven. For more information about these fascinating compounds, see Bergeron, R. J., "Cycloamyloses," *J. Chem. Educ.* **1977,** *54,* 204–207.

22.37 Isomaltose has the following structure:

6-*O*-(α-D-Glucopyranosyl)-D-glucopyranose

(1) The acid and enzymic hydrolysis experiments tell us that isomaltose has two glucose units linked by an α linkage.

(2) That isomaltose is a reducing sugar indicates that one glucose unit is present as a cyclic hemiacetal.

(3) Methylation of isomaltonic acid followed by hydrolysis gives us information about the size of the nonreducing pyranoside ring and about its point of attachment to the reducing ring. The formation of the first product (2,3,4,6-tetra-*O*-methyl-D-glucose)—a compound with an —OH at C5—tells us that the nonreducing ring is present as a pyranoside. The formation of 2,3,4,5-tetra-*O*-methyl-D-gluconic acid—a compound with an —OH at C6— shows that the nonreducing ring is linked to C6 of the reducing ring.

(4) Methylation of maltose itself tells the size of the reducing ring. That 2,3,4-tri-*O*-methyl-D-glucose is formed shows that the reducing ring is also six membered; we know this because of the free —OH at C5.

22.38 Stachyose has the following structure:

The enzymatic hydrolyses (as just indicated) give the basic structure of stachyose and raffinose. The only remaining question is the ring size of the first galactose unit of stachyose. That methylation of stachyose and subsequent hydrolysis yields 2,3,4,6-tetra-*O*-methyl-D-galactose establishes that this ring is a pyranoside.

22.39 Arbutin has the following structure:

p-Hydroxyphenyl-β-D-glucopyranoside

Compounds **X**, **Y**, and **Z** are hydroquinone, *p*-methoxyphenol, and *p*-dimethoxybenzene, respectively.

X
Hydroquinone

(*a*) Singlet δ 7.9 [2H]
(*b*) Singlet δ 6.8 [4H]

Y
p-Methoxyphenol

(*a*) Singlet δ 4.8 [1H]
(*b*) Multiplet δ 6.8 [4H]
(*c*) Singlet δ 3.9 [3H]

Z
p-Dimethoxybenzene

(*a*) Singlet δ 3.75 [6H]
(*b*) Singlet δ 6.8 [4H]

The reactions that take place are the following:

D-Glucose   Hydroquinone **X**

2,3,4,6-Tetra-*O*-methyl-D-glucose   *p*-Methoxyphenol **Y**

p-Dimethoxybenzene **Z**

22.40 Aldotetrose **B** must be D-threose because the alditol derived from it (D-threitol) is optically active (the alditol from D-erythrose, the other possible D-aldotetrose, would be meso). Due to rotational symmetry, however, the alditol from **B** (D-threitol) would produce only two ^{13}C NMR signals. Compounds **A-F** are thus in the family of aldoses stemming from D-threose. Since reduction of aldopentose **A** produces an optically inactive alditol, **A** must be D-xylose. The two diastereomeric aldohexoses **C** and **D** produced from **A** by a Kiliani-Fischer synthesis must therefore be D-idose and D-gulose, respectively. **E** and **F** are the alditols derived from **C** and **D**, respectively. Alditol **E** would produce only three ^{13}C NMR signals due to rotational symmetry while **F** would produce six signals.

22.41 There are four closely spaced upfield alkyl signals in the ^{13}C NMR spectrum (δ 26.5, δ 25.6, δ 24.9, δ 24.2), corresponding to the four methyls of the two acetonide protecting groups. (The compound is, therefore, the 1,2,5,6-bis-acetonide of mannofuranose, below.)

22.42 The final product is the acetonide of glyceraldehyde (below); two molar equivalents are formed from each molar equivalent of the 1,2,5,6-bis-acetonide of mannitol.

Challenge Problems

22.43 The β-anomer can hydrogen bond intramolecularly, as in:

In contrast, the α-anomer can only hydrogen bond intermolecularly, leading to a higher boiling point.

22.44

II

$\xrightarrow[\text{H}_2\text{O}]{\text{H}_3\text{O}^+}$

[III]

$\xrightarrow[\text{H}_2\text{O}]{\text{HO:}^-}$

IV

22.45 (a) The proton at C1. (b) Because of the single neighboring hydrogen (at C2). (c and d).

β-anomer
V

α-anomer
VI

QUIZ

22.1 Supply the appropriate structural formula or complete the partial formula for each of the following:

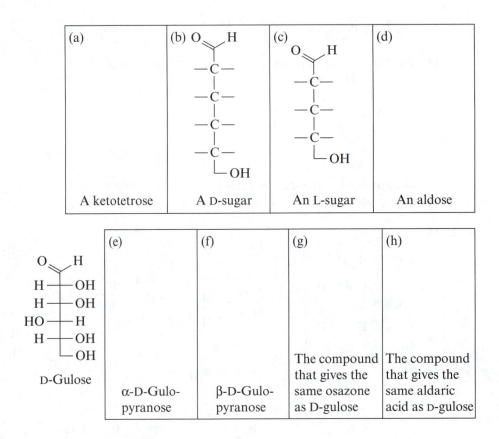

22.2 Which of the following monosaccharides yields an optically inactive alditol on $NaBH_4$ reduction?

Answer: ☐

22.3   Give the structural formula of the monosaccharide that you could use as starting material in the Kiliani-Fischer synthesis of the following compound:

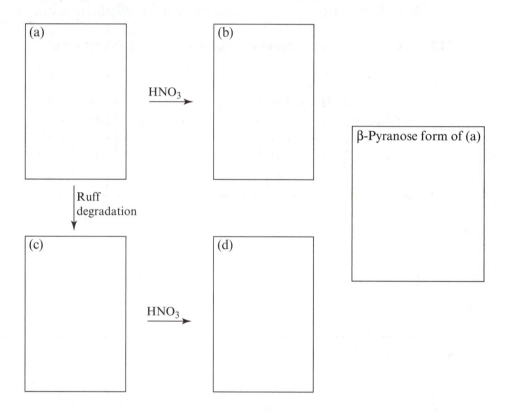

22.4   The D-aldopentose, (a), is oxidized to an aldaric acid, (b), which is optically active. Compound (a) undergoes a Ruff degradation to form an aldotetrose, (c), which undergoes oxidation to an optically inactive aldaric acid, (d). What are the structural formulas of (a), (b), (c), and (d)?

22.5   Give the structural formula of the β-pyranose form of (a) in the space just given.

22.6   Complete the following skeletal formulas and statements by filling in the blanks and circling the words that make the statements true.

The Haworth and conformational formulas of the β-cyclic hemiacetal

of

```
        O    H
         \\ /
          C
   HO ——— H
   HO ——— H
   H  ——— OH
   H  ——— OH
          OH
```

D-Mannose

are

(a) [structure] and (b) [structure]

(Fill in groups as appropriate)

This cyclic hemiacetal is (c) reducing, nonreducing; on reaction with Br_2/H_2O it gives an optically (d) active, inactive (e) aldaric, aldonic acid. On reaction with dilute HNO_3 it gives an optically (f) active, inactive (g) aldaric, aldonic acid. Reaction of the cyclic hemiacetal with (h) [   ] converts it into an optically (i) active, inactive alditol.

22.7 Outline chemical tests that would allow you to distinguish between:

(a) Glucose

```
     O    H
      \\ /
       C
  H  ——— OH
  HO ——— H
  H  ——— OH
  H  ——— OH
         OH
```

and galactose

```
     O    H
      \\ /
       C
  H  ——— OH
  HO ——— H
  HO ——— H
  H  ——— OH
         OH
```

(b) Glucose and fructose
$$\left(\begin{array}{c} \text{OH} \\ \text{O} \\ \text{HO} - \text{H} \\ \text{H} - \text{OH} \\ \text{H} - \text{OH} \\ \text{OH} \end{array}\right)$$

22.8 Hydrolysis of (+)-sucrose (ordinary table sugar) yields
(a) D-glucose
(b) D-mannose
(c) D-fructose
(d) D-galactose
(e) More than one of the above.

22.9 Select the reagent needed to perform the following transformation:

(a) CH_3OH, KOH   (b) $(CH_3CO)_2O$ (c) $(CH_3)_2SO_4$, HO^-

(d) CH_3OH, HCl   (e) CH_3OCH_3, HCl

23 LIPIDS

SOLUTIONS TO PROBLEMS

23.1 (a) There are two sets of enantiomers, giving a total of four stereoisomers:

erythro   *threo*

(±)-*threo*-9,10-Dibromohexadecanoic acids

Formation of a bromonium ion at the other face of palmitoleic acid gives a result such that the *threo* enantiomers are the only products formed (obtained as a racemic mixture).

The designations *erythro* and *threo* come from the names of the sugars called *erythrose* and *threose* (Section 22.9A).

23.2 (a,b)

Zingiberene
(a sesquiterpene)

β-Selinene
(a sesquiterpene)

Caryophyllene
(a sesquiterpene)

Squalene
(a triterpene)

23.3 (a)

Myrcene

(b)

Limonene

(c) α-Farnesene
(see Section 23.3) $\xrightarrow[\text{(2) Me}_2\text{S}]{\text{(1) O}_3}$

(d) Geraniol
(see Section 23.3) $\xrightarrow[\text{(2) Me}_2\text{S}]{\text{(1) O}_3}$

(e) Squalene
(see Section 23.3) $\xrightarrow[\text{(2) Me}_2\text{S}]{\text{(1) O}_3}$

23.4 (a)

+ CO₂

(+ further oxidation
products)

(b)

(c)

(+ rearranged products)

(d)

23.5 Br₂ or KMnO₄ at room temperature. Either reagent would give a positive result with geraniol and a negative result with menthol.

23.6 (a) (b)

23.7

23.8

5α Series

5β Series

23.9 (a)

3α-Hydroxy-5α-androstan-17-one
(androsterone)

(b)

17α-Ethynyl-17β-hydroxy-5(10)-estren-3-one
(norethynodrel)

23.10

Absolute configuration of cholesterol
(5-cholesten-3β-ol)

23.11 Estrone and estradiol are *phenols* and thus are soluble in aqueous sodium hydroxide. Extraction with aqueous sodium hydroxide separates the estrogens from the androgens.

23.12 (a)

Cholesterol   →(Br₂)   5α,6β-Dibromocholestan-3β-ol

(b)

5α,6α-Epoxycholestan-3β-ol
(prepared by epoxidation
of cholesterol; cf. Section 23.4G)   →(H₃O⁺/H₂O)   Cholestan-3β,5α,6β-triol

(c)

5α-Cholestan-3β-ol
(prepared by hydrogenation
of cholesterol; cf. Section 23.4G)   →(H₂CrO₄/acetone)   5α-Cholestan-3-one

(d)

Cholesterol

BH$_3$:THF
(cf. Sect. 23.4G)

6α-Deuterio-5α-cholestan-3β-ol

(e)

5α,6α-Epoxycholestan-3β-ol
(cf. (b) above)

HBr

6β-Bromocholestan-3β,5α-diol

23.13 (a)

(b)

(c)

23.14 (a)

$$\text{CH}_3(\text{CH}_2)_{16}\text{COOH} + \text{CH}_3\text{CH}_2\text{OH} \underset{}{\overset{\text{HA}}{\rightleftharpoons}} \text{CH}_3(\text{CH}_2)_{16}\text{COOCH}_2\text{CH}_3 + \text{H}_2\text{O}$$

$$\text{CH}_3(\text{CH}_2)_{16}\text{COOH} \xrightarrow{\text{SOCl}_2} \text{CH}_3(\text{CH}_2)_{16}\text{COCl} \xrightarrow{\text{CH}_3\text{CH}_2\text{OH}} \text{CH}_3(\text{CH}_2)_{16}\text{COOCH}_2\text{CH}_3 + \text{HCl}$$

(b)

$$\text{CH}_3(\text{CH}_2)_{16}\text{COCl} \xrightarrow{(\text{CH}_3)_3\text{COH}} \text{CH}_3(\text{CH}_2)_{16}\text{COOC}(\text{CH}_3)_3 + \text{HCl}$$

(from a)

(c)

$$\text{CH}_3(\text{CH}_2)_{16}\text{COCl} \xrightarrow[\text{(excess)}]{\text{NH}_3} \text{CH}_3(\text{CH}_2)_{16}\text{CONH}_2 + \text{NH}_4\text{Cl}$$

(d)

$$\text{CH}_3(\text{CH}_2)_{16}\text{COCl} \xrightarrow[\text{(excess)}]{(\text{CH}_3)_2\text{NH}} \text{CH}_3(\text{CH}_2)_{16}\text{CON(CH}_3)_2 + (\text{CH}_3)_2\overset{+}{\text{NH}_2}\ \text{Cl}^-$$

(e)

$$\text{CH}_3(\text{CH}_2)_{16}\text{CONH}_2 \xrightarrow[\text{(2) H}_2\text{O}]{\text{(1) LiAlH}_4,\ \text{Et}_2\text{O}} \text{CH}_3(\text{CH}_2)_{16}\text{CH}_2\text{NH}_2$$

(from c)

(f)

$$\text{CH}_3(\text{CH}_2)_{16}\text{CONH}_2 \xrightarrow[\text{HO}^-]{\text{Br}_2} \text{CH}_3(\text{CH}_2)_{15}\text{CH}_2\text{NH}_2$$

(g)

$$\text{CH}_3(\text{CH}_2)_{16}\text{COCl} \xrightarrow[\text{Et}_2\text{O},\ -78\ °\text{C}]{\text{LiAlH}(t\text{-BuO})_3} \text{CH}_3(\text{CH}_2)_{16}\text{CHO}$$

(from a)

(h)

$$\text{CH}_3(\text{CH}_2)_{16}\text{COOCH}_2\text{CH}_3 \xrightarrow[\text{(2) H}_2\text{O}]{\text{(1) LiAlH}_4,\ \text{Et}_2\text{O}} \text{HOCH}_2\text{CH}_3\ +\ \text{CH}_3(\text{CH}_2)_{16}\text{CH}_2\text{OH}$$

(from a)

$$\text{CH}_3(\text{CH}_2)_{16}\text{COCl} \longrightarrow \text{CH}_3(\text{CH}_2)_{16}\text{COOCH}_2(\text{CH}_2)_{16}\text{CH}_3$$

(i)

(j) (from g)

(k) (from h)

(l) (from k)

23.15 (a)

(b) (from a)

(c) (from a)

(d) (from a)

23.16 (a)

(b)

(c)

(d)

23.17 Elaidic acid is *trans*-9-octadecenoic acid (*trans*-octadec-9-enoic acid):

It is formed by the isomerization of oleic acid.

23.18 A reverse Diels-Alder reaction takes place.

23.19 (a)

(b) Infrared spectroscopy
(c) A peak in the 675–730 cm^{-1} region would indicate that the double bond is cis; a peak in the 960–975 cm^{-1} region would indicate that it is trans.

23.20

CH$_3$

α-Phellandrene

CH$_2$

β-Phellandrene

Note: On permanganate oxidation, the =CH$_2$ group of β-phellandrene is converted to CO$_2$ and thus is not detected in the reaction.

Roadmap Syntheses

23.21

Vaccenic acid

23.22

23.23

5α-Cholest-2-ene

A

B

Here we find that epoxidation takes place at the less hindered α face (cf. Section 23.4G). Ring opening by HBr takes place in an anti fashion to give a product with diaxial substituents.

23.24 (a)

(b) BuLi, Et₂O

(c)

(d) NC

(e) Michael addition using a basic catalyst.

23.25 First: an elimination takes place,

Then a conjugate addition occurs, followed by an aldol addition:

A cyperone

Challenge Problems

23.26

Pahutoxin

23.27   (a, b) The reaction is an intramolecular transesterification.

QUIZ

23.1   Write an appropriate formula in each box.

(a)	(b)
A naturally occurring fatty acid	A soap

(c)	(d)
A solid fat	An oil

(e)	(f)
A synthetic detergent	5α-Estran-17-one

23.2 Give a reagent that would distinguish between each of the following:

(a) Pregnane and 20-pregnanone

(b) Stearic acid and oleic acid

(c) 17α-Ethynyl-1,3,5(10)-estratriene-3,17β-diol (ethynylestradiol) and 1,3,5(10)-estratriene-3,17β-diol (estradiol)

23.3 What product would be obtained by catalytic hydrogenation of 4-androstene?

23.4 Supply the missing compounds:

23.5 The following compound is a:

(a) Monoterpene   (b) Sesquiterpene   (c) Diterpene

(d) Triterpene   (e) Tetraterpene

23.6 Mark off the isoprene units in the previous compound.

23.7 Which is a systematic name for the steroid shown here?
(a) 5α-Androstan-3α-ol
(b) 5β-Androstan-3β-ol
(c) 5α-Pregnan-3α-ol
(d) 5β-Pregnan-3β-ol
(e) 5α-Estran-3α-ol

24 AMINO ACIDS AND PROTEINS

SOLUTIONS TO PROBLEMS

24.1

(a)

(b)

(c)

predominates at the isoelectric point rather than

because of the acid-strengthening inductive effect of the α-aminium group.

(d) Since glutamic acid is a dicarboxylic acid, acid must be added (i.e., the pH must be made lower) to suppress the ionization of the second carboxyl group and thus achieve the isoelectric point. Glutamine, with only one carboxyl group, is similar to glycine or phenylalanine and has its isoelectric point at a higher pH.

24.2 The conjugate acid is highly stabilized by resonance.

24.3 (a)

DL-Leucine

$+ CO_2$

(b)

DL-Alanine

$+ CO_2 +$

(c)

NaOEt / C_6H_5—Br

$\xrightarrow[\text{heat}]{\text{NaOH}}$

$\xrightarrow[\text{heat}]{\text{HCl}}$

DL-Phenylalanine

+ CO_2

+

24.4 (a)

Phenylacetaldehyde

$\xrightarrow[\text{HCN}]{\text{NH}_3}$

$\xrightarrow{H_3O^+}$

DL-Phenylalanine

(b) CH_3SH +

$\xrightarrow{\text{base}}$ CH_3S

$\xrightarrow[\text{HCN}]{\text{NH}_3}$

$\xrightarrow{H_3O^+}$ CH_3S

DL-Methionine

24.5 Because of the presence of an electron-withdrawing 2,4-dinitrophenyl group, the labeled amino acid is relatively nonbasic and is, therefore, insoluble in dilute aqueous acid. The other amino acids (those that are not labeled) dissolve in dilute aqueous acid.

24.6 (a)

VAG

Labeled valine
(separate and identify)   Alanine   Glycine

(b)

Labeled valine   ε-Labeled lysine

+

Glycine

24.7

Phenyl isothiocyanate

MIR

Phenylthiohydantoin
derived from methionine

IR

Phenylthiohydantoin
derived from isoleucine

R

24.8 (a) Two structures are possible with the sequence ECG. Glutamic acid may be linked to cysteine through its α-carboxyl group.

or through its γ-carboxyl group,

(b) This result shows that the second structure is correct, that in glutathione the γ-carboxyl group is linked to cysteine.

24.9 We look for points of overlap to determine the amino acid sequence in each case.

(a)
```
S T
  T O
P S
-------
P S T O
```

(b)
```
A C
  C R
    R V
L A
---------
L A C R V
```

24.10 Sodium in liquid ammonia brings about reductive cleavage of the disulfide linkage of oxytocin to two thiol groups; then air oxidizes the two thiol groups back to a disulfide linkage:

24.11 Removal of the Fmoc group initially involves an elimination reaction promoted by piperidine to form the carbamic acid derivative of the amino acid and 9-methylidenefluorene, which reacts further with piperidine by nucleophilic addition to form the byproduct. Spontaneous decarboxylation of the carbamic acid generates CO_2 and the free amino acid.

24.12

Glycine + Di-*tert*-butyl carbonate $\xrightarrow{HO^-}$

Boc-G $\xrightarrow[\substack{(2)}]{(1)\ Et_3N}$ (ClCO$_2$Et)

Mixed anhydride $\xrightarrow[\substack{-CO_2,\ -EtOH}]{Valine}$

Boc-GV $\xrightarrow[\substack{(2)}]{(1)\ Et_3N}$ (ClCO$_2$Et)

Mixed anhydride $\xrightarrow{Alanine}$

Boc-GVA $\xrightarrow[\substack{25\ °C}]{CF_3CO_2H,\ CH_3CO_2H}$

$\ce{(CH2=C(CH3)2)}$ + CO_2 + GVA

24.13 (a) $2\ C_6H_5$ ... O=C-Cl + H_2N ... Lysine ... NH_2 $\xrightarrow[25\,°C]{HO^-}$

Benzyl chloro-
formate

Lysine

C_6H_5 ... $\xrightarrow{(1)\ Et_3N}$ (2) Cl–C(=O)–OEt

C_6H_5 ... $H_3\overset{+}{N}$ ... O^- $\xrightarrow{-CO_2,\ -EtOH}$

C_6H_5 ... $\xrightarrow[\text{cold}]{HBr,\ AcOH}$

$2\ C_6H_5\text{–}CH_2\text{–}Br\ +\ 2\ CO_2\ +\ H_3\overset{+}{N}$ ... NH_2

KI

24.14 The weakness of the benzyl-oxygen bond allows these groups to be removed by catalytic hydrogenolysis.

24.15 Trifluoroacetic acid protonates the carbonyl group of the ester linkage joining the resin to the peptide. Heterolysis of the ester linkage then yields the relatively stable benzyl-type carbocation at the point of attachment. Acid hydrolysis of the amide linkages (peptide bonds) requires more stringent conditions because the fragments produced are not similarly stabilized.

24.16

1. Add Ala-Fmoc.

Diisopropyl-
carbodiimide

2. Purify by washing.

3. Remove protecting group.

Piperidine in DMF

4. Purify by washing.

5. Add Phe-Fmoc.

and
Diisopropylcarbodiimide

6. Purify by washing.

Piperidine in DMF

7. Remove protecting group.

8. Purify by washing.

9. Add protected Lys.
(Side-chain amino
group protected
by Boc.)

and
Diisopropylcarbodiimide

10. Purify by washing.

Piperidine in DMF

11. Remove Fmoc
protecting group.

12. Purify by washing.

CF₃ OH

13. Detach tripeptide and
remove Lys side-chain
Boc group.

14. Isolate product.

KFA

Problems

Structure and Reactivity

24.17 (a) Isoleucine, threonine, hydroxyproline, and cystine.

(b)

$$
\begin{array}{c}
CO_2^- \\
H_3\overset{+}{N}\!\!-\!\!H \\
CH_3\!\!-\!\!H \\
CH_2 \\
CH_3
\end{array}
\quad\text{and}\quad
\begin{array}{c}
CO_2^- \\
H_3\overset{+}{N}\!\!-\!\!H \\
H\!\!-\!\!CH_3 \\
CH_2 \\
CH_3
\end{array}
$$

$$
\begin{array}{c}
CO_2^- \\
H_3\overset{+}{N}\!\!-\!\!H \\
H\!\!-\!\!OH \\
CH_3
\end{array}
\quad\text{and}\quad
\begin{array}{c}
CO_2^- \\
H_3\overset{+}{N}\!\!-\!\!H \\
HO\!\!-\!\!H \\
CH_3
\end{array}
$$

$$
\begin{array}{c}
CO_2^- \\
H_2\overset{+}{N}\!\!-\!\!H \\
\\
H\!\cdots\!OH
\end{array}
\quad\text{and}\quad
\begin{array}{c}
CO_2^- \\
H_2\overset{+}{N}\!\!-\!\!H \\
\\
HO\!\cdots\!H
\end{array}
$$

(With cystine, both chirality centers are α-carbon atoms; thus, according to the problem, both must have the L-configuration, and no isomers of this type can be written.)
(c) Diastereomers

24.18 (a)

(b)

(c) C_6H_5

24.19 (a)

$$\underset{\text{(−)-Serine}}{\overset{\displaystyle CO_2^-}{\underset{\displaystyle CH_2OH}{H_3\overset{+}{N}-\!\!\!-\!\!\!-H}}} \xrightarrow[\text{CH}_3\text{OH}]{\text{HCl}} \underset{\substack{\textbf{A} \\ (C_4H_{10}ClNO_3)}}{\overset{\displaystyle CO_2CH_3}{\underset{\displaystyle CH_2OH}{H_3\overset{+}{N}-\!\!\!-\!\!\!-H}}}\; Cl^- \xrightarrow{\text{PCl}_5} \underset{\substack{\textbf{B} \\ (C_4H_9Cl_2NO_2)}}{\overset{\displaystyle CO_2CH_3}{\underset{\displaystyle CH_2Cl}{H_3\overset{+}{N}-\!\!\!-\!\!\!-H}}}\; Cl^-$$

$$\xrightarrow[\text{(2) HO}^-]{\text{(1) H}_3\text{O}^+,\text{ H}_2\text{O, heat}} \underset{\substack{\textbf{C} \\ (C_3H_6ClNO_2)}}{\overset{\displaystyle CO_2^-}{\underset{\displaystyle CH_2Cl}{H_3\overset{+}{N}-\!\!\!-\!\!\!-H}}} \xrightarrow[\text{dil H}_3\text{O}^+]{\text{Na-Hg}} \underset{\text{L-(+)-Alanine}}{\overset{\displaystyle CO_2^-}{\underset{\displaystyle CH_3}{H_3\overset{+}{N}-\!\!\!-\!\!\!-H}}}$$

(b) **B** $\xrightarrow{\text{HO}^-}$ $\underset{\substack{\textbf{D} \\ (C_4H_8ClNO_2)}}{\overset{\displaystyle CO_2CH_3}{\underset{\displaystyle CH_2Cl}{H_2N-\!\!\!-\!\!\!-H}}}$ $\xrightarrow{\text{NaSH}}$ $\underset{\substack{\textbf{E} \\ (C_4H_9NO_2S)}}{\overset{\displaystyle CO_2CH_3}{\underset{\displaystyle CH_2SH}{H_2N-\!\!\!-\!\!\!-H}}}$

$$\xrightarrow[\text{(2) HO}^-]{\text{(1) H}_3\text{O}^+,\text{ H}_2\text{O, heat}} \underset{\text{L-(+)-Cysteine}}{\overset{\displaystyle CO_2^-}{\underset{\displaystyle CH_2SH}{H_3\overset{+}{N}-\!\!\!-\!\!\!-H}}}$$

(c) $\underset{\text{L-(−)-Asparagine}}{\overset{\displaystyle CO_2^-}{\underset{\displaystyle \underset{\displaystyle \overset{\|}{O}}{CH_2CNH_2}}{H_3\overset{+}{N}-\!\!\!-\!\!\!-H}}}$ $\xrightarrow[\substack{\text{Hofmann} \\ \text{rearrangement}}]{\text{NaOBr, HO}^-}$ $\underset{\substack{\textbf{F} \\ (C_3H_7N_2O_2)}}{\overset{\displaystyle CO_2^-}{\underset{\displaystyle CH_2NH_2}{H_2N-\!\!\!-\!\!\!-H}}}$ $\xleftarrow{\text{NH}_3}$ $\underset{\substack{\textbf{C} \\ \text{[from part (a)]}}}{\overset{\displaystyle CO_2^-}{\underset{\displaystyle CH_2Cl}{H_3\overset{+}{N}-\!\!\!-\!\!\!-H}}}$

24.20 (a)

$$\xrightarrow[\substack{\text{reflux 6 h} \\ \text{(66\% yield)}}]{\text{concd HCl}} \underset{\text{DL-Glutamic acid}}{\overset{}{\underset{\displaystyle \overset{+}{N}H_3}{HO-\!\!C(=O)-CH_2-CH_2-CH-C(=O)-O^-}}} + \; CH_3C(=O)OH \; + \; 2\; CH_3CH_2OH \; + \; ^+NH_4 \; + \; CO_2$$

(b)

24.21 At pH 2–3 the γ-carboxyl groups of polyglutamic acid are uncharged. (They are present as –CO$_2$H groups.) At pH 5 the γ-carboxyl groups ionize and become negatively charged. (They become γ-CO$_2^-$ groups.) The repulsive forces between these negatively charged groups cause an unwinding of the α helix and the formation of the random coil.

Peptide Sequencing

24.22 We look for points of overlap:

```
           F S
         P G F
       P P     S P F
     R P             F R
   R P P G F S P F R
```

Bradykinin

24.23 1. This experiment shows that valine is the N-terminal amino acid and that valine is attached to leucine. (Lysine labeled at the ε-amino group is to be expected if lysine is not the N-terminal amino acid and if it is linked in the polypeptide through its α-amino group.)

2. This experiment shows that alanine is the C-terminal amino acid and that it is linked to glutamic acid.

At this point, then, we have the following information about the structure of the heptapeptide.

```
V L (A, K, F) E A
```

Sequence
unknown

3. (a) This experiment shows that the dipeptide, **A**, is: L K

(b) The carboxypeptidase reaction shows that the C-terminal amino acid of the tripeptide, **B**, is glutamic acid; the DNP labeling experiment shows that the N-terminal amino acid is phenylalanine. Thus, the tripeptide **B** is : F A E

Putting these pieces together in the only way possible, we arrive at the following amino acid sequence for the heptapeptide.

```
V L
  L K
    F A E
        E A
_____
V L K F A E A
```

Challenge Problem

24.24 The observation that the ^{1}H NMR spectrum taken at room temperature shows two different signals for the methyl groups suggests that they are in different environments. This would be true if rotation about the carbon-nitrogen bond was not taking place.

δ 8.05   H   CH$_3$ δ 2.95

C — N

O   CH$_3$ δ 2.80

We assign the δ 2.80 signal to the methyl group that is on the same side as the electronegative oxygen atom.

The fact that the methyl signals appear as doublets (and that the formyl proton signal is a multiplet) indicates that long-range coupling is taking place between the methyl protons and the formyl proton.

That the two doublets are not simply the result of spin-spin coupling is indicated by the observation that the distance that separates one doublet from the other changes when the applied magnetic field strength is lowered. [*Remember!* The magnitude of a chemical shift is proportional to the strength of the applied magnetic field, while the magnitude of a coupling constant is not.]

That raising the temperature (to 111 °C) causes the doublets to coalesce into a single signal indicates that at higher temperatures the molecules have enough energy to surmount the energy barrier of the carbon-nitrogen bond. Above 111 °C, rotation is taking place so rapidly that the spectrometer is unable to discriminate between the two methyl groups.

QUIZ

24.1 Write the structural formula of the principal ionic species present in aqueous solutions at pH 2, 7, and 12 of isoleucine (2-amino-3-methylpentanoic acid).

At pH = 2

(a)

At pH = 7

(b)

At pH = 12

(c)

24.2 A hexapeptide gave the following products:

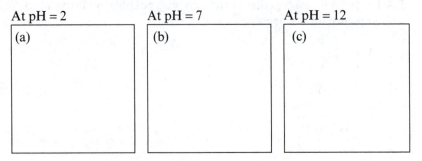

= Proline (P)

Hexapeptide $\xrightarrow{\text{3}N \text{ HCl, 100 °C}}$ 2 G, 1 L, 1 F, 1 P, 1 Y

Hexapeptide $\xrightarrow{\text{1}N \text{ HCl, 80 °C}}$ FGY + GFG + PLG + LGF

The structure of the hexapeptide (using abbreviations such as G, L etc.) is

25 NUCLEIC ACIDS AND PROTEIN SYNTHESIS

SOLUTIONS TO PROBLEMS

25.1 Adenine:

Guanine:

Cytosine:

Thymine (R = CH₃) or Uracil (R = H):

25.2 (a) The nucleosides have an *N*-glycosidic linkage that (like an *O*-glycosidic linkage) is rapidly hydrolyzed by aqueous acid but is one that is stable in aqueous base.

25.3

25.4 (a) The isopropylidene group is part of a cyclic acetal and is thus susceptible to hydrolysis by mild acid.

(b) It can be installed by treating the nucleoside with acetone and a trace of acid and by simultaneously removing the water that is produced.

25.5 (a) Cordycepin is

(3'-Deoxyadenosine)

(b)

25.6 (a) 3×10^9 base pairs $\times \dfrac{34\text{Å}}{10 \text{ base pairs}} \times \dfrac{10^{-10}\text{m}}{\text{Å}} \cong 1 \text{ m}$

(b) $6 \times 10^{-12} \dfrac{\text{g}}{\text{ovum}} \times 6.5 \times 10^9 \text{ ova} = 4 \times 10^{-2}\text{g}$

25.7 (a)

Lactim form   Thymine
of guanine

(b) Thymine would pair with adenine and thus adenine would be introduced into the complementary strand where guanine should occur.

25.8 (a) A diazonium salt and a heterocyclic analog of a phenol.

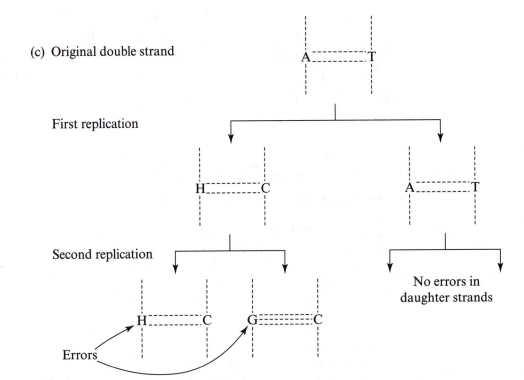

(b)

Hypoxanthine   Cytosine

(c) Original double strand

First replication

Second replication

Errors

No errors in daughter strands

25.9

Uracil
(in mRNA)

Adenine
(in DNA)

25.10 (a)

ACC	CCC	AAA	AUG	UCG	*m*RNA

(b)

T	P	K	M	S	Amino acids

(c)

UGG	GGC	UUU	UAC	AGC	Anticodons

25.11

R	·	I	·	C	·	Y	·	V	Amino acids

(a)

AGA	AUA	UGC	UGG	GUA	*m*RNA

(b)

TCT	TAT	ACG	ACC	CAT	DNA

(c)

UCU	UAU	ACG	ACC	CAU	Anticodons

Problems

Nucleic Acid Structure

25.12

25.13

Mechanisms

25.14 Ionization of the benzoyl group at the anomeric carbon (C1) is assisted by the carbonyl oxygen of the C2 benzoyl group, resulting in a stabilized cation blocked from attack on the α-face.

25.15

Mitomycin A

Leuco-aziridinomitosene

Protonation and ring opening

Resonance-stabilized cation intermediate

Alkylation by N^2 of a deoxyguanosine in DNA

Deprotonation :A⁻

A monoadduct with DNA

Tautomerization

1-Dihydromitosene A

Second alkylation by N^2 of another deoxyguanosine in DNA

A cross-linked adduct with DNA

25.16

A

SPECIAL TOPIC
^{13}C NMR Spectroscopy

A.1 Analysis of the molecular formula $C_5H_{10}O$ indicates an IHD = 1. Structural possibilities are: a C=C, a C=O, or a ring.

  The δ 211.0 signal must be due to a carbonyl group. The remaining C_4H_{10} is represented by only two signals in the alkyl group region, suggesting symmetry in the molecule and two unique carbons.

  Two CH_3CH_2 groups must be present, with the CH_3 giving rise to the signal at $\delta \sim 10$ and the CH_2 giving the signal at $\delta \sim 37$.

  Hence, the structure is

A.2 Qualifying structures are:

  and

  Each will give four ^{13}C NMR signals, one of which is due to the carbonyl group (δ 211.0).

A.3 (a) 4 signals   (b) 6 signals   (c) 6 signals

A.4 (a) 5 signals   (b) 7 signals   (c) 8 signals
   (d) 6 signals   (e) 4 signals

A.5 (a) 4 signals   (b) 6 signals

   (c) 4 signals

B.1 Splitting tree diagram for a quartet. The signal from the observed hydrogen (^{a}H) is split into four peaks of 1:3:3:1 intensity by three equivalent hydrogens (^{b}H). The same method of analysis we used for the triplet pattern applies here, wherein each successively lower level in the tree diagram represents splitting by another one of the coupled hydrogens. Again, because in this case J_{ab} is the same for splitting of the ^{a}H the signal by all three ^{b}H hydrogens, internal legs of the diagram overlap, and intensities amplify each time this occurs. The result is a pattern of 1:3:3:1 intensities, as we would observe for a quartet in an actual spectrum. The possible magnetic orientations shown under the tree diagram for the three ^{b}H hydrogens indicate that all three of the adjacent hydrogens may be aligned with the applied field, or two may be aligned with the field and one against (in three equal energy combinations), or two against and one with the field (in three equal energy combinations), or all three may be aligned against the applied field.

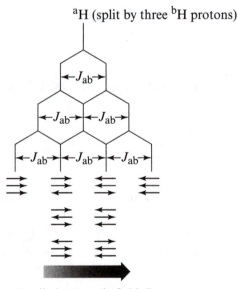

^{a}H (split by three ^{b}H protons)

Applied magnetic field, **B$_0$**

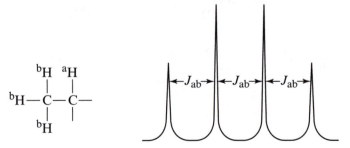

C

SPECIAL TOPIC
Chain-Growth Polymers

C.1 (a)

Atactic polystyrene
$(Ph = C_6H_5)$

Syndiotactic polystyrene
$(Ph = C_6H_5)$

Isotactic polystyrene
$(Ph = C_6H_5)$

(b) The solution of isotactic polystyrene would be optically active.

D

SPECIAL TOPIC
Electrocyclic and Cycloaddition Reactions

SOLUTIONS TO PROBLEMS

D.1 According to the Woodward-Hoffmann rule for electrocyclic reactions of $4n$ π-electron systems (Section D.2A), the photochemical cyclization of *cis, trans*-2,4-hexadiene should proceed with *disrotatory motion*. Thus, it should yield *trans*-3,4-dimethylcyclobutene:

cis, trans-2,4-Hexadiene   *trans*-3,4-Dimethylcyclobutene

D.2 (a)

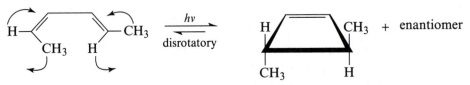

ψ_2 of a hexadiene
(Section D.2A)

(b) This is a thermal electrocyclic reaction of a $4n$ π-electron system; it should, *and does*, proceed with conrotatory motion.

D.3

trans, trans-2,4-Hexadiene   *cis*-3,4-Dimethylcyclobutene

cis, trans-2,4-Hexadiene

Here we find that two consecutive electrocyclic reactions (the first photochemical, the second thermal), provide a stereospecific synthesis of *cis, trans*-2,4-hexadiene from *trans, trans*-2,4-hexadiene.

D.4 (a) This is a photochemical electrocyclic reaction of an eight π-electron system—a $4n$ π-electron system where $n = 2$. It should, therefore, proceed with disrotatory motion.

cis-7,8-Dimethyl-1,3,5-cyclooctatriene

(b) This is a thermal electrocyclic reaction of the eight π-electron system. It should proceed with conrotatory motion.

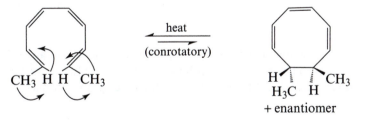

cis-7,8-Dimethyl-1,3,5-cyclooctatriene

D.5 (a) This is conrotatory motion, and since this is a 4n π-electron system (where $n = 1$) it should occur under the influence of heat.

$$H_3CO_2C \xrightarrow[\text{(conrotatory)}]{\text{heat}} H-\cdots CO_2CH_3$$

(b) This is conrotatory motion, and since this is also a 4n π-electron system (where $n = 2$) it should occur under the influence of heat.

$$\xrightarrow[\text{(conrotatory)}]{\text{heat}}$$

+ enantiomer

(c) This is disrotatory motion. This, too, is a 4n π-electron system (where $n = 1$); thus it should occur under the influence of light.

D.6 (a) This is a $(4n + 2)$ π-electron system (where $n = 1$); a thermal reaction should take place with disrotatory motion:

(b) This is also a $(4n + 2)$ π-electron system; a photochemical reaction should take place with conrotatory motion.

D.7 Here we need a conrotatory ring opening of *trans*-5,6-dimethyl-1,3-cyclohexadiene (to produce *trans,cis,trans*-2,4,6-octatriene); then we need a disrotatory cyclization to produce *cis*-5,6-dimethyl-1,3-cyclohexadiene.

trans-5,6-Dimethyl-1,3-
cyclohexadiene

trans,cis,trans-2,4,6-
Octatriene

cis-5,6-Dimethyl-1,3-
cyclohexadiene

Since both reactions involve $(4n + 2)$ π-electron systems, we apply light to accomplish the first step and heat to accomplish the second. It would also be possible to use heat to produce *trans,cis,cis*-2,4,6-octatriene and then use light to produce the desired product.

D.8 The first electrocyclic reaction is a thermal, conrotatory ring opening of a $4n$ π-electron system. The second electrocyclic reaction is a thermal, disrotatory ring closure of a $(4n + 2)$ π-electron system.

D.9 (a)   (b)   (c)   (d)

Chorismate
(a key metabolite in plants
and microorganisms)

D.10 (a)   (b)

D.11 (a) There are two possible products that can result from a concerted cycloaddition. They are formed when *cis*-2-butene molecules come together in the following ways:

(b) There are two possible products that can be obtained from *trans*-2-butene as well.

D.12 This is an intramolecular [2 + 2] cycloaddition.

D.13 (a)

(b)

+
Enantiomer

E

SPECIAL TOPIC
Step-Growth Polymers

SOLUTIONS TO PROBLEMS

E.1 (a) The reaction of cyclohexane with O_2 (cat.) gives cyclohexanol and cyclohexanone, which with HNO_3 (cat.) give adipic acid (hexanedioic acid, HOOC(CH$_2$)$_4$COOH) $+$ N_2O.

(b) Adipic acid $+$ 2 NH_3 $\rightarrow$ the diammonium salt $+NH_4$ $^-O_2C(CH_2)_4CO_2^-$ $+NH_4$, which on heat (-2 H$_2$O) gives the diamide H_2N—CO(CH$_2$)$_4$CO—NH$_2$; at 350 °C with catalyst gives the dinitrile $NC(CH_2)_4CN$; with 4 H$_2$/catalyst gives hexamethylenediamine $H_2N(CH_2)_6NH_2$.

(c) 1,3-Butadiene $+$ Cl$_2$ $\rightarrow$ ClCH$_2$CH=CHCH$_2$Cl, then 2 NaCN $\rightarrow$ NCCH$_2$CH=CHCH$_2$CN, then H$_2$/Ni $\rightarrow$ NC(CH$_2$)$_4$CN, then 4 H$_2$/catalyst $\rightarrow$ H$_2$N(CH$_2$)$_6$NH$_2$.

(d) Tetrahydrofuran $+$ 2 HCl $\rightarrow$ Cl(CH$_2$)$_4$Cl, then 2 NaCN $\rightarrow$ NC(CH$_2$)$_4$CN, then 4 H$_2$/catalyst $\rightarrow$ H$_2$N(CH$_2$)$_6$NH$_2$.

671

E.2 (a) HO—CH₂CH₂—OH + :B⁻ ⇌ ⁻O—CH₂CH₂—OH + HB

[R = CH₃– or HO—CH₂CH₂—〜]

CH₃O⁻ + HB ⇌ CH₃OH + :B⁻

(b)

[R = CH₃– or HO—CH₂CH₂—〜]

E.3 (a)

(b) By high-pressure catalytic hydrogenation

E.4

E.5

Lexan

E.6 (a) The resin is probably formed in the following way. Base converts the bisphenol A to a double phenoxide ion that attacks a carbon atom of the epoxide ring of each epichlorohydrin:

(b) The excess of epichlorohydrin limits the molecular weight and ensures that the resin has epoxy ends.

(c) Adding the hardener brings about cross linking by reacting at the terminal epoxide groups of the resin:

E.7 (a)

(b) To ensure that the polyester chain has —CH_2OH end groups.

E.8 Because the para position is occupied by a methyl group, cross-linking does not occur and the resulting polymer remains thermoplastic (See Section E.4.)

E.9

Bakelite

SPECIAL TOPIC
Thiols, Sulfur Ylides, and Disulfides

SOLUTIONS TO PROBLEMS

F.1 (a)
$$\text{cyclohexanone} =\!O + CH_2\!=\!S(CH_3)_2 \longrightarrow \text{epoxide} + CH_3SCH_3$$

(b)
$$\text{isopropylidene} =\!O + CH_2\!=\!S(CH_3)_2 \longrightarrow \text{epoxide} + CH_3SCH_3$$

F.2 (a) benzyl-S=C(NH$_2$)(NH$_2$)$^+$ Br$^-$
(b) benzyl-SH

(c) benzyl-S—S-benzyl
(d) benzyl-S$^-$ Na$^+$

(e) benzyl-S-benzyl

F.3
$$\text{allyl-Br} + S\!=\!C(NH_2)(NH_2) \xrightarrow[\text{(2) } HO^-, H_2O]{\text{(1) } \text{EtOH}} \text{allyl-SH}$$

$$\xrightarrow{H_2O_2} \text{allyl-S—S-allyl}$$

F.4
$$\text{allyl-OH} \xrightarrow{Br_2} Br\text{-CH}_2\text{CHBr-OH} \xrightarrow{NaSH} HS\text{-CH}_2\text{CH(SH)-OH}$$

677

F.5 (a) Cl—CH₂CH₂—C(=O)—CH₂CH₂CH₂—C(=O)—O—CH₂CH₃

(This step is the Friedel-Crafts alkylation of an alkene.)

(b) SOCl₂

(c) 2 C₆H₅—CH₂—SH and KOH

(d) H₃O⁺

(e)

G

SPECIAL TOPIC
Thiol Esters and Lipid Biosynthesis

SOLUTION TO PROBLEM

G.1

Farnesol

Bisabolene

SPECIAL TOPIC
Alkaloids

SOLUTIONS TO PROBLEMS

H.1 (a) The first step is similar to a crossed Claisen condensation (see Section 19.2B):

(b) This step involves hydrolysis of an amide (lactam) and can be carried out with either acid or base. Here we use acid.

(c) This step is the decarboxylation of a β-keto acid; it requires only the application of heat and takes place during the acid hydrolysis of step (b).

(d) This is the reduction of a ketone to a secondary alcohol. A variety of reducing agents can be used, including sodium borohydride, for example.

(e) Here we convert the secondary alcohol to an alkyl bromide with hydrogen bromide; this reagent also gives a hydrobromide salt of the aliphatic amine.

(f) Treating the salt with base produces the secondary amine; it then acts as a nucleophile and attacks the carbon atom bearing the bromine. This reaction leads to the formation of a five-membered ring and (±) nicotine.

H.2 (a) The chirality center adjacent to the ester carbonyl group is racemized by base (probably through the formation of an enolate anion that can undergo inversion of configuration; cf. Section 18.3A).

(b)

H.3 (a)

Tropine   (±) Tropic acid

(b) Tropine is a meso compound; it has a plane of symmetry that passes through the ⟩CHOH group, the ⟩NCH₃ group, and between the two —CH₂— groups of the five-membered ring.

(c)

ψ-Tropine

H.4

Tropine $\xrightarrow{-H_2O}$ $C_8H_{13}N$ $\xrightarrow{CH_3I}$ $C_9H_{16}NI$ $\xrightarrow[\text{(2) heat}]{\text{(1) Ag}_2O/H_2O}$

$C_9H_{15}N$ $\xrightarrow{CH_3I}$ $C_{10}H_{18}NI$ $\xrightarrow[\text{(2) heat}]{\text{(1) Ag}_2O/H_2O}$

H.5 One possible sequence of steps is the following:

Tropinone

H.6

$C_{20}H_{25}NO_5$

Dihydropapaverine

Papaverine

H.7 A Diels-Alder reaction was carried out using 1,3-butadiene as the diene component.

H.8 Acetic anhydride acetylates both —OH groups.

Heroin

H.9 (a) A Mannich-type reaction (see Section 19.8).

(b)

Gramine

A

APPENDIX
Empirical and Molecular Formulas

In the early and middle 19th century methods for determining formulas for organic compounds were devised by J. J. Berzelius, J. B. A. Dumas, Justus Liebig and Stanislao Cannizzaro. Although the experimental procedures for these analyses have been refined, the basic methods for determining the elemental composition of an organic compound today are not substantially different from those used in the nineteenth century. A carefully weighed quantity of the compound to be analyzed is oxidized completely to carbon dioxide and water. The weights of carbon dioxide and water are carefully measured and used to find the percentages of carbon and hydrogen in the compound. The percentage of nitrogen is usually determined by measuring the volume of nitrogen (N_2) produced in a separate procedure.

Special techniques for determining the percentage composition of other elements typically found in organic compounds have also been developed, but the direct determination of the percentage of oxygen is difficult. However, if the percentage composition of all the other elements is known, then the percentage of oxygen can be determined by difference. The following examples will illustrate how these calculations can be carried out.

EXAMPLE A

A new organic compound is found to have the following elemental analysis.

Carbon	67.95%
Hydrogen	5.69
Nitrogen	26.20
Total:	99.84%

Since the total of these percentages is very close to 100% (within experimental error), we can assume that no other element is present. For the purpose of our calculation it is convenient to assume that we have a 100-g sample. If we did, it would contain the following:

67.95 g of carbon
 5.69 g of hydrogen
26.20 g of nitrogen

In other words, we use percentages *by weight* to give us the ratios *by weight* of the elements in the substance. To write a formula for the substance, however, we need *ratios by moles.*

We now divide each of these weight-ratio numbers by the atomic weight of the particular element and obtain the number of moles of each element, respectively, in 100 g of the compound. This operation gives us the ratios *by moles* of the elements in the substance:

$$C \quad \frac{67.95 \text{ g}}{12.01 \text{ g mol}^{-1}} = 5.66 \text{ mol}$$

$$H \quad \frac{5.69 \text{ g}}{1.008 \text{ g mol}^{-1}} = 5.64 \text{ mol}$$

$$N \quad \frac{26.20 \text{ g}}{14.01 \text{ g mol}^{-1}} = 1.87 \text{ mol}$$

One possible formula for the compound, therefore, is $C_{5.66}H_{5.64}N_{1.87}$.

By convention, however, we use *whole* numbers in formulas. Therefore, we convert these fractional numbers of moles to whole numbers by dividing each by 1.87, the smallest number.

$$C \quad \frac{5.66}{1.87} = 3.03 \text{ which is} \sim 3$$

$$H \quad \frac{5.64}{1.87} = 3.02 \text{ which is} \sim 3$$

$$N \quad \frac{1.87}{1.87} = 1.00$$

Thus, within experimental error, the ratios by moles are 3 C to 3 H to 1 N, and C_3H_3N is the *empirical formula*. By empirical formula, we mean the formula in which the subscripts are the smallest integers that give the ratio of atoms in the compound. In contrast, a *molecular* formula discloses the complete composition of one molecule. The molecular formula of this particular compound could be C_3H_3N or some whole number multiple of C_3H_3N; that is, $C_6H_6N_2$, $C_9H_9N_3$, $C_{12}H_{12}N_4$, and so on. If, in a separate determination, we find that the molecular weight of the compound is 108 ± 3, we can be certain that the *molecular formula* of the compound is $C_6H_6N_2$.

FORMULA	MOLECULAR WEIGHT
C_3H_3N	53.06
$C_6H_6N_2$	106.13 (which is within the range 108 ± 3)
$C_9H_9N_3$	159.19
$C_{12}H_{12}N_4$	212.26

The most accurate method for determining molecular weights is by high-resolution mass spectrometry (Section 9.14A). A variety of other methods based on freezing point depression, boiling point elevation, osmotic pressure, and vapor density can also be used to determine molecular weights.

EXAMPLE B

Histidine, an amino acid isolated from protein, has the following elemental analysis:

Carbon	46.38%
Hydrogen	5.90
Nitrogen	27.01
Total:	79.29
Difference	20.71 (assumed to be oxygen)
	100.00%

Since no elements, other than carbon, hydrogen, and nitrogen, are found to be present in histidine, the difference is assumed to be oxygen. Again, we assume a 100-g sample and divide the weight of each element by its gram-atomic weight. This gives us the ratio of moles (A).

$$\qquad\qquad\text{(A)} \qquad\qquad \text{(B)} \qquad\qquad\qquad \text{(C)}$$

$$\text{C} \quad \frac{46.38}{12.01} = 3.86 \quad \frac{3.86}{1.29} = 2.99 \times 2 = 5.98 \sim 6 \text{ carbon atoms}$$

$$\text{H} \quad \frac{5.90}{1.008} = 5.85 \quad \frac{5.85}{1.29} = 4.53 \times 2 = 9.06 \sim 9 \text{ hydrogen atoms}$$

$$\text{N} \quad \frac{27.01}{14.01} = 1.93 \quad \frac{1.93}{1.29} = 1.50 \times 2 = 3.00 = 3 \text{ nitrogen atoms}$$

$$\text{O} \quad \frac{20.71}{16.00} = 1.29 \quad \frac{1.29}{1.29} = 1.00 \times 2 = 2.00 = 2 \text{ oxygen atoms}$$

Dividing each of the moles (A) by the smallest of them does not give a set of numbers (B) that is close to a set of whole numbers. Multiplying each of the numbers in column (B) by 2 does, however, as seen in column (C). The empirical formula of histidine is, therefore, $C_6H_9N_3O_2$.

In a separate determination, the molecular weight of histidine was found to be 158 ± 5. The empirical formula weight of $C_6H_9N_3O_2$ (155.15) is within this range; thus, the molecular formula for histidine is the same as the empirical formula.

PROBLEMS

A.1 What is the empirical formula of each of the following compounds?

(a) Hydrazine, N_2H_4

(b) Benzene, C_6H_6

(c) Dioxane, $C_4H_8O_2$

(d) Nicotine, $C_{10}H_{14}N_2$

(e) Cyclodecane, $C_{10}H_{20}$

(f) Acetylene, C_2H_2

A.2 The empirical formulas and molecular weights of several compounds are given next. In each case, calculate the molecular formula for the compound.

	EMPIRICAL FORMULA	MOLECULAR WEIGHT
(a)	CH_2O	179 ± 5
(b)	CHN	80 ± 5
(c)	CCl_2	410 ± 10

A.3 The widely used antibiotic, penicillin G, gave the following elemental analysis: C, 57.45%; H, 5.40%; N, 8.45%; S, 9.61%. The molecular weight of penicillin G is 330 ± 10. Assume that no other elements except oxygen are present and calculate the empirical and molecular formulas for penicillin G.

ADDITIONAL PROBLEMS

A.4 Calculate the percentage composition of each of the following compounds.

(a) $C_6H_{12}O_6$

(b) $CH_3CH_2NO_2$

(c) $CH_3CH_2CBr_3$

A.5 An organometallic compound called *ferrocene* contains 30.02% iron. What is the minimum molecular weight of ferrocene?

A.6 A gaseous compound gave the following analysis: C, 40.04%; H, 6.69%. At standard temperature and pressure, 1.00 g of the gas occupied a volume of 746 mL. What is the molecular formula of the compound?

A.7 A gaseous hydrocarbon has a density of 1.251 g L^{-1} at standard temperature and pressure. When subjected to complete combustion, a 1.000-L sample of the hydrocarbon gave 3.926 g of carbon dioxide and 1.608 g of water. What is the molecular formula for the hydrocarbon?

A.8 Nicotinamide, a vitamin that prevents the occurrence of pellagra, gave the following analysis: C, 59.10%; H, 4.92%; N, 22.91%. The molecular weight of nicotinamide was shown in a separate determination to be 120 ± 5. What is the molecular formula for nicotinamide?

A.9 The antibiotic chloramphenicol gave the following analysis: C, 40.88%; H, 3.74%; Cl, 21.95%; N, 8.67%. The molecular weight was found to be 300 ± 30. What is the molecular formula for chloramphenicol?

SOLUTIONS TO PROBLEMS OF APPENDIX A

A.1 (a) NH_2 (b) CH   (c) C_2H_4O (d) C_5H_7N (e) CH_2 (f) CH

A.2

EMPIRICAL FORMULA	EMPIRICAL FORMULA WEIGHT	$\left(\dfrac{\text{MOLECULAR WEIGHT}}{\text{EMP. FORM. WT.}}\right)$	MOLECULAR FORMULA
(a) CH_2O	30	$\dfrac{179}{30} \cong 6$	$C_6H_{12}O_6$
(b) CHN	27	$\dfrac{80}{27} \cong 3$	$C_3H_3N_3$
(c) CCl_2	83	$\dfrac{410}{83} \cong 5$	C_5Cl_{10}

A.3 If we assume that we have a 100-g sample, the amounts of the elements are

	WEIGHT	Moles (A)	(B)
C	57.45	$\dfrac{57.45}{12.01} = 4.78$	$\dfrac{4.78}{0.300} = 15.9 \cong 16$
H	5.40	$\dfrac{5.40}{1.008} = 5.36$	$\dfrac{5.36}{0.300} = 17.9 \cong 18$
N	8.45	$\dfrac{8.45}{14.01} = 0.603$	$\dfrac{0.603}{0.300} = 2.01 \cong 2$
S	9.61	$\dfrac{9.61}{32.06} = 0.300$	$\dfrac{0.300}{0.300} = 1.00 = 1$
O*	19.09	$\dfrac{19.09}{16.00} = 1.19$	$\dfrac{1.19}{0.300} = 3.97 \cong 4$
	$\overline{100.00}$		

(* by difference from 100)

The empirical formula is thus $C_{16}H_{18}N_2SO_4$. The empirical formula weight (334.4) is within the range given for the molecular weight (330 ± 10). Thus, the molecular formula for penicillin G is the same as the empirical formula.

A.4 (a) To calculate the percentage composition from the molecular formula, first determine the weight of each element in 1 mol of the compound. For $C_6H_{12}O_6$,

$$C_6 = 6 \times 12.01 = 72.06 \qquad \frac{72.06}{180.2} = 0.400 = 40.0\%$$

$$H_{12} = 12 \times 1.008 = 12.10 \qquad \frac{12.10}{180.2} = 0.0671 = 6.7\%$$

$$O_6 = \underline{6 \times 16.00 = 96.00} \qquad \frac{96.00}{180.2} = 0.533 = 53.3\%$$

$$\text{MW} \qquad 180.16$$

(MW = molecular weight)

Then determine the percentage of each element using the formula

$$\text{Percentage of A} = \frac{\text{Weight of A}}{\text{Molecular Weight}} \times 100$$

(b) 　　$C_2 = 2 \times 12.01 = 24.02$　　$\dfrac{24.02}{75.07} = 0.320 = 32.0\%$

　　　　$H_5 = 5 \times 1.008 = \;\;5.04$　　$\dfrac{5.04}{75.07} = 0.067 = 6.7\%$

　　　　$N = 1 \times 14.01 = 14.01$　　$\dfrac{14.01}{75.07} = 0.187 = 18.7\%$

　　　　$O_2 = 2 \times 16.00 = \underline{32.00}$　　$\dfrac{32.00}{75.07} = 0.426 = 42.6\%$

　　　　　　　Total $= 75.07$

(c) 　　$C_3 = 3 \times 12.01 = \;\;36.03$　　$\dfrac{36.03}{280.77} = 0.128 = 12.8\%$

　　　　$H_5 = 5 \times 1.008 = \;\;5.04$　　$\dfrac{5.04}{280.77} = 0.018 = 1.8\%$

　　　　$Br_3 = 3 \times 79.90 = \underline{239.70}$　　$\dfrac{239.70}{280.77} = 0.854 = 85.4\%$

　　　　　　　Total $= 280.77$

A.5   If the compound contains iron, each molecule must contain at least one atom of iron, and 1 mol of the compound must contain at least 55.85 g of iron. Therefore,

$$\text{MW of ferrocene} = 55.85 \frac{\text{g of Fe}}{\text{mol}} \times \frac{1.000 \text{ g}}{0.3002 \text{ g of Fe}}$$

$$= 186.0 \frac{\text{g}}{\text{mol}}$$

A.6   First, we must determine the empirical formula. Assuming that the difference between the percentages given and 100% is due to oxygen, we calculate:

C	40.04	$\dfrac{40.04}{12.01} = 3.33$	$\dfrac{3.33}{3.33} = 1$
H	6.69	$\dfrac{6.69}{1.008} = 6.64$	$\dfrac{6.64}{3.33} \cong 2$
O	53.27	$\dfrac{53.27}{16.00} = 3.33$	$\dfrac{3.33}{3.33} = 1$
	$\overline{100.00}$		

The empirical formula is thus CH_2O.

To determine the molecular formula, we must first determine the molecular weight. At standard temperature and pressure, the volume of 1 mol of an ideal gas is 22.4 L. Assuming ideal behavior,

$$\frac{1.00 \text{ g}}{0.746 \text{ L}} = \frac{MW}{22.4 \text{ L}} \quad \text{where MW} = \text{molecular weight}$$

$$MW = \frac{(1.00)(22.4)}{0.746} = 30.0 \text{ g}$$

The empirical formula weight (30.0) equals the molecular weight; thus, the molecular formula is the same as the empirical formula.

A.7 As in Problem A.6, the molecular weight is found by the equation

$$\frac{1.251 \text{ g}}{1.00 \text{ L}} = \frac{MW}{22.4 \text{ L}}$$

$$MW = (1.251)(22.4)$$

$$MW = 28.02$$

To determine the empirical formula, we must determine the amount of carbon in 3.926 g of carbon dioxide, and the amount of hydrogen in 1.608 g of water.

$$C \quad \left(3.926 \text{ g CO}_2\right)\left(\frac{12.01 \text{ g C}}{44.01 \text{ g CO}_2}\right) = 1.071 \text{ g carbon}$$

$$H \quad \left(1.608 \text{ g H}_2\text{O}\right)\left(\frac{2.016 \text{ g H}}{18.016 \text{ g H}_2\text{O}}\right) = \frac{0.180 \text{ g hydrogen}}{1.251 \text{ g sample}}$$

The weight of C and H in a 1.251-g sample is 1.251 g. Therefore, there are no other elements present.

To determine the empirical formula, we proceed as in Problem A.6 except that the sample size is 1.251 g instead of 100 g.

$$C \quad \frac{1.071}{12.01} = 0.0892 \qquad \frac{0.0892}{0.0892} = 1$$

$$H \quad \frac{0.180}{1.008} = 0.179 \qquad \frac{0.179}{0.0892} = 2$$

The empirical formula is thus CH_2. The empirical formula weight (14) is one-half the molecular weight. Thus, the molecular formula is C_2H_4.

A.8 Use the procedure of Problem A.3.

$$C \quad 59.10 \quad \frac{59.10}{12.01} = 4.92 \qquad \frac{4.92}{0.817} = 6.02 \cong 6$$

$$H \quad 4.92 \quad \frac{4.92}{1.008} = 4.88 \qquad \frac{4.88}{0.817} = 5.97 \cong 6$$

$$N \quad 22.91 \quad \frac{22.91}{14.01} = 1.64 \quad \frac{1.64}{0.817} = 2$$

$$O \quad \underline{13.07} \quad \frac{13.07}{16.00} = 0.817 \quad \frac{0.817}{0.817} = 1$$
$$\qquad\quad 100.00$$

The empirical formula is thus $C_6H_6N_2O$. The empirical formula weight is 122.13, which is equal to the molecular weight within experimental error. The molecular formula is thus the same as the empirical formula.

A.9

$$C \quad 40.88 \quad \frac{40.88}{12.01} = 3.40 \quad \frac{3.40}{0.619} = 5.5 \quad 5.5 \times 2 = 11$$

$$H \quad 3.74 \quad \frac{3.74}{1.008} = 3.71 \quad \frac{3.71}{0.619} = 6 \quad 6 \times 2 = 12$$

$$Cl \quad 21.95 \quad \frac{21.95}{35.45} = 0.619 \quad \frac{0.619}{0.619} = 1 \quad 1 \times 2 = 2$$

$$N \quad 8.67 \quad \frac{8.67}{14.01} = 0.619 \quad \frac{0.619}{0.619} = 1 \quad 1 \times 2 = 2$$

$$O \quad \underline{24.76} \quad \frac{24.76}{16.00} = 1.55 \quad \frac{1.55}{0.619} = 2.5 \quad 2.5 \times 2 = 5$$
$$\qquad\quad 100.00$$

The empirical formula is thus $C_{11}H_{12}Cl_2N_2O_5$. The empirical formula weight (323) is equal to the molecular weight; therefore, the molecular formula is the same as the empirical formula.

B APPENDIX
Answers to Quizzes

EXERCISE 1

1.1 (d) **1.2** (d) **1.3** (e) **1.4** (d) **1.5** (c) **1.6**

1.7 $CH_3CHCH_2CH_3$ and $CH_3-\overset{\underset{|}{CH_3}}{\underset{|}{CH_3}}{C}-CH_3$

1.8

1.9 (a) sp^2 (b) sp^3 (c) 0 (d) trigonal planar (e) 0 D

1.10 (a) + 1 (b) 0 (c) − 1

1.11

1.12

1.13

EXERCISE 2

2.1 (e)   **2.2** (a)   **2.3** (b)

2.4 (a)   (b)   (c)

(d)   (e)   (f) CH_3

(g)

2.5 or

2.6 (a)   (b)   (c)

(d) CH_3 (e)

2.7 (a) Isopropyl phenyl ether

(b) Ethylmethylphenylamine

(c) Isopropylamine

EXERCISE 3

3.1 (a)   **3.2** (c)   **3.3** (b)   **3.4** (e)   **3.5** (b)

3.6 (b)

3.7 $H_2SO_4 + NaF \longrightarrow NaHSO_4 + HF$

3.8 $(CH_3)_2NH$

3.9 (a) CH_3CH_2Li (b) D_2O

3.10 (a) CH_3Li (b) $CH_3\overset{\underset{|}{CH_3}}{C}HCH_2OH$ (c) $CH_3\overset{\underset{|}{CH_3}}{C}HCH_2OLi$

EXERCISE 4

4.1 (c)   **4.2** (c)   **4.3** (b)   **4.4** (a)   **4.5** (b)

4.6 (a)   **4.7** (a)

4.8 (a)   (b) or

(c)

4.9 H_2, Pt, pressure   or H_2, Ni, pressure

4.10

EXERCISE 5

5.1 (a)   **5.2** (b)   **5.3** (b)   **5.4** (e)   **5.5** (b)

5.6 and

5.7   **5.8**
(*E* and *Z*)

5.9 (d)   **5.10** (e)

EXERCISE 6

6.1 (b)   **6.2** (a)

6.3

6.4 (b)

6.5 A =   B =

C =

EXERCISE 7

7.1 (c)   **7.2** (d)   **7.3** (a)

7.4 (a) Li, $C_2H_5NH_2$, −78 °C, then NH_4Cl

(b) $H_2/Ni_2B(P-2)$ or $H_2/Pd/CaCO_3$ (Lindlar's catalyst)

(c) H_2/Ni, pressure   or H_2/Pt, pressure (using at least 2 molar equivalents of H_2)

(d) C_2H_5ONa/C_2H_5OH, heat

(e) $(CH_3)_3COK/(CH_3)_3COH$, heat

7.5

7.6 (a)   (b) ═⁻Na⁺   (c) ═H

(d) ═⁻Na⁺   (e)

7.7 A =   B =

C =   D =

EXERCISE 8

8.1 (e)   **8.2** (c)   **8.3** (e)   **8.4** (a)   **8.5** (d)   **8.6** (c)

8.7 (c)   **8.8** (b)

EXERCISE 9

9.1 (a)   (b)   (c)

(d)   (e)

9.2 (c)   **9.3** (a)   **9.4** (b)   **9.5** (c)   **9.6** (c,d,e)   **9.7**

EXERCISE 10

10.1 (d)   **10.2** (b)   **10.3** (c)   **10.4** (b)

10.5

10.6 Six

EXERCISE 11

11.1 (d)   **11.2** (a)   **11.3** (e)

11.4   **A** =   **B** =

C =   **D** =

EXERCISE 12

12.1 (b)   **12.2** (a)

12.3 A = ——≡Li   or   ——≡MgBr

 B = NaH

 C = CH_3I

12.4 A =

 B = PCC or Swern oxidation

 C =

 D =

12.5 A =

or

 (if excess CH_3MgBr is used)

EXERCISE 13

13.1 (d)   **13.2** (c)   **13.3** (c)   **13.4** (c)   **13.5** (b)

EXERCISE 14

14.1 (e)   **14.2** (a)   **14.3** (b)   **14.4** (b)

14.5

14.6 Azulene

EXERCISE 15

15.1 (a)   **15.2** (a)   **15.3** (b)

15.4   (a)   **A** = SO$_3$/H$_2$SO$_4$ **B** =

     C = H$_2$O, H$_2$SO$_4$, heat   **D** =

   (b)   **A** = SOCl$_2$ or PCl$_5$ **B** =   +   AlCl$_3$

     C = Zn(Hg), HCl, reflux   **D** = Br$_2$/FeBr$_3$
            or Wolff–Kishner reduction

EXERCISE 16

   16.1   (d)   **16.2**   (b)   **16.3**   (b)

   16.4   (a)   **A** =   **B** = NaCN

         C   (1)   DIBAL-H, hexane, −78 °C   (2) H$_2$O

     (b)   **A** = PCC or Swern oxidation   **B** = HO⁀OH, H$_3$O$^+$ **C** = Mg

     (c)   **A** = (C$_6$H$_5$)$_3$P   **B** = $^+$P(C$_6$H$_5$)$_3$ Br$^-$ **C** =

     (d)   **A** = (1) CH$_3$MgBr   **B** = HCN   **C** = (1) LiAlH$_4$, Et$_2$O
             (2) H$_2$CrO$_4$ or Swern oxidation   (2) H$_2$O

16.5

The *gem*-diol formed in the alkaline hydrolysis step readily loses water to form the aldehyde.

16.6 The general formula for an oxime is

Both carbon and nitrogen are sp^2 hybridized; the electron pair on nitrogen occupies one sp^2 orbital. Aldoximes and ketoximes can exist in either of these two stereoisomeric forms:

This type of stereoisomerism is also observed in the case of other compounds that possess the C=N group, for example, phenylhydrazones and imines.

EXERCISE 17

17.1 (b)   **17.2** (d)   **17.3** (d)

17.4 **A** = 3-Chlorobutanoic acid

B = Methyl 4-nitrobenzoate or methyl *p*-nitrobenzoate

C = *N*-Methylaniline

17.5 (a) **A** = (1) KMnO$_4$, HO$^-$, heat   **B** = SOCl$_2$ or PCl$_5$
(2) H$_3$O$^+$

(b) **A** =

B = CH$_3$—$\overset{18}{\text{OH}}$

C =

(c) A = [structure: cyclopentane with C(=O)NH₂]   B = [structure: cyclopentane with C≡N]   C = [structure: cyclopentane with CHO]

17.6 (b)

EXERCISE 18

18.1 (a)

18.2 A = [structure: pyrrolidine with N–H]   B = [structure: allyl bromide]

C = [structure: pyrrolidinium enamine with allyl group, Br⁻]

18.3 (c)   **18.4** (e)   **18.5** (b)

EXERCISE 19

19.1 (c)   **19.2** (e)   **19.3** (b)

19.4 (a) A = EtO–C(=O)–CH(Et)–C(=O)–OEt   B = EtO–C(=O)–C⁻(Et)–C(=O)–OEt K⁺

C = EtO–C(=O)–C(Et)₂–C(=O)–OEt   D = HO–C(=O)–C(Et)₂–C(=O)–OH

E = [structure: 2-ethylbutanoic acid]

(b) **A** = [structure: benzaldehyde, with O and H]

B = [structure: phenyl-substituted chain with four ester groups, OEt OEt and OEt OEt]

C = [structure: phenyl-substituted chain with four carboxylic acid groups, OH groups]

(c) **A** = [structure: tetrahydropyranone enolate with Li⁺]

B = [structure: ethyl iodide]

(d) **A** = [structure: pyrrolidine, N–H]

B = [structure: allyl bromide]

C = [structure: pyrrolidinium iminium with allyl-substituted cyclohexane ring, Br⁻]

19.5 (a) [structure: tetralone with OH and methyl]

(b) [structure: dihydronaphthalenone with methyl]

(c) CH₃MgI, Et₂O

(d) [structure: tetralone with two methyl substituents]

+ enantiomer and
simple addition product

(e) Zn(Hg)/HCl or Wolff–Kishner reduction

19.6 (e)

19.7 (a) (b) (c) HCN

19.8 (a) (b) (c) (E and Z)

(d) LiAlH$_4$, Et$_2$O   (e) H$_2$, Ni, pressure   (f) CH$_3$OH (excess), HA

(g)   (h)

(i) (1) OEt, LDA, −78 °C
    (2) H$_2$O

EXERCISE 20

20.1 (d)   **20.2** (e)

20.3 (a) 2., 3., 4.   (b) 3., 4.   (c) 4.

20.4 (a) A = HNO$_3$/H$_2$SO$_4$ B =   C = NaNO$_2$, HCl, 0–5 °C

D = CuCN   E = LiAlH$_4$, Et$_2$O   F =

(b) A = NaN$_3$ B =   C =

D =   E = H$_3$PO$_2$

20.5 (a) (2)   (b) (2)   (c) (1)   (d) (1)   (e) (2)   (f) (2)

EXERCISE 21

21.1 (a) +2

(b) 0

(c) 0

21.2 + $\xrightarrow{\text{Ru}=}$?

21.3 (c)

21.4 (c)

21.5 (b) (e)

21.6 (c)

EXERCISE 22

22.1 (a)
(b)
(c)

(d)
(e)
(f)

(g)
(h)

22.2 C

22.3

22.4 (a)   (b)   (c)   (d)

22.5

22.6 (a)   (b)   (c) reducing

(d) active   (e) aldonic   (f) active   (g) aldaric   (h) NaBH$_4$

(i) active

22.7 (a) Galactose $\xrightarrow{\text{NaBH}_4}$ optically *inactive* alditol; glucose → optically active alditol

Galactose $\xrightarrow[\text{HNO}_3]{\text{dil}}$ optically *inactive* aldaric acid; glucose → optically active aldaric acid

(b) HIO$_4$ oxidation $\longrightarrow$ different products:

Fructose $\longrightarrow$ 2 mol + CO$_2$ + 3

Glucose $\longrightarrow$ 1 mol + 5

22.8 (e)   **22.9** (d)

EXERCISE 23

23.1 (a) OH or C_{16} or C_{18} (b) ONa   (c)

(d)   (e) SO_3Na (f)

23.2 (a) I_2/HO^- (iodoform test)   (b) Br_2 (c) Ethynylestradiol shows IR absorption at ~ 3300 cm^{-1} for the terminal alkyne hydrogen.

23.3 5α-Androstane

23.4 (a) H   (b) :$^-$Na$^+$ (c) Cl

(d) KCN   (e) OH   (f) H_2/Pd

23.5 (b) Sesquiterpene

23.6

23.7 (e)

EXERCISE 24

24.1 (a)   (b)   (c)

24.2 PLGFGY

C

APPENDIX
Molecular Model Set Exercises

The exercises in this appendix are designed to help you gain an understanding of the three-dimensional nature of molecules. You are encouraged to perform these exercises with a model set as described.

These exercises should be performed as part of the study of the chapters shown below.

Chapter in Text	Accompanying Exercises
4	1, 3, 4, 5, 6, 8, 10, 11, 12, 14, 15, 16, 17, 18, 20, 21
5	2, 7, 9, 13, 24, 25, 26, 27
7	9, 19, 22, 28
22	29
24	30
13	31
14	23, 27

The following molecular model set exercises were originally developed by Ronald Starkey.

Refer to the instruction booklet that accompanies your model set for details of molecular model assembly.

Exercise 1 (Chapter 4)

Assemble a molecular model of methane, CH_4. Note that the hydrogen atoms describe the apexes of a regular tetrahedron with the carbon atom at the center of the tetrahedron. Demonstrate by attempted superposition that two models of methane are identical.

Replace any one hydrogen atom on each of the two methane models with a halogen to form two molecules of CH_3X. Are the two structures identical? Does it make a difference which of the four hydrogen atoms on a methane molecule you replace? How many different configurations of CH_3X are possible?

Repeat the same considerations for two disubstituted methanes with two identical substituents (CH_2X_2), and then with two different substituents (CH_2XY). Two colors of atom-centers could be used for the two different substituents.

Exercise 2 (Chapter 5)

Construct a model of a trisubstituted methane molecule (CHXYZ). Four different colored atom-centers are attached to a central tetrahedral carbon atom-center. Note that the carbon now has four different substituents. Compare this model with a second model of CHXYZ. Are the two structures identical (superposable)?

Interchange any two substituents on one of the carbon atoms. Are the two CHXYZ molecules identical now? Does the fact that interchange of any two substituents on the carbon interconverts the stereoisomers indicate that there are only two possible configurations of a tetrahedral carbon atom?

Compare the two models that were not identical. What is the relationship between them? Do they have a mirror-image relationship? That is, are they related as an object and its mirror image?

Exercise 3 (Chapter 4)

Make a model of ethane, CH_3CH_3. Does each of the carbon atoms retain a tetrahedral configuration? Can the carbon atoms be rotated with respect to each other without breaking the carbon-carbon bond?

Rotate about the carbon-carbon bond until the carbon-hydrogen bonds of one carbon atom are aligned with those of the other carbon atom. This is the eclipsed conformation. When the C—H bond of one carbon atom bisects the H—C—H angle of the other carbon atom the conformation is called staggered. Remember, conformations are arrangements of atoms in a molecule that can be interconverted by bond rotations.

In which of the two conformations of ethane you made are the hydrogen atoms of one carbon closer to those of the other carbon?

Exercise 4 (Chapter 4)

Prepare a second model of ethane. Replace one hydrogen, any one, on each ethane model with a substituent such as a halogen, to form two models of CH_3CH_2X. Are the structures identical? If not, can they be made identical by rotation about the C—C bond? With one of the models, demonstrate that there are three equivalent staggered conformations (see Exercise 3) of CH_3CH_2X. How many equivalent eclipsed conformations are possible?

Exercise 5 (Chapter 4)

Assemble a model of a 1,2-disubstituted ethane molecule, CH_2XCH_2X. Note how the orientation of and the distance between the X groups changes with rotation about the carbon-carbon bond. The arrangement in which the X substituents are at maximum separation is the *anti*-staggered conformation. The other staggered conformations are called *gauche*. How many *gauche* conformations are possible? Are they energetically equivalent? Are they identical?

Exercise 6 (Chapter 4)

Construct two models of butane, ‿⁀‿. Note that the structures can be viewed as dimethyl-substituted ethanes. Show that rotations about the C2, C3 bond of butane produce eclipsed, *anti*-staggered, and *gauche*-staggered conformations. Measure the distance between C1 and C4 in the conformations just mentioned. [The scale of the Darling Framework Molecular Model Set, for example, is: 2.0 inches in a model corresponds to approximately 1.0 Å (0.1 nm) on a molecular scale.] In which eclipsed conformation are the C1 and C4 atoms closest to each other? How many eclipsed conformations are possible?

Exercise 7 (Chapter 5)

Using two models of butane, verify that the two hydrogen atoms on C2 are not stereo-chemically equivalent. Replacement of one hydrogen leads to a product that is not identical to that obtained by replacement of the other C2 hydrogen atom. Both replacement products have the same condensed formula, $CH_3CHXCH_2CH_3$. What is the relationship of the two products?

Exercise 8 (Chapter 4)

Make a model of hexane, . Extend the six-carbon chain as far as it will go. This puts C1 and C6 at maximum separation. Notice that this *straight-chain* structure maintains the tetrahedral bond angles at each carbon atom and therefore the carbon chain adopts a zigzag arrangement. Does this extended chain adopt staggered or eclipsed conformations of the hydrogen atoms? How could you describe the relationship of C1 and C4?

Exercise 9 (Chapters 5 and 7)

Prepare models of the four isomeric butenes, C_4H_8. Note that the restricted rotation about the double bond is responsible for the cis-trans stereoisomerism. Verify this by observing that breaking the π bond of *cis*-2-butene allows rotation and thus conversion to *trans*-2-butene. Is any of the four isomeric butenes chiral (nonsuperposable with its mirror image)? Indicate pairs of butene isomers that are structural (constitutional) isomers. Indicate pairs that are diastereomers. How does the distance between the C1 and C4 atoms in *trans*-2-butene compare with that of the *anti* conformation of butane? Compare the C1 to C4 distance in *cis*-2-butene with that in the conformation of butane in which the methyls are eclipsed.

1-Butene   *cis*-2-Butene   *trans*-2-Butene   2-Methylpropene

Exercise 10 (Chapter 4)

Make a model of cyclopropane. Take care not to break your models due to the angle strain of the carbon-carbon bonds of the cyclopropane ring. It should be apparent that the ring carbon atoms must be coplanar. What is the relationship of the hydrogen atoms on adjacent carbon atoms? Are they staggered, eclipsed, or skewed?

Exercise 11 (Chapter 4)

A model of cyclobutane can be assembled in a conformation that has the four carbon atoms coplanar. How many eclipsed hydrogen atoms are there in the conformation? Torsional strain (strain caused by repulsions between the aligned electron pairs of eclipsed bonds) can be relieved at the expense of increased angle strain by a slight folding of the ring. The deviation of one ring carbon from the plane of the other three carbon atoms is about 25°. This folding compresses the C—C—C bond angle to about 88°. Rotate the ring carbon bonds of the planar conformation to obtain the folded conformation. Are the hydrogen atoms on adjacent carbon atoms eclipsed or skewed? Considering both structural and stereoisomeric forms, how many dimethylcyclobutane structures are possible? Do deviations of the ring from planarity have to be considered when determining the number of possible dimethylcyclobutane structures?

Exercise 12 (Chapter 4)

Cyclopentane is a more flexible ring system than cyclobutane or cyclopropane. A model of cyclopentane in a conformation with all the ring carbon atoms coplanar exhibits minimal deviation of the C—C—C bond angles from the normal tetrahedral bond angle. How many eclipsed hydrogen interactions are there in this planar conformation? If one of the ring carbon atoms is pushed slightly above (or below) the plane of the other carbon atoms, a model of the envelope conformation is obtained. Does the envelope conformation relieve some of the torsional strain? How many eclipsed hydrogen interactions are there in the envelope conformation?

Cyclopentane

Exercise 13 (Chapter 5)

Make a model of 1,2-dimethylcyclopentane. How many stereoisomers are possible for this compound? Identify each of the possible structures as either cis or trans. Is it apparent that cis-trans isomerism is possible in this compound because of restricted rotation? Are any of the stereoisomers chiral? What are the relationships of the 1,2-dimethylcyclopentane stereoisomers?

Exercise 14 (Chapter 4)

Assemble the six-membered ring compound cyclohexane. Is the ring flat or puckered? Place the ring in a chair conformation and then in a boat conformation. Demonstrate that the chair and boat are indeed conformations of cyclohexane—that is, they may be interconverted by rotations about the carbon-carbon bonds of the ring.

Chair form   Boat form

Note that in the chair conformation carbon atoms 2, 3, 5, and 6 are in the same plane and carbon atoms 1 and 4 are below and above the plane, respectively. In the boat conformation, carbon atoms 1 and 4 are both above (they could also both be below) the plane described by carbon atoms 2, 3, 5, and 6. Is it apparent why the boat is sometimes associated with the flexible form? Are the hydrogen atoms in the chair conformation staggered or eclipsed? Are any hydrogen atoms eclipsed in the boat conformation? Do carbon atoms 1 and 4 have an *anti* or *gauche* relationship in the chair conformation? (*Hint*: Look down the C2, C3 bond).

A twist conformation of cyclohexane may be obtained by slightly twisting carbon atoms 2 and 5 of the boat conformation as shown.

Boat form   Twist form

Note that the C2, C3 and the C5, C6 sigma bonds no longer retain their parallel orientation in the twist conformation. If the ring system is twisted too far, another boat conformation results. Compare the nonbonded (van der Waals repulsion) interactions and the torsional strain present in the boat, twist, and chair conformations of cyclohexane. Is it apparent why the relative order of thermodynamic stabilities is: chair > twist > boat?

Exercise 15 (Chapter 4)

Construct a model of methylcyclohexane. How many chair conformations are possible? How does the orientation of the methyl group change in each chair conformation?

Identify carbon atoms in the chair conformation of methylcyclohexane that have intramolecular interactions corresponding to those found in the *gauche* and *anti* conformations of butane. Which of the chair conformations has the greatest number of *gauche* interactions? How many more? If we assume, as in the case for butane, that the *anti* interaction is 3.8 kJ mol^{-1} more favorable than *gauche*, then what is the relative stability of the two chair conformations of methylcyclohexane? *Hint*: Identify the relative number of *gauche* interactions in the two conformations.

Exercise 16 (Chapter 4)

Compare models of the chair conformations of monosubstituted cyclohexanes in which the substituent alkyl groups are methyl, ethyl, isopropyl, and *tert*-butyl.

Rationalize the relative stability of axial and equatorial conformations of the alkyl group given in the table for each compound. The chair conformation with the alkyl group equatorial is more stable by the amount shown.

ALKYL GROUP	$\Delta G°$ (kJ mol^{-1}) EQUATORIAL $\rightleftharpoons$ AXIAL
CH_3	7.3
CH_2CH_3	7.5
$CH(CH_3)_2$	9.2
$C(CH_3)_3$	21 (approximate)

Exercise 17 (Chapter 4)

Make a model of 1,2-dimethylcyclohexane. Answer the questions posed in Exercise 13 with regard to 1,2-dimethylcyclohexane.

Exercise 18 (Chapter 4)

Compare models of the neutral and charged molecules shown next. Identify the structures that are isoelectronic, that is, those that have the same electronic structure. How do those structures that are isoelectronic compare in their molecular geometry?

CH_3CH_3 CH_3NH_2 CH_3OH

$CH_3\bar{C}H_2$ $CH_3\overset{+}{N}H_3$ $CH_3\overset{+}{O}H_2$

$CH_3\bar{N}H$

Exercise 19 (Chapter 7)

Prepare a model of cyclohexene. Note that chair and boat conformations are no longer possible, as carbon atoms 1, 2, 3, and 6 lie in a plane. Are cis and trans stereoisomers possible for the double bond? Attempt to assemble a model of *trans*-cyclohexene. Can it be done? Are cis and trans stereoisomers possible for 2,3-dimethylcyclohexene? For 3,4-dimethylcyclohexene?

Cyclohexene

Assemble a model of *trans*-cyclooctene. Observe the twisting of the π-bond system. Would you expect the cis stereoisomer to be more stable than *trans*-cyclooctene? Is *cis*-cyclooctene chiral? Is *trans*-cyclooctene chiral?

Exercise 20 (Chapter 4)

Construct models of *cis*-decalin (*cis*-bicyclo[4.4.0]decane) and *trans*-decalin. Observe how it is possible to convert one conformation of *cis*-decalin in which both rings are in chair conformations to another all-chair conformation. This interconversion is not possible in the case of the *trans*-decalin isomer. Suggest a reason for the difference in the behavior of the cis and trans isomers. *Hint*: What would happen to carbon atoms 7 and 10 of *trans*-decalin if the other ring (indicated by carbon atoms numbered 1–6) is converted to the alternative chair conformation. Is the situation the same for *cis*-decalin?

trans-Decalin   *cis*-Decalin

Exercise 21 (Chapter 4)

Assemble a model of norbornane (bicyclo[2.2.1]heptane). Observe the two cyclopentane ring systems in the molecule. The structure may also be viewed as having a methylene (CH_2) bridge between carbon atoms 1 and 4 of cyclohexane. Describe the conformation of the cyclohexane ring system in norbornane. How many eclipsing interactions are present?

Norbornane

Using a model of twistane, identify the cyclohexane ring systems held in twist conformations. In adamantane, find the chair conformation cyclohexane systems. How many are present? Evaluate the torsional and angle strain in adamantane. Which of the three compounds in this exercise are chiral?

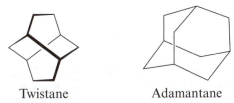

Twistane   Adamantane

Exercise 22 (Chapter 7)

An hypothesis known as Bredt's Rule states that a double bond to a bridgehead of a small-ring bridged bicyclic compound is not possible. The basis of this rule can be seen if you attempt to make a model of bicyclo[2.2.1]hept-1-ene, **A.** One approach to the assembly of this model is to try to bridge the number 1 and number 4 carbon atoms of cyclohexene with a methylene (CH_2) unit. Compare this bridging with the ease of installing a CH_2

bridge between the 1 and 4 carbon atoms of cyclohexane to form a model of norbornane (see Exercise 21). Explain the differences in ease of assembly of these two models.

A   **B**

Bridgehead double bonds can be accommodated in larger ring-bridged bicyclic compounds such as bicyclo[3.2.2]non-1-ene, **B**. Although this compound has been prepared in the laboratory, it is an extremely reactive alkene.

Exercise 23 (Chapter 14)

Not all cyclic structures with alternating double and single bonds are aromatic. Cyclooctatetraene shows none of the aromatic characteristics of benzene. From examination of molecular models of cyclooctatetraene and benzene, explain why there is π-electron delocalization in benzene but not in cyclooctatetraene. *Hint*: Can the carbon atoms of the eight-membered ring readily adopt a planar arrangement?

Benzene   Cyclooctatetraene

Note that benzene can be represented in several different ways with most molecular model sets. In this exercise, the Kekulé representation with alternating double and single bonds is appropriate. Alternative representations of benzene, such as a form depicting molecular orbital lobes, are shown in your model set instruction booklet.

Exercise 24 (Chapter 5)

Consider the $CH_3CHXCHYCH_3$ system. Assemble all possible stereoisomers of this structure. How many are there? Indicate the relationship among them. Are they all chiral?
Repeat the analysis with the $CH_3CHXCHXCH_3$ system.

Exercise 25 (Chapter 5)

The $CH_3CHXCHXCH_3$ molecule can exist as the stereoisomers shown here. In the eclipsed conformation (meso) shown on the left (**E**), the molecule has a plane of symmetry that bisects the C2, C3 bond. This is a more energetic conformation than any of the three staggered conformations, but it is the only conformation of this configurational stereoisomer that has a plane of symmetry. Can you consider a molecule achiral if only one conformation, and in this case not even the most stable conformation, has a plane of symmetry? Are any of the staggered conformations achiral (superposable on its mirror image)? Make a model of the staggered conformation shown here (**S**) and make another model that is the mirror image of it. Are these two structures different conformations of the same configurational stereoisomer (e.g., are they conformers that can be inter-converted by bond rotations), or are they configurational stereoisomers? Based on your answer to the last question, suggest an explanation for the fact that the molecule is not optically active.

Exercise 26 (Chapter 5)

Not all molecular chirality is a result of a tetrahedral chirality center, such as CHXYZ. Cumulated dienes (1,2-dienes, or allenes) are capable of generating molecular chirality. Identify, using models, which of the following cumulated dienes are chiral.

Are the following compounds chiral? How are they structurally related to cumulated dienes?

Is the cumulated triene **F** chiral? Explain the presence or absence of molecular chirality. More than one stereoisomer is possible for triene **F**. What are the structures, and what is the relationship between those structures?

Exercise 27 (Chapters 5 and 14)

Substituted biphenyl systems can produce molecular chirality if the rotation about the bond connecting the two rings is restricted. Which of the three biphenyl compounds indicated here are chiral and would be expected to be optically active? Build models of **J, K,** and **L** to help determine your answers.

J a = f = CH₃   **K** a = b = CH₃   **L** a = f = CH₃

b = e = $^+$N(CH₃)₃   e = f = $\overset{+}{N}$(CH₃)₃   b = e = H

Exercise 28 (Chapter 7)

Assemble a simple model of ethyne (acetylene). The linear geometry of the molecule should be readily apparent. Now, use appropriate pieces of your model set to depict the σ and both the π bonds of the triple bond system using *sp* hybrid carbon atoms and pieces that represent orbitals. Based on attempts to assemble cycloalkynes, predict the smallest cycloalkyne that is stable.

Exercise 29 (Chapter 22)

Construct a model of β-D-glucopyranose. Note that in one of the chair conformations all the hydroxyl groups and the CH₂OH group are in an equatorial orientation. Convert the structure of β-D-glucopyranose to α-D-glucopyranose, to β-D-mannopyranose, and to β-D-galactopyranose. Indicate the number of large ring substituents (OH or CH₂OH) that are axial in the more favorable chair conformation of each of these sugars. Is it reasonable that the β-anomer is more stable than the α-anomer of D-glucopyranose?

Make a model of β-L-glucopyranose. What is the relationship between the D and L configurations? Which is more stable?

β-D-Glucopyranose

D-(+)-Glucose   D-(+)-Mannose   D-(+)-Galactose

Exercise 30 (Chapter 24)

Assemble a model of tripeptide **A** shown here. Restricted rotation of the C—N bond in the amide linkage results from resonance contribution of the nitrogen nonbonding electron pair. Note the planarity of the six atoms associated with the amide portions of the molecule caused by this resonance contribution. Which bonds along the peptide chain are free to rotate? The amide linkage can either be cisoid or transoid. How does the length (from the N-terminal nitrogen atom to the C-terminal carbon atom) of the tripeptide chain that is transoid compare with one that is cisoid? Which is more "linear"? Convert a model of tripeptide **A** in the transoid arrangement to a model of tripeptide **B**. Which tripeptide has a longer chain?

Tripeptide **A**	R = CH$_3$	(L-Alanine)
Tripeptide **B**	R = CH$_2$OH	(L-Serine)

Exercise 31 (Chapter 13)

Make models of the π molecular orbitals for the following compounds. Use the phase representation of each contributing atomic orbital shown in your molecular model set instruction booklet. Compare each model with π molecular orbital diagrams shown in the textbook.

(a) π_1 and π_2 of ethene (CH$_2$=CH$_2$, or ═)

(b) π_1 through π_4 of 1,3-butadiene (CH$_2$=CH—CH=CH$_2$, or)

(c) π_1, π_2, and π_3 of the allyl (propenyl) radical (CH$_2$=CH—ĊH$_2$, or)

EXERCISE 32

Use your model set to construct several of the interesting representative natural product structures shown here.

Progesterone   Caryophyllene   Longifolene

Morphine   Strychnine

MOLECULAR MODEL SET EXERCISES — SOLUTIONS

Solution 1   Replacement of any hydrogen atom of methane leads to the same monosubstituted product CH$_3$X. Therefore, there is only one configuration of a monosubstituted methane. There is only one possible configuration for a disubstituted methane of either the CH$_2$X$_2$ or CH$_2$XY type.

Solution 2   Interchange of any two substituents converts the configuration of a tetrahedral chirality center to that of its enantiomer. There are only two possible configurations. If the models are not identical, they will have a mirror-image relationship.

Solution 3   The tetrahedral carbon atoms may be rotated without breaking the carbon-carbon bond. There is no change in the carbon-carbon bond orbital overlap during rotation. The eclipsed conformation places the hydrogen atoms closer together than they are in the staggered conformation.

Eclipsed   Staggered
conformation   conformation

Solution 4   All monosubstituted ethanes (CH$_3$CH$_2$X) may be made into identical structures by rotations about the C—C bond. The following structures are three energetically equivalent staggered conformations.

The three equivalent eclipsed conformations are

Solution 5   The two *gauche* conformations are energetically equivalent, but not identical (superposable) since they are conformational enantiomers. They bear a mirror-image relationship and are interconvertible by rotation about the carbon-carbon bond.

anti Conformation   *gauche* Conformations

Solution 6   There are three eclipsed conformations. The methyl groups (C1 and C4) are closest together in the methyl-methyl eclipsed conformation. The carbon-carbon internuclear distances between C1 and C4 are shown in the following table. The number of conformations of each type and the molecular distances in angstroms (Å) are shown.

CONFORMATION	NUMBER	DISTANCES (Å)
Eclipsed (CH_3, CH_3)	1	2.5
Gauche	2	2.8
Eclipsed (H, CH_3)	2	3.3
Anti	1	3.7

Solution 7   The enantiomers formed from replacement of the C2 hydrogen atoms of butane are

Solution 8   The extended chain assumes a staggered arrangement. The relationship of C1 and C4 is *anti*.

Solution 9   None of the isomeric butenes is chiral. They all have a plane of symmetry. All the isomeric butenes are related as constitutional (or structural) isomers except *cis*-2-butene and *trans*-2-butene, which are diastereomers.

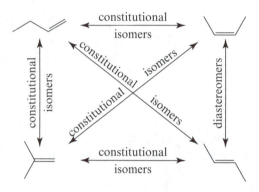

Molecular Model Set C1 to C4 Distances:

COMPOUND	DISTANCES (Å)
cis-2-Butene	2.0
trans-2-Butene	3.7
Butane (*gauche*)	2.8
Butane (*anti*)	3.7

Solution 10   The hydrogen atoms are all eclipsed in cyclopropane.

Solution 11   All the hydrogen atoms are eclipsed in the planar conformation of cyclobutane. The folded ring system has skewed hydrogen interactions. There are six possible isomers of dimethyl-cyclobutane. Since the ring is not held in one particular folded conformation, deviations of the ring planarity need not be considered in determining the number of possible dimethyl structures.

Solution 12   In the planar conformation of cyclopentane, all five pairs of methylene hydrogen atoms are eclipsed. That produces 10 eclipsed hydrogen interactions. Some torsional strain is relieved in the envelope conformation since there are only six eclipsed hydrogen interactions.

Solution 13   The three configurational stereoisomers of 1,2-dimethylcyclopentane are shown here. Both trans stereoisomers are chiral, while the cis configuration is an achiral meso compound.

Solution 14 The puckered ring of the chair and the boat conformations can be interconverted by rotation about the carbon-carbon bonds. The chair is more rigid than the boat conformation. All hydrogen atoms in the chair conformation have a staggered arrangement. In the boat conformation, there are eclipsed relationships between the hydrogen atoms on C2 and C3, and also between those on C5 and C6. Carbon atoms that are 1,4 to each other in the chair conformation have a *gauche* relationship. An evaluation of the three conformations confirms the relative stability: chair > twist > boat. The boat conformation has considerable eclipsing strain and nonbonded (van der Waals repulsion) interactions, the twist conformation has slight eclipsing strain, and the chair conformation has a minimum of eclipsing and nonbonded interactions.

Solution 15 Interconversion of the two chair conformations of methylcyclohexane changes the methyl group from an axial to a less crowded equatorial orientation, or the methyl that is equatorial to the more crowded axial position.

Axial methyl   Equatorial methyl

The conformation with the axial methyl group has two *gauche* (1,3-diaxial) interactions that are not present in the equatorial methyl conformation. These *gauche* interactions are axial methyl to C3 and axial methyl to C5. The methyl to C3 and methyl to C5 relationships with methyl groups in an equatorial orientation are anti.

Solution 16 The $\Delta G°$ value reflects the relative energies of the two chair conformations for each structure. The crowding of the alkyl group in an axial orientation becomes greater as the bulk of the group increases. The increased size of the substituent has little effect on the steric interactions of the conformation that has the alkyl group equatorial. The *gauche* (1,3-diaxial) interactions are responsible for the increased strain for the axial conformation. Since the ethyl and isopropyl groups can rotate to minimize the nonbonded interactions, their effective size is less than their actual size. The *tert*-butyl group cannot relieve the steric interactions by rotation and thus there is a considerably greater difference in potential energy between the axial and equatorial conformations.

Solution 17 All four stereoisomers of 1,2-dimethylcyclohexane are chiral. The *cis*-1,2-dimethylcyclohexane conformations have equal energy and are readily interconverted, as shown here.

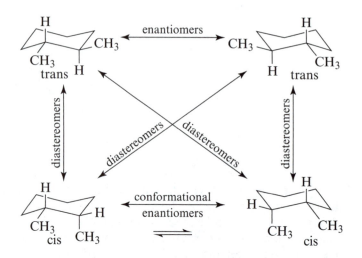

Solution 18 The structures that are isoelectronic have the same geometry. Isoelectronic structures are

$$CH_3CH_3 \qquad \text{and} \qquad CH_3\overset{+}{N}H_3$$
$$CH_3NH_2 \qquad CH_3\bar{C}H_2 \qquad \text{and} \qquad CH_3\overset{+}{O}H_2$$

Structure $CH_3\bar{N}H$ would be isoelectronic to CH_3OH.

Solution 19 Cis-trans stereoisomers are possible only for 3,4-dimethylcyclohexene. The ring size and geometry of the double bond prohibit a trans configuration of the double bond. Two configurational isomers (they are enantiomers) are possible for 2,3-dimethylcyclohexene.

cis-Cyclooctene is more stable because it has less strain than the *trans*-cyclooctene structure. The relative stability of cycloalkene stereoisomers in rings larger than cyclodecene generally favors trans. The *trans*-cyclooctene structure is chiral.

trans-Cyclooctene
(one enantiomer)

Solution 20 The ring fusion in *trans*-decalin is equatorial, equatorial. That is, one ring is attached to the other as 1,2-diequatorial substituents would be. Interconversion of the chair conformations of one ring (carbon atoms 1 through 6) in *trans*-decalin would require the other ring to adopt a 1,2-diaxial orientation. Carbon atoms 7 and 10 would both become axial substituents to the other ring. The four carbon atoms of the *substituent* ring (carbon atoms 7 through 10) cannot bridge the diaxial distance. In *cis*-decalin both conformations have an axial, equatorial ring fusion. Four carbon atoms can easily bridge the axial, equatorial distance.

Solution 21 The cyclohexane ring in norbornane is held in a boat conformation, and therefore has four hydrogen eclipsing interactions. All the six-membered ring systems in twistane are in twist conformations. All four of the six-membered ring systems in adamantane are chair conformations.

Solution 22 Bridging the 1 and 4 carbon atoms of cyclohexane is relatively easy since in the boat conformation the flagpole hydrogen atoms (on C1 and C4) are fairly close and their C—H bonds are directed toward one another. With cyclohexene, the geometry of the double bond and its inability to rotate freely make it impossible to bridge the C1, C4 distance with a single methylene group. Note, however, that a cyclohexene ring can accommodate a methylene bridge between C3 and C6. This bridged bicyclic system (bicyclo[2.2.1]hept-2-ene) does not have a bridgehead double bond.

Bicyclo[2.2.1]hept-2-ene

Solution 23 The 120° geometry of the double bond is ideal for incorporation into a planar six-membered ring, as the internal angle of a regular hexagon is 120°. Cyclooctatetraene cannot adopt a planar ring system without considerable angle strain. The eight-membered ring adopts a "tub" conformation that minimizes angle strain and does not allow significant *p*-orbital overlap other than that of the four double bonds in the system. Thus, cyclooctatetraene has four isolated double bonds and is not a delocalized π-electron system.

Cyclooctatetraene (tub conformation)

Solution 24 In the $CH_3CHXCHYCH_3$ system, there are four stereoisomers, all of which are chiral.

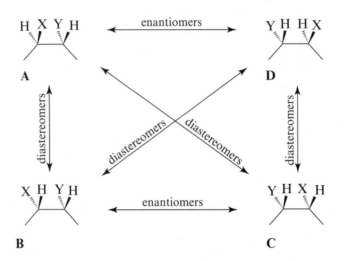

In the $CH_3CHXCHXCH_3$ system, there are three stereoisomers, two of which are chiral. The third stereoisomer (**E**) (shown below) is an achiral meso structure.

Solution 25 If at least one conformation of a molecule in which free rotation is possible has a plane of symmetry, the molecule is achiral. For a molecule with the configurations specified, there are two achiral conformations: the eclipsed conformation **E** shown in the exercise and staggered conformation **F**.

A model of **F** is identical with its mirror image. It is achiral, although it does not have a plane of symmetry, due to the presence of a *center* of symmetry that is located between C2 and C3. A center of symmetry, like a plane of symmetry, is a reflection symmetry element. A center of symmetry involves reflection through a point; a plane of symmetry requires reflection about a plane. A model of the mirror image of **S** (structure **T**) is not identical to **S**, but is a conformational enantiomer of **S**. They can be made identical by rotation about the C2, C3 bond. Since **S** and **T** are conformational enantiomers, the two will be present in equal amounts in a solution of this configurational stereoisomer. Both conformation **S** and conformation **T** are chiral and therefore should rotate the plane of plane polarized light. Since they are enantiomeric, the rotations of light will be equal in magnitude but *opposite* in direction. The net result is a racemic form of conformational enantiomers, and thus optically inactive. A similar argument can be made for any other chiral conformation and this configuration of $CH_3CHXCHXCH_3$.

Chemical interchange of two groups at either chirality center in meso compound **E** leads to a pair of enantiomers (**G** and **H**).

Solution 26   Structures **B** and **C** are chiral. Structure **A** has a plane of symmetry and is therefore achiral. Compounds **D** and **E** are both chiral. The relative orientation of the terminal groups in **D** and **E** is perpendicular, as is the case in the cumulated dienes.

Cumulated triene **F** is achiral. It has a plane of symmetry passing through all six carbon atoms. Structure **F** has a trans configuration. The cis diastereomer is the only other possible stereoisomer.

Solution 27   Structure **J** can be isolated as a chiral stereoisomer because of the large steric barrier to rotation about the bond connecting the rings. Biphenyl **K** has a plane of symmetry and is therefore achiral. The symmetry plane of **K** is shown here. Any chiral conformation of **L** can easily be converted to its enantiomer by rotation. It is only when a ≠ b and f ≠ e and rotation is restricted by bulky groups that chiral (optically active) stereoisomers can be isolated.

A plane of symmetry

Solution 28   A representation of the molecular orbitals in ethyne is given in Section 1.14 of the text. The smallest stable cycloalkyne is the nine-membered ring cyclononyne.

Solution 29   As shown here, the alternative chair conformation of β-D-glucopyranose has all large substituents in an axial orientation. The structures α-D-glucopyranose, β-D-mannopyranose, and β-D-galactopyranose all have one large axial substitutent in the most favorable conformation. β-L-Glucopyranose is the enantiomer (mirror image) of β-D-glucopyranose. Enantiomers are of equal thermodynamic stability.

β-D-Glucopyranose

α-D-Glucopyranose

β-D-Galactopyranose

β-D-Mannopyranose

β-L-Glucopyranose

Solution 30 The peptide chain bonds not free to rotate are those indicated by the bold lines in the structure shown here. The transoid arrangement produces a more linear tripeptide chain. The length of the tripeptide chain does not change if you change the substituent **R** groups.

Solution 31 The models of the π molecular orbitals for ethene are shown in the Orbital Symmetry section of the Darling Framework Molecular Model Set instruction booklet. A representation of these orbitals can be found in the text in Section 1.13.

The π molecular orbitals for 1,3-butadiene are shown in the text in Section 13.6C. A model of the π molecular orbitals of 1,3-butadiene is also shown in the Orbital Symmetry section of the Darling Framework Molecular Model instruction booklet. The phases of the contributing atomic orbitals to the molecular orbitals of the allyl radical can be found in the text in Section 13.2. The π molecular orbital of the allyl radical has a node at C2.